T0269571

CAMBRIDGE MONOGRAPHS
ON MECHANICS AND APPLIED MATHEMATICS

GENERAL EDITORS

G. K. BATCHELOR, PH.D., F.R.S.
Professor of Applied Mathematics at the University of Cambridge

J. W. MILES, PH.D.
Professor of Applied Mathematics, University of California, La Jolla

BUOYANCY EFFECTS IN FLUIDS

BUOYANCY EFFECTS
IN FLUIDS

BY

J. S. TURNER

Professor of Geophysical Fluid Dynamics
Research School of Earth Sciences
Australian National University

CAMBRIDGE UNIVERSITY PRESS

CAMBRIDGE

LONDON   NEW YORK   NEW ROCHELLE

MELBOURNE   SYDNEY

CAMBRIDGE UNIVERSITY PRESS
Cambridge, New York, Melbourne, Madrid, Cape Town, Singapore,
São Paulo, Delhi, Dubai, Tokyo, Mexico City

Cambridge University Press
The Edinburgh Building, Cambridge CB2 8RU, UK

Published in the United States of America by
Cambridge University Press, New York

www.cambridge.org
Information on this title: www.cambridge.org/9780521297264

First published 1973
First paperback edition 1979

A catalogue record for this publication is available from the British Library

Library of Congress catalogue card number: 72-76085

ISBN 978-0-521-08623-3 Hardback
ISBN 978-0-521-29726-4 Paperback

TO MY FATHER

CONTENTS

PREFACE

Buoyancy forces arise as a result of variations of density in a fluid subject to gravity, and produce a wide range of phenomena of importance in many branches of fluid mechanics. Progress in this field has been made largely through the desire to solve very practical problems, arising for instance in meteorology or in hydraulic engineering. This emphasis on particular applications has meant that parallel developments have often been made in different disciplines without much cross reference to related work, and some results, well understood in one context, are less familiar in another where they might be used to advantage. In this book I have attempted to write a coherent account of the various fluid motions which can be driven or influenced by the presence of small density differences. It is intended as a general introduction to the subject and its literature, in which the physical understanding of the phenomena is emphasized, rather than the applications on the one hand or detailed mathematical theory on the other.

The selection of subject matter must always be a personal one, however, and my own research interests have certainly influenced the topics chosen and the amount of space given to each of them. I have worked with laboratory models of small scale processes in the ocean and atmosphere, and so laboratory and geophysical examples come most readily to mind, but comparisons have also been made with results from various fields of engineering where possible. I have been particularly interested in two problems, those of buoyant convection, and mixing in stably stratified fluids. The sections on these topics cover a larger fraction of the available material (though much of this in the latter field is still rather speculative), while other subjects which have a much firmer foundation in the published literature are treated only to the extent needed to provide a background for the understanding of more complicated processes. An outline of the general plan of the book is given in the opening section of chapter 1.

Some phenomena which might be implied by the title chosen for this book have been arbitrarily omitted; I have taken this to mean effects which are due *primarily* to *small* density variations in an ordinary fluid such as water. Thus waves at a free surface are not considered in detail, nor are bubbles and particles in a liquid, except insofar as they produce small changes in the mean density. For the most part we will be concerned with miscible fluids, so that surface tension phenomena receive little attention. Flows in porous media are mentioned only in passing, in the context of numerical experiments in thermal convection, and not much is said about conditionally unstable motions (such as convection in clouds, when the release of latent heat dominates the buoyancy). The effects of compressibility are ignored too, after some discussion of the conditions under which a compressible fluid may be treated as incompressible. Perhaps the most important restriction on the scope is that no explicit results are given for rotating stratified fluids. This implies that the natural phenomena of interest are small enough or fast enough for the Coriolis forces due to the earth's rotation to be negligible compared with the buoyancy and inertia forces.

Many people have helped in various ways during the preparation of this book. I am grateful to Professor O. M. Phillips for suggesting that I should write it, to Professor G. K. Batchelor for his encouragement both as head of my department and as editor of this series, and to the British Admiralty and the U.S. Office of Naval Research for supporting my research during the years I have been learning about stratified fluids. T. H. Ellison, H. B. Fischer, P. B. Rhines and J. D. Woods, as well as many colleagues and students in Cambridge and Woods Hole, have made helpful comments on parts of the manuscript; and I owe a special debt to H. E. Huppert and S. A. Thorpe who read the whole of an earlier draft and did much to clarify the presentation. I acknowledge with thanks Mrs Glynis Coulson and Mrs Susan Gray who did the typing, P. F. Linden who assisted with the proofs, and last but not least, my wife and family for their forbearance during the months when my attention was even more divided than usual.

I am indebted to those who have permitted the reproduction of their photographs and diagrams, and who in many cases have supplied original prints. Permission to reproduce figures was received from the editors of the following journals:

Journal of Fluid Mechanics
Tellus
Quarterly Journal of the Royal Meteorological Society
Journal of Geophysical Research
Philosophical Transactions of the Royal Society
Proceedings of the Royal Society
Journal of the Atmospheric Sciences
Journal of Applied Meteorology
Deep-Sea Research
Nature
International Journal of Heat and Mass Transfer
Advances in Geophysics
Science
Physics of Fluids.

Cambridge   J. S. TURNER

June 1972

Note to the 1979 reprint

The basic concepts underlying this subject have not changed greatly in the six years since this book first appeared, though there has been a rapid growth in the literature related to the applications. It did not seem appropriate at present to attempt a fully revised edition, and I have made only minor corrections of points which have come to my attention, mainly of typographic errors, before reprinting. A short, selective, annotated bibliography of recent papers and review articles has been added, to allow readers to bring themselves up to date on specific topics discussed in the text.

Canberra   J. S. T.

March 1979

CHAPTER 1

INTRODUCTION AND PRELIMINARIES

1.1. The topics to be discussed

It seems useful to begin by outlining the range of subjects covered in this book, to give a broad picture of the way in which the several parts of the field have developed, and at the same time some explanation of the theme which has been used to connect them. The phenomena studied all depend on gravity acting on small density differences in a non-rotating fluid. Often the undisturbed fluid has a density distribution which varies in the vertical but is constant in horizontal planes; this will be called a stratified system whether the density changes smoothly or discontinuously. Special attention will be given to the problems of buoyant convection (arising from an unstable density distribution) and to various mechanisms of mixing when the stratification is stable.

Chapters 2 and 3 summarize relevant results on internal waves, and these were also historically the first phenomena to be studied. The original applications of the methods of perfect fluid theory to motion under gravity were to the problems of small amplitude surface waves and tides (subjects which will not be discussed here). These were soon extended to the case of two layers of uniform density with a density discontinuity between them. Some of the basic results had already been obtained by 1850 (notably by Stokes 1847), and they were applied to phenomena such as the drag experienced by a ship when it creates a wave on an interface close to the surface (Ekman 1904), and to internal seiches in lakes. The later developments in the theory of internal waves owe more to meteorology than to the study of the sea or lakes, probably because of the directly visible and often spectacular effects which are caused by gravity waves in the atmosphere. The studies of waves in a density gradient, waves in the lee of obstacles and the effect on these of wind velocity changes with height, all originated in this context. Some of these results have been extended to the case where the waves have a large amplitude. The recent work on interactions between waves has again been discussed with the oceanographic application in mind (Phillips 1966a).

Various finite amplitude flow phenomena in a stratified fluid are also considered in chapter 3.† Many of these occur in the context of hydraulic engineering, for example in the prediction of the velocity of intrusion of saline water into a lock filled with fresh water when the gate separating the two is opened, or the conditions under which one layer can be withdrawn from a stratified fluid without removing an adjacent layer. An elementary discussion of small scale fronts in the atmosphere also comes under this heading. The phenomenon of blocking, and the jet-like motions which can arise in slow flows of a stratified fluid are mentioned briefly, with some discussion of the effects of viscosity and diffusion.

Instabilities of various shear flows of a stratified fluid are treated in chapter 4. Some parts of this subject have been well understood for a long time, but other results are of more recent origin. A classification of the mechanisms of generation of turbulence is also given here, as a logical prelude to the discussion of turbulent flow in a stratified medium. The subject of forced and free convection in a shear flow over a rough plane again owes much to the meteorological work, which will be used as the basis for chapter 5. The behaviour of turbulent wakes in a stratified fluid seems to fit in naturally here.

In the next two chapters we consider gravitationally unstable flows, i.e. the various mechanisms of buoyant convection. The historical order of development is reversed here; the models of convection which emphasize the buoyant elements themselves are treated first, followed by studies of convection between horizontal planes and some discussion of the relation between the two. The impetus for much of this work has come from the problem of heat transfer from the ground to the lower atmosphere, but some attention is also given to other geometries, and to current work on the numerical simulation of turbulent motions. The discussion of convection is extended in chapter 8 to the case where two properties

† A more thorough historical review of this part of the subject has been given recently by Hinwood (1970).

with different molecular diffusitivities are present simultaneously in a fluid. When these have opposing effects on the vertical density gradient, convection in well-mixed layers can be driven by an unstable buoyancy flux (for example by a flux of heat from below), while a net density difference is preserved across the interfaces between them. These effects are believed to have important implications for vertical mixing in the ocean.

In the last two chapters, many of the ideas developed earlier are used to discuss the processes responsible for mixing in large bodies of stratified fluid, particularly in the ocean and atmosphere. It is shown that there are many mechanisms which will cause a smoothly stratified fluid to break up into a series of steps, and so a basic problem is the understanding of mixing across a density interface. Such mixing can be driven either directly by a boundary source of mechanical energy, or by energy propagated into the interior by internal waves.

It is impossible to arrange a subject with so many different strands in an entirely satisfactory order, but I hope that enough cross-references have been given to allow readers to choose a different sequence to suit their own interests. Some basic ideas and approximations which are common to the whole book are outlined in the following sections of this chapter, but it will be obvious that neither this introduction nor the later chapters can be comprehensive or complete. Most theoretical results will be quoted without proof, and a background knowledge of fluid mechanics of homogeneous media is assumed. (See, for example, Batchelor 1967.)

1.2. Equilibrium and departures from it

The only external force field considered in this book is that of gravity, which exerts a body force $\rho\mathbf{g}$ per unit volume on each element of fluid (where ρ is the local density and $\mathbf{g}$ is the acceleration due to gravity). The effects to be described result from variations of ρ from point to point in the fluid, which will nearly always be regarded as incompressible. The nature of the fluid is unimportant; most of the ideas may be applied to liquids, in which density variations are due to differences of temperature or the concentrations of solutes or sediment, or to gases, in which there may be differences of

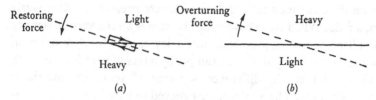

Fig. 1.1. Displacements from hydrostatic equilibrium:
(a) stable, and (b) unstable density distributions.

temperature or composition. The compressibility of gases becomes
significant in deep layers, but for many purposes these too can be
treated as incompressible by using potential temperature and
potential density (as defined below) in place of the actual density.

A body of homogeneous, inviscid incompressible fluid at rest is in
a state of neutral equilibrium. At every point the weight of a fluid
element is then exactly balanced by the pressure exerted on it by
neighbouring fluid, and this continues to hold true if the elements
are displaced to another position of rest. When ρ varies the hydrostatic equation

$$p = p_0 - g \int_0^z \rho \, dz \qquad (1.2.1)$$

shows that the fluid (in the absence of diffusion) is in equilibrium
only when the density as well as the pressure is constant in every
horizontal plane. This equilibrium stratification is stable when the
heavier fluid lies below, since tilting of a density surface will produce
a restoring force: the resulting motion can overshoot the equilibrium
position and oscillate about it, thus giving rise to internal waves.
When light fluid lies below heavier, the equilibrium is unstable and
small displacements of density surfaces from the horizontal will
grow and lead to convective motions (see fig. 1.1).

The corresponding state of neutral static stability in a compressible
fluid is that for which the entropy is constant with depth. A change
of pressure results in a change in temperature in an adiabatic
process; and this must be taken into account when comparing displaced fluid with its surroundings. The potential temperature θ and
potential density ρ_θ are defined to be the temperature and density
when the fluid is compressed adiabatically to a standard pressure p_0.
In a perfect gas for which $p = R\rho T$ (a good approximation for the

atmosphere), R being the gas constant, they are related to the absolute temperature T and the local properties ρ and p by

$$T = \theta \left(\frac{p}{p_0}\right)^{(\gamma-1)/\gamma}, \quad \rho = \rho_\theta \left(\frac{p}{p_0}\right)^{1/\gamma}, \quad (1.2.2)$$

and it follows that

$$-\frac{1}{\rho_\theta}\frac{\partial \rho_\theta}{\partial z} = \frac{1}{\theta}\frac{\partial \theta}{\partial z} = \frac{1}{T}\frac{\partial T}{\partial z} - \frac{\gamma-1}{\gamma}\cdot\frac{1}{p}\frac{\partial p}{\partial z}, \quad (1.2.3)$$

where γ is the ratio of specific heats $\gamma = C_p/C_v = C_p/(C_p - R)$, and is about 1.4 for air. An isothermal atmosphere has $T = $ constant, and it then follows from (1.2.1) and (1.2.3) that

$$p = p_0 e^{-gz/RT} \quad \text{and} \quad \rho = \rho_0 e^{-gz/RT}, \quad (1.2.4)$$

where ρ_0 is a reference density. The length $H_s = RT/g$ in which the density falls off by a factor e is called the scale height (it is about 8 km in the earth's atmosphere). In the *isothermal* atmosphere

$$\frac{1}{\rho_\theta}\frac{\partial \rho_\theta}{\partial z} = \frac{\gamma-1}{\gamma}\frac{1}{\rho}\frac{\partial \rho}{\partial z}. \quad (1.2.5)$$

An *adiabatic* atmosphere is one in which θ is constant, and the absolute temperature gradient is then

$$\frac{\partial T}{\partial z} = -\frac{\gamma-1}{\gamma}\frac{g}{R} = -\frac{g}{C_p} = -\Gamma \quad (1.2.6)$$

or $\partial \ln T/\partial \ln p = (\gamma - 1)/\gamma$ using pressure as the vertical coordinate.

If the actual 'lapse rate', or decrease of temperature with height, equals this ($\Gamma = 10\,°C/km$ in the earth's atmosphere), a displaced fluid element will always have the same density as its new surroundings and the equilibrium will be neutral; if the temperature decreases less rapidly than Γ the situation will be stable. An isothermal atmosphere will be very stable in this sense, and an *inversion*, in which the absolute temperature increases with height over some interval, even more so. Thorough mixing of an arbitrarily stratified compressible fluid results in the formation of an adiabatic atmosphere. When the air is moist and lifting produces saturation, the release of latent heat of condensation will heat the rising air, and the rate of cooling of an unmixed parcel will be reduced to the *saturated adiabatic* lapse rate. An atmosphere stable to dry convection may be unstable when there is moisture present, and one must be careful to specify which adiabatic process is to be used for reference when assessing the static stability. Strictly speaking, the potential density should be used also for liquids. The compressibility of small volumes of water, say in the laboratory, may be neglected, but it becomes significant for the deep ocean (see Phillips 1966 a, p. 13). More restrictions must be put on the velocity and amplitudes of the *motion* in a compressible fluid before it can be regarded as incompressible, and these will be considered in the next section.

1.3. The equations of motion, and various approximations

The characteristic differences between the motion of a homogeneous and a heterogeneous fluid can most easily be explained by writing down the equations of motion in several forms: and it will also be useful to have these here for reference in later chapters. (For a detailed derivation of these equations, see for example Yih (1965).) From now on the fluid will be assumed incompressible and non-diffusive unless it is explicitly stated otherwise, and this means that

$$\frac{D\rho}{Dt} = 0, \tag{1.3.1}$$

where D/Dt denotes differentiation following the motion. The continuity equation in vector notation is

$$\nabla \cdot \mathbf{u} = 0, \tag{1.3.2}$$

where $\mathbf{u} = (u, v, w)$ is the velocity. The momentum (Navier–Stokes) equations with the force of gravity included can be written (with $\mathbf{g} = (0, 0, -g)$, the x and y axes being in the horizontal plane and z vertically upwards) as

$$\rho \frac{D\mathbf{u}}{Dt} = \rho \left(\frac{\partial \mathbf{u}}{\partial t} + (\mathbf{u} \cdot \nabla \mathbf{u}) \right) = -\nabla p + \rho \mathbf{g} + \mu \nabla^2 \mathbf{u}. \tag{1.3.3}$$

The last term is the result of molecular viscosity μ (assumed constant here), and if this is neglected one obtains the Euler equations of motion.

If p and ρ are now expanded about the values p_0 and ρ_0 in a reference state of hydrostatic equilibrium for which $\nabla p_0 = \rho_0 \mathbf{g}$

(i.e. one sets $p = p_0 + p'$ and $\rho = \rho_0 + \rho'$), the Euler equations can be written in terms of the deviations p' and ρ' from this state as

$$\rho \frac{Du}{Dt} = -\nabla p' + \rho' g. \tag{1.3.4}$$

Thus as already implied in the elementary discussion of static stability, only differences of density ρ' from some standard value are relevant in determining the effect of gravity. In a two-layer system, for example, the layer with the standard density ρ_0 may be regarded as weightless, and that with density $\rho = \rho_0 + \rho'$ as if it were acted on by a reduced gravitational acceleration $g\rho'/\rho$. (See Prandtl 1952, p. 368.)

Taking the curl of the Navier-Stokes equation (1.3.3) leads to an equation for the vorticity $\boldsymbol{\zeta} = \nabla \times \mathbf{u}$, namely

$$\frac{D\boldsymbol{\zeta}}{Dt} = \boldsymbol{\zeta} \cdot \nabla \mathbf{u} + \nu \nabla^2 \boldsymbol{\zeta} + \nabla p \times \nabla \left(\frac{1}{\rho}\right), \tag{1.3.5}$$

where $\nu = \mu/\rho_0$ is the kinematic viscosity (which is also taken to be a constant, implying the neglect of ρ' compared to ρ_0 in this term). The first three terms are the same as for a fluid of constant density; a change of vorticity of a fluid element can again be brought about by a stretching of vortex lines, or by the diffusion of vorticity from boundaries. The last term contains the essential difference between a stratified fluid and a uniform one. Vorticity will be created whenever a non-homogeneous fluid is displaced from a state in which ∇p and $\nabla \rho$ are parallel (the only condition for which the vector product is zero). In the simplest case where p effectively depends on gravity alone (and it is often true that contributions to p' due to other accelerations are negligible compared to the hydrostatic part) displacements of density surfaces away from the horizontal will produce vorticity. This will oscillate in magnitude and direction in stable stratification (so that internal waves are *rotational* phenomena), and it will tend to increase monotonically during the development of convection.

The creation of vorticity also implies the creation of circulation defined by

$$\Gamma = \int_C \mathbf{u} \cdot d\mathbf{l} = \iint_S \boldsymbol{\zeta} \cdot d\mathbf{A}, \tag{1.3.6}$$

where dl is a line element of a closed curve C and d$\mathbf{A}$ is an element of a surface S bounded by C. The result for an inviscid fluid, obtained originally by Bjerknes, can be written in the two forms

$$\frac{d\Gamma}{dt} = -\int_C \frac{1}{\rho} \nabla p \cdot d\mathbf{l}$$

$$= \iint_S \left(\nabla p \times \nabla \left(\frac{1}{\rho} \right) \right) \cdot d\mathbf{A}. \qquad (1.3.7)$$

This is a generalization of Kelvin's theorem, which states that in an inviscid fluid of constant density the circulation round a closed curve moving with the fluid remains constant. When ρ is variable along the path of integration, circulation is generated unless density and pressure surfaces coincide. For example, a circuit taken just below and just above an interface separating layers of different density shows that a shear must be developed at the interface when it is tilted from a state of rest, so that the pressure varies along the path (fig. 1.1). If the surface S is chosen to be one of constant density, the integrand of (1.3.7) vanishes, and a motion started from rest can be treated as a two-dimensional irrotational flow within each density surface, the vortex lines being imbedded in these surfaces. This last property can be exploited to derive an integrated form of the Euler equations, which correspond to Bernoulli's equation in a uniform fluid, and in which the Bernoulli constant is replaced by a function of density alone. These are given and used explicitly in chapter 3.

Two widely used approximate forms of the Euler equations (1.3.4) must also be introduced. The first simplification is that of *linearization*, the neglect of the non-linear convection terms like $u \, \partial u / \partial x$ in comparison with $\partial u / \partial t$. This procedure is justified when the motions and velocities are of small amplitude. The parameter ϵ which must be kept small is different in different situations (see chapter 3), but typically it is a ratio of a vertical displacement to a horizontal lengthscale. The product terms are of order ϵ^2 and can be omitted to give the first order equations

$$\rho \frac{\partial \mathbf{u}}{\partial t} = -\nabla p' + \rho' \mathbf{g}. \qquad (1.3.8)$$

The importance of the linear equations lies in the fact that any small oscillation described by them can be resolved into a set of 'normal modes', in each of which the particle motions are simple harmonic and independent of all other modes. (See Lamb 1932, ch. 8.) Note that no assumption has been made here about the magnitude of the density variation, and some linear problems can be solved without further approximation (see §2.2.1).

In the second approximation to be mentioned here, on the other hand, the density variation ρ' is assumed to be small compared to ρ_0. Rewriting (1.3.4) in the form

$$\left(1 + \frac{\rho'}{\rho_0}\right)\frac{D\mathbf{u}}{Dt} = -\frac{1}{\rho_0}\nabla p' + \frac{\rho'}{\rho_0}\mathbf{g}, \qquad (1.3.9)$$

we see that the density ratio ρ'/ρ_0 appears twice, in the first (inertia) term and in the buoyancy term. When ρ'/ρ_0 is small, it produces only a small correction to the inertia compared to a fluid of density ρ_0, but it is of primary importance in the buoyancy term. The approximation introduced by Boussinesq (1903) consists essentially of neglecting variations of density in so far as they affect inertia, but retaining them in the buoyancy terms, where they occur in the combination $g' = g\rho'/\rho_0$. When viscosity and diffusion are included, variations of fluid properties are also neglected in this approximation.

There are other restrictions necessary in compressible fluids which are often included by the name 'Boussinesq approximation'; these are discussed fully by Spiegel and Veronis (1960) and only the results will be quoted here. First one must replace density by potential density, as already shown in §1.2. The limitation of small density deviations from a standard ρ_0 implies two things; the vertical scale of the *mean* motion must be much smaller than the scale height H_s (1.2.4), and the fluctuating density changes due to local pressure variations must also be negligible. The latter is the most important of the extra conditions; it implies that the fluid can be treated as incompressible (so that (1.3.2) is an adequate approximation to the continuity equation), and it therefore excludes sound and shock waves. Finally, the ratio of the length to the timescales of any variation in an unsteady flow should be much smaller than the velocity of sound, to ensure that information about pressure changes is transmitted effectively instantaneously, as it is in an incompressible fluid.

The Boussinesq approximation may be made either independently of the linear (small amplitude) assumption or in combination with it. The *linearized* Boussinesq equations for an inviscid liquid are

$$\frac{\partial \mathbf{u}}{\partial t} = -\frac{1}{\rho_0}\nabla p' + \frac{\rho'}{\rho_0}\mathbf{g}, \qquad (1.3.10)$$

together with (1.3.1) and (1.3.2). The linearized form of (1.3.1) is just

$$\frac{\partial \rho'}{\partial t} + w\frac{\partial \rho_0}{\partial z} = 0, \qquad (1.3.11)$$

expressing the fact that changes of density at a point are due to bodily displacement of the mean density structure. The second (product) term is not small here, since $\partial \rho_0/\partial z$ is a finite quantity.

It is important to note immediately, however, that there are some circumstances where it is inconsistent to use the Boussinesq approximation, even when the density differences are everywhere small. In the theory of internal solitary waves, for example, (§3.1.2) certain non-linear terms are retained which are of the same order as those neglected in obtaining (1.3.9), and the whole phenomenon depends on the consistent inclusion of all terms to this order.

Under other circumstances the Boussinesq approximation works surprisingly well in fluids with large density variations (even in the atmosphere), but this depends very much on the type of flow considered. Most meteorological flows with a vertical scale less than about 1 km (small compared to H_s) can be treated in this way, and this includes wave motions which extend to great heights, provided the vertical excursions of individual fluid particles are not too large. The assumptions are most seriously violated by convective motions extending through a height of the order H_s or more since fluid then can traverse the whole depth. However another simplifying result is available for incompressible fluids which can in some problems remove the inertial effect of the density variation entirely from the equations, and allow one to calculate it *afterwards* from the solution of the corresponding Boussinesq problem. The simplest example is a steady flow in which gravity effects are absent (Yih 1958). In this case a transformation of (1.3.4) to remove ρ from the convective terms $\rho u(\partial u/\partial x)$ shows that the actual velocity is a factor $(\rho_0/\rho)^{\frac{1}{2}}$ greater than that calculated using a constant ρ_0. With a fixed pressure gradient, this just implies that the kinetic energy per unit volume, ρu^2 say, is constant throughout such a flow. This same form of result is obtained in §2.2.1 for small amplitude waves of a particular form in an exponential density gradient (a problem which certainly involves gravity), and Drazin (1969) has shown that this remains true for internal gravity waves even when the amplitude is large. It has not been strictly proved for other cases, but scaling u with $\rho^{-\frac{1}{2}}$ will often give a good approximation for other incompressible flows in regions of large density variation. No such general simplification is possible for non-linear motions of compressible fluids over many scale heights. (See Claus (1964) and §3.1.4.)

1.4. Basic parameters of heterogeneous flows

Several quantities and concepts which recur in different contexts throughout this subject will now be introduced in an elementary way (leaving a more detailed discussion of their physical signifi-cance to the appropriate later chapter). Consider first the motion of an element of inviscid fluid displaced a small distance η vertically from its equilibrium position in a stable environment. The vertical component of (1.3.10) (neglecting the small pressure fluctuation), together with (1.3.11) gives

$$\frac{\partial^2 \eta}{\partial t^2} = \frac{g}{\rho}\frac{\partial \rho_0}{\partial z}\eta. \tag{1.4.1}$$

The element will thus oscillate in simple harmonic motion with angular frequency

$$N = \left(-\frac{g}{\rho}\frac{\partial \rho_0}{\partial z}\right)^{\frac{1}{2}}. \tag{1.4.2}$$

This is the frequency associated with the names of Brunt (in meteorology, where of course the potential density gradient must be used) and Väisälä (especially in oceanography), but it will be referred to here by the less cumbersome and more descriptive name of *buoyancy frequency*. The corresponding periods $2\pi/N$ are typically a few minutes in the atmosphere and the oceanic thermocline, and up to many hours in the deep ocean. Notice that N is constant when the density varies exponentially with height (if $\rho = \rho_0 e^{-z/H}$ then $N^2 = g/H$), and a variation of this kind is often assumed to simplify the analysis. In the limit of small density differences the corresponding gradient is linear, so the latter assumption is useful when the Boussinesq equations are appropriate.

In a shear flow the vertical gradient of the horizontal velocity also has the dimensions of frequency, and the non-dimensional ratio

$$Ri = N^2 \bigg/ \left(\frac{\partial u}{\partial z}\right)^2 = -g\frac{\partial \rho}{\partial z}\bigg/ \rho\left(\frac{\partial u}{\partial z}\right)^2 \qquad (1.4.3)$$

is called the *gradient Richardson number*. (This was named in honour of L. F. Richardson, though it is not exactly the form he used – see Brunt 1952.) There are other related ratios called by this name, for example, the flux Richardson number Rf defined in chapter 5, which is the ratio of the rate of removal of energy by buoyancy forces to its production by the shear, and the particular definition implied in each case must be kept clearly in mind. Another ratio of this kind has a fundamental significance as an *overall* parameter describing a whole flow (Batchelor 1953 a). When the Boussinesq and hydrostatic approximations are made and the motion is steady, the ratio of the buoyancy to the inertia terms in (1.3.9) is the *only* dimensionless number needed to specify an inviscid flow. Using the scales of velocity U and length L imposed by the boundary conditions, this can be written

$$Ri_0 = g'L/U^2. \qquad (1.4.4)$$

A subscript will always be used to denote the overall or finite difference form of the Richardson number.

In hydraulic engineering it is more common to use, instead of Ri_0, the inverse square root of Ri_0 i.e.

$$F = U/(g'L)^{\frac{1}{2}}, \qquad (1.4.5)$$

which is called the internal or densimetric *Froude number*. (Sometimes this is written with a subscript i to denote 'internal', but this will be dropped here.) This usage has arisen because of the correspondence with the ordinary Froude number (defined with g

replacing g' in the above), which compares a characteristic flow velocity with the velocity of long waves on a free surface. An entirely analogous interpretation of F involving internal waves is given in chapter 3. The use of the corresponding gradient parameter $(\partial u/\partial z)/N (= F_g$ say) instead of Ri (cf. (1.4.3)) has much to recommend it, but both Ri and F will be retained since common usage seems to demand them.

Other parameters of course can become important when extra physical effects are taken into account. The most obvious is the *Reynolds number* $Re = UL/\nu$ (where ν is the kinematic viscosity) a measure of the balance between inertial and viscous terms in (1.3.3). When the length and velocity scales, and therefore Re, are large, it is justifiable in a homogeneous fluid to treat the motion as inviscid, except perhaps near boundaries. Even in laboratory experiments, especially the kind of free turbulent flows considered in chapter 6, it is often possible to ignore the viscous effects and still to make reliable comparisons between a model and the prototype in nature. In a strongly stratified fluid, however, there is another possibility which must be kept in mind: the density gradient can suppress vertical motions, and so introduce a much smaller internal lengthscale into the problem. The Reynolds number defined using this new scale need no longer be large, and viscosity (and also the effects of molecular diffusion) again become relevant. (See § 10.2.3.)

When molecular diffusion is taken into account explicitly, various non-dimensional ratios may be defined, the simplest of which is the *Prandtl number* $Pr = \nu/\kappa$. The other parameters depend on the ones already defined, and they are not so fundamental to the whole range of problems under discussion, so their introduction will be left to the appropriate place in later chapters.

CHAPTER 2

LINEAR INTERNAL WAVES

Natural bodies of fluid such as the atmosphere, the oceans and lakes are characteristically stably stratified: that is, their mean (potential) density decreases as one goes upwards, in most regions and for most of the time. When they are disturbed in any way, internal waves are generated. These ubiquitous motions take many forms, and they must be invoked to explain phenomena ranging from the temperature fluctuations in the deep ocean to the formation of clouds in the lee of a mountain. In this chapter we summarize the results which can be obtained using linear theory (i.e. when the amplitudes are assumed to be small), and in §3.1 extend some of them to describe waves of large amplitude.

Many of the elementary properties of infinitesimal wave motions in stratified fluids can be introduced conveniently by considering waves at an interface between two superposed layers, and so this case is treated first in §2.1. These waves are analogous to waves on a free water surface, and therefore seem very familiar. It should be emphasized at the outset, however, that they are *not* the most general wave motions which can occur in a continuously stratified fluid. Energy can propagate through such a fluid at an angle to the horizontal, not just along surfaces of constant density, and our intuition based on surface waves is of little help here. The more general theory, and a comparison between the two descriptions, is given in §2.2.

2.1. Waves at a boundary between homogeneous layers

Consider two superposed layers, each of which has constant density ρ_0 and $\rho_1 = \rho_0 + \rho'$ say and is initially at rest. Only two-dimensional waves will be allowed, with motions confined to the x, z plane (z positive upwards) and crests consisting of parallel ridges running in the y-direction. Waves of this kind are easily set up in a laboratory channel (see fig. 2.1 pl. 1)†, and more general three-dimensional disturbances can be treated by superimposing such motions (because the equations are linear when the motions are kept small). The fluid will be assumed inviscid in these introductory sections, and viscous effects added where they are needed to explain particular phenomena. For the derivation of the results quoted here, reference should be made to Lamb (1932, ch. 9).

The linearized equations (1.3.8) can be applied to each of the layers separately, with ρ constant. *Within* the layers the motion is irrotational and so potential theory can be used to describe the flows above and below the interface. The equations satisfied by the velocity potentials (which are defined by $u = -\partial\phi/\partial x$, $w = -\partial\phi/\partial z$ and because of the continuity equation satisfy $\nabla^2\phi = 0$ in each layer) are obtained by integrating the component equations of (1.3.8). They are

$$\frac{p_0}{\rho_0} = \frac{\partial\phi_0}{\partial t},$$

and

$$\frac{p_1}{\rho_1} = \frac{\partial\phi_1}{\partial t} - g\frac{\rho'}{\rho_1}z. \tag{2.1.1}$$

These equations apply whatever the (small) motion in the two layers, but we can simplify the discussion by considering just one component of a Fourier decomposition, a sinusoidal travelling wave which displaces the interface an amount

$$\eta = a \exp i(kx - \omega t), \tag{2.1.2}$$

(where k is a wavenumber and ω is the angular frequency). Clearly η and ϕ must be related by $\partial\eta/\partial t = -\partial\phi/\partial z$, since these are alternative expressions for the vertical velocity. The form of solution also depends on the boundary condition away from the interface, and two important cases are treated separately below.

2.1.1. *Progressive waves in deep water*

Suppose first that the layers are both very deep compared with the wavelength, so that the disturbances vanish at $kz = \pm\infty$, where the

† The plates are to be found between p. 96 and p. 97.

origin of z is at the interface. Velocity potentials satisfying (2.1.1) and the boundary conditions (2.1.2) and at infinity are

$$\left.\begin{aligned} \phi_0 &= A_0 e^{-kz} \exp \mathrm{i}(kx - \omega t), \\ \phi_1 &= A_1 e^{kz} \exp \mathrm{i}(kx - \omega t). \end{aligned}\right\} \qquad (2.1.3)$$

The displacement η must be the same in the two layers, so

$$kA_0 = -kA_1 = \mathrm{i}\omega a. \qquad (2.1.4)$$

A second condition relating the constants A_0 and A_1 is that the pressure must be continuous at the interface (provided no extra forces are introduced by surface tension). For these to be consistent, there must be a special relation between ω and k (the dispersion relation) which can be expressed in terms of the phase velocity $c = \omega/k$ and the wavelength $\lambda = 2\pi/k$ as

$$c^2 = \frac{g}{k}\frac{\rho'}{\rho_0 + \rho_1} = \frac{g\lambda}{2\pi}\frac{\rho_1 - \rho_0}{\rho_1 + \rho_0}. \qquad (2.1.5)$$

Such waves are called *dispersive*, because motions generated with given frequency but different wavelengths move at different velocities and will separate as they travel away from the source. For example if the oceanic thermocline is disturbed in an arbitrary way, the longest waves generated will always be the first to reach a distant observing point.

As with all gravity waves, the energy is divided between the kinetic and potential forms, the two contributions being equal when averaged over a wavelength. The phase velocity c (which will sometimes be written c_p) is reduced by the factor

$$[\rho'/(\rho_0 + \rho_1)]^{\frac{1}{2}}$$

compared with that for surface waves. In typical situations in the ocean, for example, where $\rho'/\rho_0 \approx 10^{-3}$, internal waves travel with only a few per cent of the velocity of surface waves. Energy is propagated along the interface with the group velocity

$$c_g = \frac{\mathrm{d}\omega}{\mathrm{d}k} \qquad (2.1.6)$$

which in this case is just $c_g = \frac{1}{2}c$, a relation which also holds for surface waves on deep water. Fluid particles move in paths which

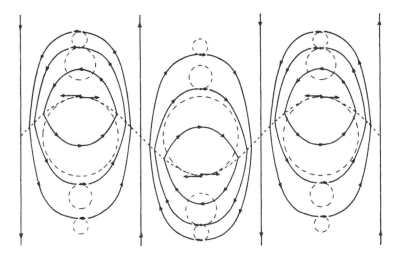

Fig. 2.2. Streamlines and orbits in a progressive internal wave travelling from left to right along the interface between two fluids. From Defant (1961).

(in the linear approximation) are circular, with amplitude decreasing exponentially with distance from the interface. (See fig. 2.2.)

The solutions (2.1.3) are consistent with the general rotational property of stratified motions (§ 1.3). Although the normal velocity has been made continuous, (2.1.4) shows that the tangential velocity changes sign across the interface. The potential flows in two layers are separated by a vortex sheet, with a maximum shear at the crests and troughs of the waves. In a real fluid, of course, such discontinuities will be spread out by viscosity into a vortex layer of finite thickness. With miscible fluids in the two layers (the case we will be most concerned with) diffusion of the property producing the density difference (i.e. heat or salt) will already have produced a continuous density distribution rather than strictly a step, and this will modify the vorticity distribution associated with the wave even without viscosity (cf. § 4.3.3). There will always be a maximum shear where the density gradient is a maximum. (See, for example, Phillips 1966a, p. 168.)

2.1.2. *Waves between layers of finite thickness*

More general boundary conditions can be put on (2.1.1), with (2.1.2) again applied at the interface. When the fluids are confined between rigid horizontal planes at $z = h_0$ and $z = -h_1$ so that $w = 0$ at these depths, the dispersion relation becomes

$$c^2 = \frac{g\rho'}{k}(\rho_0 \coth kh_0 + \rho_1 \coth kh_1)^{-1}. \qquad (2.1.7)$$

Various special cases are recovered using the limits $\coth kh \to 1$ as $kh \to \infty$ (which when kh_0 and kh_1 are both large gives (2.1.5)), and $\coth kh \to (kh)^{-1}$ as $kh \to 0$. When kh_0 is large and kh_1 is small, the phase velocity is

$$c^2 = \frac{\rho'}{\rho_1}gh_1. \qquad (2.1.8)$$

Small amplitude *long* waves on a layer of denser fluid underlying a deep light fluid are thus *non-dispersive*. Their phase velocity depends on the layer depth and the densities (through a reduced acceleration) but not on the wavenumber, and $c_g = c$. In this limit the vertical acceleration may be neglected (a result which continues to be valid for *finite* amplitude long waves – see §3.1.1), and the horizontal velocity is the same for all particles on a vertical line. This implies that a line of dye just moves back and forth as the wave passes by; the amplitude of the vertical velocity increases linearly from zero with distance above the solid bottom.

When there is a free surface above the upper layer the boundary condition to be applied there is more complicated; this case will be considered only to the extent needed to show how the internal wave motions may be affected. Because of the extra degree of freedom, the dispersion relation is now a quadratic in c^2; for each k, two different wave modes may exist. If one now adds the Boussinesq approximation (for the first time in this chapter restricting the density variation), it can be shown that the extra root is approximately

$$c_1^2 = \frac{g}{k}\tanh k(h_0 + h_1), \qquad (2.1.9)$$

and c_2^2 is of the form (2.1.7). The first wave (the 'barotropic' mode) is identical with a surface wave on a layer of constant density with depth $(h_0 + h_1)$. It has particle velocities decreasing with depth in the same way, with no discontinuity of tangential velocity at the internal boundary, and this remains true of the *surface* mode for an arbitrary distribution of density with depth.

The second mode with phase velocity c_2 is the internal wave, with largest amplitude and a discontinuity in u at the interface. Just below the free surface is a level where the motions vanish, and the phase of the displacements at the surface is opposite to that at the interface. The horizontal velocity u at the surface can be large, and regions of maximum convergence of u lie above the nodes of the interface displacement. The resulting gathering of contaminants (or some other process which changes the reflectivity at a convergence) can produce visible lines or 'slicks' which are an indication of the presence of two-dimensional internal waves (La Fond 1962, and see fig. 2.3 pl. IV). The vertical displacement of the free surface is smaller than that at the interface by a factor of the order $-\rho'/\rho_0$; for many geophysical applications this is negligible, and the surface can be treated as a rigid lid as far as the internal waves are concerned.

This last statement can, however, be turned round the other way: because of the much smaller potential energy changes needed to produce an internal wave, *imposed* disturbances at the surface can be very effective generators of internal waves. The phenomenon of 'dead water', an increased resistance to the motion of a ship when it is steaming on a thin light surface layer overlying denser water, can be explained in this way, as was shown in a classic series of experiments by Ekman (1904). So long as the speed is less than $c = (g'h_0)^{\frac{1}{2}}$ there will be an extra drag due to an internal wave which moves with the ship, but if the speed can be increased beyond this value, only the surface mode can be generated and the drag is much reduced.

2.1.3. Standing waves

The properties of internal standing waves, in which sections of an interface oscillate vertically between fixed nodes, can be derived by superimposed identical wavetrains travelling in opposite directions, and few specific results need be quoted here. There is more to be said about the finite amplitude case (§3.1.1) but we should note one result which is used in §4.3.2: the maximum shear occurs now at the nodes i.e. the position of maximum slope (instead of at the crests and troughs as for progressive waves). The second wavetrain can be formed by reflection of a travelling wave from a solid barrier, and so standing waves (called surface or internal seiches) are of interest especially in closed basins such as lakes. One is thus led to study the free modes of vibration for various density distributions and basin shapes, solving an eigenvalue problem to determine the normal modes and associated resonant periods.

In the simplest case of longitudinal oscillations of two layers contained in a rectangular basin which is narrow but much longer than its depth, the long wave approximation is appropriate for the first few modes, the relevant wavelength for the seiche with one nodal line being twice the length of the lake. The two-layer approximation is often a good one here, since typical observations in lakes (or the ocean) show a well-mixed warm surface layer separated from a colder homogeneous region by a shallow *thermocline*, or layer of high gradient. The periods of all the internal seiches of this kind are (cf. (2.1.7))

$$T = \frac{2l}{n}\left(g\frac{\rho'}{\rho_0}\frac{h_0 h_1}{h_0 + h_1}\right)^{-\frac{1}{2}}, \qquad (2.1.10)$$

where n is the number of nodal lines. The motions are predominantly horizontal and of opposite sign above and below the interface, and again the vertical displacements at the surface are smaller than those at the interface by a factor ρ'/ρ_0, with opposite senses of tilt at these two levels.

A seiche can be set up by wind blowing along a lake, which piles surface water up at the leeward end. At the same time the bottom layer comes closer to the surface to windward, driven by a horizontal pressure gradient in the opposite direction to that in the surface layer, and it may even reach the surface there (see fig. 2.4). This quasi-steady state (which when the wind drops leads to the seiches) is complicated by the return circulations set up in each of the layers. It should be noted too that seiches set up in this way can be of such large amplitude that non-linear theory must be used to give a full description of them (cf. §3.2.2); a particularly good example in Loch Ness has been studied recently by Thorpe, Hall and Crofts (1972).

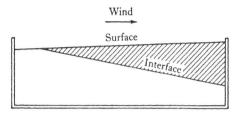

Fig. 2.4. The mechanism of generation of an internal seiche.

Proudman (1953) has shown how such calculations can be extended to basins of variable width, and to those in which the transverse oscillations must also be considered. Cases where there are
several superposed layers, rather than just two, are also little different in principle, with another period and associated mode of
oscillation in the vertical becoming possible each time a layer is
added. For example, Mortimer (1952) obtained good agreement
between observed and computed periods of internal oscillations in
Lake Windermere by approximating to the density distributions
using three layers. As discussed in chapters 8 and 10, step-like
distributions are often observed even in large bodies of stratified
fluid (such as the ocean), and the multi-layer wave theory will also
be relevant there. When the lowest modes dominate, of course, the
motions will be little affected by the presence of thin layers, though
the microstructure will influence measurements of temperature
fluctuations made at fixed depths (Phillips 1971, Garrett and
Munk 1971).

When the density distribution becomes truly continuous, an
infinite set of modes becomes possible and this leads us to the subject of the next section, where a simple example is given. A method
of practical computation of modes for an arbitrary density distribution was developed by Fjeldstad (1933), but neither this nor more
recent work on this problem will be described here.

2.2. Waves in a continuously stratified fluid

Infinitesimal motions in a continuously stratified incompressible
fluid, otherwise at rest, are described by (1.3.1), (1.3.2) and (1.3.8),
in which arbitrarily large variations of mean density with height are

retained, but the non-linear terms are neglected because they are products of two small quantities. For two-dimensional motions, the component equations can be written as

$$
\left.
\begin{aligned}
\frac{\partial u}{\partial t}+\frac{1}{\rho}\frac{\partial p'}{\partial x} &= 0, \\[1ex]
\frac{\partial w}{\partial t}+\frac{1}{\rho}\frac{\partial p'}{\partial z}+g' &= 0, \\[1ex]
\frac{\partial u}{\partial x}+\frac{\partial w}{\partial z} &= 0, \\[1ex]
\frac{\partial g'}{\partial t}-N^2 w &= 0,
\end{aligned}
\right\}
\qquad (2.2.1)
$$

where $\rho = \rho(z)$ (now assumed to be continuous), and N, defined by

$$
N^2 = -(g/\rho)(\partial\rho/\partial z),
$$

is the buoyancy frequency. The fourth equation indicates that density variations at a fixed point are due entirely to the vertical displacement, and are 90° out of phase with the vertical velocity.

Two complementary approaches to this problem are instructive and we will first consider the one which emphasizes the similarity with the problem of waves in layered fluids.

2.2.1. *Description in terms of modes*

Various forms of equations for velocity components or a stream-function can be obtained from the above set by the successive elimination of the variables; one such reduction will be made in §2.2.2 after adding the restriction of the Boussinesq approximation, but this is not a necessary restriction. It is already clear that wave-like solutions of (2.2.1) are possible, with u, w, p', and g' depending on time through a factor $e^{i\omega t}$. An especially simple form of solution has been obtained for the case of an exponential density distribution, say

$$
\rho = \rho_s e^{-z/H} \qquad (2.2.2)
$$

(which gives $N^2 = g/H$, a constant), since exponential factors can

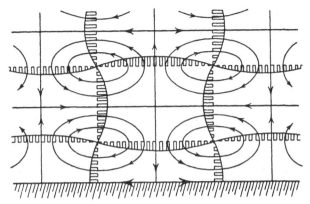

Fig. 2.5. Displacements and streamlines in a cellular standing internal gravity wave. (From Prandtl 1952.)

appear in each of the other terms. Only the velocity components are written down here following Prandtl (1952):

$$u = a\omega\, e^{z/2H} \cos kx \left(\frac{1}{2Hk} \cos mz - \frac{m}{k} \sin mz \right) e^{i\omega t},$$

$$w = a\omega\, e^{z/2H} \sin kx \cos mz\, e^{i\omega t},$$

$$(2.2.3)$$

where a is the (small) amplitude, and the real part is implied in each case. The associated pattern of displacements is drawn in fig. 2.5.

These expressions represent a cellular standing wave, with horizontal and vertical wavelengths $\lambda_h = 2\pi/k$ and $\lambda_v = 2\pi/m$ which are twice the distance between the nodal lines (i.e. twice the width or height of the cells shown). The close correspondence with the modal description of waves on layers is evident; this pattern can be fitted in a closed rectangular region whose boundaries coincide with the planes of no normal motion, and there can be an arbitrary number of cells in the horizontal and vertical. The cells are identical in size, but the amplitude increases upwards in such a way that the energy (proportional to $\rho(u^2 + w^2)$) is the same in each. This is a particular example of the result quoted in §1.3; the velocity is increased by a factor $(\rho_s/\rho)^{\frac{1}{2}}$ compared with that calculated by assuming small variations of density (small z/H). There is a definite frequency associated with each mode, namely

$$\omega = N \left(\frac{k^2}{k^2 + m^2 + 1/4H^2} \right)^{\frac{1}{2}}. \qquad (2.2.3a)$$

In the limit where H or the scale height H_s are large, the same form
of result applies to gravity waves in an isothermal compressible
atmosphere (another case of an exponential density variation), but
in the compressible case $N^2 = [(\gamma - 1)/\gamma](g/H_s)$ (see (1.2.5)).

Internal standing waves with a variety of modal structures have
been produced in a laboratory tank by Thorpe (1968a), and several
of these are illustrated in fig. 2.6 pl. 1. The bands of colour are
merely markers in a continuous density gradient, which bring out
clearly the property common to this and stepped structures, namely
the vertical oscillation of density surfaces (in phase over the whole of
a cell for standing waves).

2.2.2. Description in terms of rays

We now turn to the second way of describing internal wave motions
in a fluid with continuous stratification. Though this can be shown
to be entirely equivalent to the mode analysis, it is especially illu-
minating when considering waves produced by a small localized
disturbance, propagating through an environment with gradually
varying properties. As shown first by Görtler (1943) and redis-
covered by Mowbray and Rarity (1967), the oscillatory motions can
then be concentrated in bands bounded by 'rays' originating at
the edges of the source. Fig. 2.7 pl. II, taken from the latter paper,
shows the effect in a striking way; we will return to a detailed discussion of these experiments later.

The equations (2.2.1) will be used here in the Boussinesq approximation, setting $\rho = \rho_0$, a constant. The elimination of the pressure
from the first two by cross differentiation then gives

$$\frac{\partial}{\partial t}\left(\frac{\partial u}{\partial z} - \frac{\partial w}{\partial x}\right) - \frac{\partial g'}{\partial x} = 0. \qquad (2.2.4)$$

This form shows that only horizontal gradients of buoyancy are
responsible for changing the y-component of vorticity

$$\zeta_y = \frac{\partial u}{\partial z} - \frac{\partial w}{\partial x},$$

in agreement with the general result (1.3.5). Further differentiation and combination of the equations can be used to eliminate g' and u to give a single equation for w,

$$\frac{\partial^2}{\partial t^2}\left(\frac{\partial^2 w}{\partial x^2}+\frac{\partial^2 w}{\partial z^2}\right) + N^2(z)\frac{\partial^2 w}{\partial x^2} = 0, \qquad (2.2.5)$$

in terms of which it is convenient to continue the analysis. (Of course corresponding equations can be derived for the other variables, whose solutions must have the same time and space dependence.) Assuming a plane progressive wave solution of the form

$$w = \hat{w}(z)\exp \mathrm{i}(kx - \omega t)$$

and substituting into (2.2.5) gives for $\hat{w}$

$$\frac{\mathrm{d}^2\hat{w}}{\mathrm{d}z^2} + \left(\frac{N^2}{\omega^2} - 1\right)k^2\hat{w} = 0. \qquad (2.2.6)$$

Instead of applying the boundary conditions round the edge of a closed region (which was the procedure used in the 'mode' approach to derive (2.2.3) for example), consider the implications of (2.2.6) when a small local disturbance is applied to the stratified fluid. The solutions of this equation only have a wave-like character when $\omega < N$; energy can then propagate away from the region where it is generated. If $\omega > N$ the disturbances remain local and fall off exponentially away from the source, and waves cannot exist. Thus the buoyancy frequency N can be interpreted as the upper limit of frequency for which wave motions can exist in a stratified fluid. In a fluid with variable N, waves will be trapped within the layer where N exceeds the imposed frequency (see (2.2.11) and the following discussion), but sharp interfaces can sustain waves of arbitrary frequency, since N is then very large.

Now let us specialize further to the case where N^2 is constant (i.e. to a linear density gradient in the Boussinesq approximation), and $\hat{w} = \overline{w}\mathrm{e}^{\mathrm{i}mz}$ is also periodic in z. It follows that the dispersion relation is (c.f. (2.2.3))

$$\omega = N\left(\frac{k^2}{k^2+m^2}\right)^{\frac{1}{2}}, \qquad (2.2.7)$$

or $$\omega = N\cos\theta, \qquad (2.2.8)$$

where θ is the angle ($|\theta| \leqslant \frac{1}{2}\pi$) between the horizontal and the resultant wavenumber vector $\mathbf{k}$ (whose components are k and m).

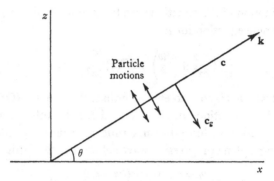

Fig. 2.8. The relation between the phase and group velocities and the particle motions for plane waves propagating through a continuously stratified fluid.

For a given stratification, therefore, waves with fixed $\omega < N$ propagate at a fixed angle to the horizontal independent of the wavelength, and this result is accurately verified by experiments of the kind shown in fig. 2.7 pl. II. The horizontal phase velocity $c_p = \omega/k$ clearly does still depend on the wavelength and hence on the scale of the phenomenon of interest.

The continuity equation implies that

$$i\mathbf{k} \cdot \bar{\mathbf{u}} = 0, \qquad (2.2.9)$$

so the particle velocities are all in lines perpendicular to $\mathbf{k}$ (and parallel to the direction of the wavefronts – see fig. 2.8). For a motion of this kind the non-linear terms $\mathbf{u} \cdot \nabla \mathbf{u}$ in the equations of motion are *identically* zero, and thus a single wave component is a solution of the full equations of motion. There is now no restriction on amplitude, but nevertheless the linear theory remains applicable.

Several particular cases are of interest. When $m = 0$, $\omega = N$ and the wavefronts and particle motions are vertical; this is in accord with the original definition of N as the natural frequency of a vertically displaced particle. A similar argument can be used to show that $N \cos \theta$ is the natural frequency of a particle displaced at an angle $(\frac{1}{2}\pi - \theta)$ to the horizontal. This then gives a physical reason why special directions are so important; when a frequency is imposed externally the fluid picks out that direction of oscillation which allows it to match the forcing motion. The other limit of low frequency (i.e. a small steady perturbation) must be such that $k^2 \ll m^2$. The motion is necessarily horizontal, and in phase over all x. In two dimensions, a solid body moved slowly in the x-direction will carry with it the layer of fluid bounded by tangent planes above and below, since there can be no vertical flow to carry fluid around it. This result (which is of course obtainable directly from the equations of motion in this limit) is analogous to the formation of 'Taylor columns' in rotating fluids, and will be referred to again in §3.3.2. (See also Veronis (1967).)

The group velocity is again significant in determining the direction of energy transfer; for waves propagating in two dimensions it has components (cf. (2.1.6))

$$\mathbf{c}_g = \left(\frac{\partial \omega}{\partial k}, \frac{\partial \omega}{\partial m}\right). \qquad (2.2.10)$$

Since ω depends only on the direction of $\mathbf{k}$ and not on its magnitude (2.2.8) the group velocity must be normal to $\mathbf{k}$ and hence to the phase velocity. Energy is transmitted away from a source along rays which coincide with the directions of the particle motions; these are characteristics of the differential equation derived from (2.2.1) and lie at an angle $\sin^{-1} \omega/N$ to the horizontal (see fig. 2.8). The magnitude of $\mathbf{c}_g$ is $|\mathbf{c}_g| = (N/|\mathbf{k}|)\sin\theta$ (using the notation of (2.2.8) and fig. 2.8), with components

$$\left.\begin{aligned} u_g &= N\frac{m^2}{(k^2+m^2)^{\frac{3}{2}}} = \frac{N}{|\mathbf{k}|}\sin^2\theta, \\ w_g &= -N\frac{km}{(k^2+m^2)^{\frac{3}{2}}} = -\frac{N}{|\mathbf{k}|}\sin\theta\cos\theta. \end{aligned}\right\} \qquad (2.2.11)$$

The horizontal components of $\mathbf{c}$ and $\mathbf{c}_g$ are thus directed in the same sense, while the vertical components are in opposite senses.

These ideas can be applied to fluids in which the density gradient varies slowly from point to point. One uses a method similar to that of geometrical optics, tracing the paths of rays whose directions are defined locally by (2.2.8). There are two conditions under which a group of waves can no longer propagate vertically; when either $\theta = 0$ or $\frac{1}{2}\pi$ the vertical group velocity (2.2.11) becomes zero. As $\theta \to 0$, the wavefronts are turned towards the vertical and the energy is reflected at the level where $\omega = N$, with a cusp in the ray path (fig. 2.9). Successive reflections of this kind ensure that the wave

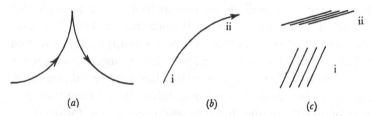

(a) (b) (c)

Fig. 2.9. Reflection and absorption of waves in a continuously stratified fluid. Paths of rays (a) in a weakening gradient, reflection where $\theta = 0$, $\omega = N$, (b) at a critical level, absorption when $\theta \to \frac{1}{2}\pi$, (c) the behaviour of a wave group in case (b).

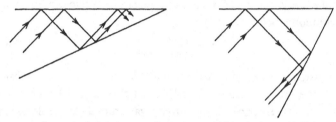

Fig. 2.10. Reflection conditions at solid boundaries. The angle which rays make with the horizontal stays constant.

energy is retained in the region where $N > \omega$ (in the ocean thermocline for example). As $\theta \to \frac{1}{2}\pi$ the wavefronts become nearly horizontal and $c_p \to 0$; this case is of special importance in a moving fluid, and will be discussed further in §2.3.3.

The reflection conditions at a solid boundary should also be mentioned here. The angle the rays make to the *horizontal* $(\frac{1}{2}\pi - \theta)$ must be the same for the incident and reflected rays, since this depends only on frequency (2.2.8) regardless of the angle of the boundary β. If $\beta > \frac{1}{2}\pi - \theta$ energy will be reflected back, with the horizontal component of group velocity in the opposite direction to the incident ray, but if $\beta < \frac{1}{2}\pi - \theta$ it will continue in the same direction after reflection (fig. 2.10). For a given frequency, a shallow enough wedge acts as a perfect absorber of internal waves, since energy cannot escape backwards. From the geometry of adjoining rays, it is clear that the wavelength and velocity do change on reflection (except at vertical or horizontal boundaries) and in a wedge the wavelength will decrease and the amplitude grow. This problem has been treated in detail by Wunsch (1969), including the singular case where $\beta = \frac{1}{2}\pi - \theta$. Eventually the waves must break and the energy is absorbed by viscous dissipation (see §4.3.4).

2.2.3. Laboratory experiments on waves in bounded regions

It is now possible to explain most, but not all, of the features of experiments such as that pictured in fig. 2.7 pl. II. In this section these and related observations will be described in more detail, with the additions needed to give a more complete understanding. To produce this picture Mowbray and Rarity (1967) used an elegant optical (Schlieren) technique to detect horizontal density gradients caused by the propagation of two-dimensional waves away from an oscillated cylinder. Similar patterns are seen when aluminium particles suspended in the flow are illuminated from the top (see fig. 2.11 pl. II); in that case the orientation of the particles, and the light reflected from them, is a measure of the local shear. Experiments using either of these techniques verify the main predictions of the ray theory. In a uniform density gradient, the dominant particle motions are in straight lines and confined to bands lying between rays emanating from the edges of the wavemaker, at an angle $\sin^{-1} \omega/N$ to the horizontal. Reflections from the vertical and horizontal boundaries can be seen in fig. 2.11, with a steepening of the ray paths at the top and bottom (because the condition that there can be no diffusion through a boundary reduces the density gradient to zero there, and the 'cusp' condition for reflection is approached).

One also sees in both these experiments an optical manifestation of crests and troughs (i.e. brighter and darker lines) moving across the bands, in a direction normal to the particle motion. These are clear evidence for a phase velocity in the predicted direction, but cannot be explained purely in terms of motions in phase with the plunger, which would cause the whole of a band to move as a solid plug with a shear layer at its edge. There are also seen to be smaller oscillatory motions in the regions outside the bands, and both of these effects can be described by adding a wave component which is 90° out of phase. This also allows the boundary conditions to be fitted at the wavemaker and round the whole perimeter of a closed region.

At first sight it seems inconsistent to attempt a description of a

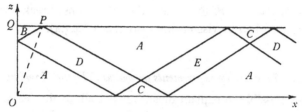

Fig. 2.13. The flow regions associated with the oscillation of a segment PQ of the boundary in a closed box (see text).

boundary value problem in terms of the ray theory, but the fact that it is possible underlines the duality of the two approaches. The mode theory emphasizes the free oscillations of a region with given fixed boundaries, whereas ray theory is appropriate when the motion is forced by oscillating part of the boundary (though the rest of the boundary may be fixed). Both features are brought out in the photograph of fig. 2.12 pl. II due to Thorpe (1968a). He was generating various modes in a container by oscillating a paddle at one end, but a band of strong motions propagating away from the edge of the wavemaker in accordance with ray theory is also clearly seen. Such problems with mixed boundary conditions have been solved completely in several cases: as noted above it has been done for a wedge (Wunsch 1969, Hurley 1970), for waves in a channel passing a thin vertical barrier (Larsen 1969a and Sandstrom 1969), and for the emission of internal waves by vibrating slender cylinders (Hurley 1969). Baines (1969) has treated the case where a section PQ of the upper boundary at one end of a semi-infinite channel is oscillated vertically (see fig. 2.13); his results will be described briefly here since they correspond closely to the experiment shown in fig. 2.11 pl. II in which the wavemaker was a solid piston.

Baines showed how the boundary conditions round the solid walls plus that of outward propagating energy towards infinity can be satisfied using a sum of internal wave modes plus a barotropic motion, all with frequency ω. The net effect of the summation varies strongly with position however. In the regions D and E (fig. 2.13) the in-phase component is a block motion along the characteristics, i.e. described by ray theory; there is vertical motion under the piston in B, and horizontal motion in C. In each of these regions there is also an out-of-phase component, which has a logarithmic infinity at the edge of D and E because of the discontinuity in the forcing motion (which of course is removed by viscosity in a real fluid). When these two components are added, a line of no motion is found to move across the bands from left to right during a cycle, and it is this which can be observed visually and interpreted as the motion of a crest. Outside the regions covered by the beam and its reflections (i.e. in A) only the out-of-phase motion is present. Streak photography of neutrally buoyant particles has confirmed these predictions for the various regions at different phases of the cycle. A special case is worth mentioning; when the frequency is such that the characteristic from the corner of the piston at P goes to O, only the regions B and C are then present. All the internal wave modes cancel, and the barotropic mode remains, with vertical motion in B and horizontal in C, and a strong shear layer between. This motion can conveniently be produced in a channel of finite length by having a second plunger at the other end, oscillating in antiphase.

Transient problems of wave generation have also received some attention. Rarity (1967) has studied the waves produced by a two-dimensional body moving through a stratified fluid in a straight line at an arbitrary angle to the horizontal. Experiments confirming the essential predictions of this theory have been reported by Stevenson (1968), who has also considered the three-dimensional problem. Larsen (1969b) has observed the free oscillations of solid spheres displaced in a fluid having a constant density gradient. It is relevant to mention this here, since he can explain the measured damping of the oscillation entirely in terms of the radiated waves, with only a minor contribution to the drag coming from the viscous boundary layer.

2.3. Waves in a moving stratified fluid

2.3.1. *Velocity constant with height*

We will now discuss the additional effects which become possible when infinitesimal two-dimensional waves are propagating in a *moving* fluid with continuous stratification (though some recent developments in this subject must be left to §3.1.4, where non-linear processes are considered). When the fluid velocity is constant with height, the first extension is a relatively simple one. Any travelling wave mode in a region bounded by a flat bottom can have a constant velocity superimposed on it; if this is chosen to be equal and opposite to the horizontal phase velocity, then another kind of stationary wave results, with fluid moving through the crests over the whole region $-\infty < x <\infty$. (The cellular wave solution (2.2.3), for example, can be extended in this way.) Of more practical importance is the generation of a stationary pattern of *lee waves* by an obstacle placed in the flow (e.g. a mountain at the bottom of the atmosphere). The choice of the proper sum of wave-like and decaying motions which satisfy the imposed boundary conditions is the problem here, and results are given below for an inviscid, Boussinesq liquid having constant N^2. Some subtle points which arise in the complex variable analysis used to obtain them will be ignored here.

Consider first a flow with velocity U confined between fixed planes a distance H apart in the vertical, containing stationary waves produced by an obstacle of (small) height h and width b placed at the origin. The phase velocity of the waves must be upstream relative to the fluid, and such that $c_p = - U$. The group velocity also has an upstream component; moreover it follows from (2.2.7) and (2.2.11) that the horizontal component of c_g is less than c_p for all possible modes trapped between $z = 0$, $z = H$, and so energy is being swept downstream. Thus the waves are indeed 'lee waves' in this situation, since no energy is available to sustain them upstream.

When the flow of interest is steady relative to fixed axes, the time derivatives in (2.2.5) may be replaced by the spatial derivatives (i.e. one can set $\mathrm{D}/\mathrm{D}t \equiv U \partial/\partial x$) which gives

$$U^2\frac{\partial^2}{\partial x^2}\left(\frac{\partial^2 w}{\partial x^2}+\frac{\partial^2 w}{\partial z^2}\right)+N^2\frac{\partial^2 w}{\partial x^2} = 0. \qquad (2.3.1)$$

The linearization here depends on h/H being small, and this also implies that the boundary condition at the obstacle, which is $w = U\mathrm{d}h/\mathrm{d}x$, can be applied at $z = 0$, not on the raised obstacle itself. The other boundary conditions are $w = 0$ at $z = H$, and $w = 0$ as $x \rightarrow -\infty$ for all z; no assumption is made about w as $z \rightarrow +\infty$. From (2.3.1) one can derive an equation equivalent to

(2.2.6) for the amplitude $\hat{w}(z, k)$ of a wave with horizontal wave-number k, namely

$$\frac{\partial^2 \hat{w}}{\partial z^2} + \left(\frac{N^2}{U^2} - k^2\right) \hat{w} = 0. \tag{2.3.2}$$

It is clear that any motion with $k = 0$ will also satisfy (2.3.1); these modes, which are called 'columnar' because the motions are uniform in the x-direction, do not enter in the steady linear theory, but they will be referred to again in §3.1.3.

The kinds of waves which are possible in this flow can best be understood by examining the case where $\hat{w}$ is set equal to zero at $z = 0$ as well as at $z = H$ (i.e. the free wave modes in the channel with the obstacle removed). The vertical velocity has non-decaying (sinusoidal) solutions satisfying these simpler top and bottom boundary conditions only for a discrete set of $k = k_n$ such that

$$(k_0^2 - k_n^2)^{\frac{1}{2}} H = n\pi, \tag{2.3.3}$$

where $k_0 = N/U$ and n is an integer. There is a finite number of separate wavetrains (i.e. k_n), which can exist with a given flow and stratification when there are rigid boundaries above and below. The behaviour of this system thus depends largely on the magnitude of the parameter $Ri_0 = F^{-2} = k_0^2 H^2 = N^2 H^2 / U^2$, which has the form of an overall Richardson number (§1.4). If this is very small (e.g. U is too large for a given N), no stationary lee waves can form, since all such waves are then swept downstream. As Ri_0 increases above π^2 (or F falls below π^{-1} and the flowspeed becomes *subcritical* for the particular mode in question) first one, then two wavenumbers can satisfy (2.2.3), and so on.

These conclusions are not changed when a shallow obstacle is placed in the flow and the forced motions are of interest, but the information about its shape has not yet been used; this enters through the application of the proper boundary condition on (2.3.2), which fixes the amplitudes of all the components. As we will see again in chapter 3, for an infinitesimal obstacle the amplitude is proportional to hb, i.e. to its cross-sectional area, rather than just its height. When b is increased to finite values, the wave amplitude will be largest when $k_n b$ is of order unity. This condition implies that the width of the obstacle is comparable with the wavelength of one of the possible lee wave modes, thus ensuring that the forcing frequency is close to a natural frequency of the system. This can happen in the atmosphere for commonly found values of U and N^2, and some mountain ranges with just the right width to give this kind of resonant excitation of the lowest mode k_1 frequently have prominent trains of lee waves associated with them.

The atmosphere, of course, differs from the above model in that it is virtually unconfined at the top, rather than having a rigid upper lid, but the arguments can be extended to that case without difficulty. (Historically important solutions were obtained for instance by Queney (1941, 1948) and Lyra (1943) for exponential density distributions and special obstacle shapes, but these will not be discussed in detail.) Both exponentially decaying and oscillatory types of solution in z can be relevant here. The latter must always be chosen to have a form which corresponds to the *upward* propagation of energy, and if for practical reasons the calculations is cut off at some fixed upper level, this outward radiation condition must always be applied there. The continual radiation of energy upwards (which spreads it through a large volume) means that the amplitude of the waves decreases with increasing x at large distances downstream, even without friction, in contrast to the case of a rigid upper lid where the energy is reflected back and remains in the flow. In the unbounded case, the limit of (2.3.3) as $H \to \infty$ implies that lee waves should be possible for any wind speed (since $Ri_0 \to \infty$ provided $k_0 \neq 0$), but in practice lee waves are not observed if $k_0 b$ is small. Another overall Richardson number based on the dimensions of the obstacle instead of H becomes relevant and the square root of this, $\kappa = Nh/U$, is the parameter used in the non-linear theory of §3.1.4. Detailed flow patterns will be described in that section.

2.3.2. Lee waves with varying properties in the vertical

If the horizontal velocity u is now allowed to be a function of height z instead of constant, an extra term $w\,\partial u/\partial z$ appears in the horizontal momentum equation (2.2.1 a), representing the vertical convection of mean horizontal momentum by the perturbed flow. Proceeding as before, it follows that the equation for the amplitude $\hat{w}$ (replacing (2.3.2)) is

$$\frac{\partial^2 \hat{w}}{\partial z^2} + \left(\frac{N^2}{u^2} - \frac{1}{u}\frac{\partial^2 u}{\partial z^2} - k^2 \right) \hat{w} = 0,$$

or

$$\left. \frac{\partial^2 \hat{w}}{\partial z^2} + (l^2(z) - k^2) \hat{w} = 0 \right\} \qquad (2.3.4)$$

with

$$l^2(z) = \frac{N^2}{u^2} - \frac{1}{u}\frac{\partial^2 u}{\partial z^2}.$$

Provided $l^2 > 0$, the behaviour is little changed by the addition of the extra term; in a flow with $\partial u/\partial z$ decreasing upwards, this just acts to increase the restoring force on a displaced particle (i.e. like an increased density gradient). In cases of practical interest the effect of allowing N also to be variable is small and so $l(z)$ can be assumed to contain this variation as well.

Much of the detailed work based on (2.3.4) has assumed for the convenience of calculation that the stratified fluid (e.g. the earth's atmosphere) can be approximated by a series of layers in which l^2 is step-wise constant, reducing the problem again to one with constant coefficients. Scorer (1949) considered two layers, the lower having depth h and $l = l_1$ and the upper very deep with $l = l_2$, and connected them by requiring both components of velocity to be continuous at the boundary. A mode analysis then showed that lee waves are possible provided

$$l_1^2 - l_2^2 > \pi^2/4h^2, \qquad (2.3.5)$$

i.e. provided l decreases upwards sufficiently rapidly. The waves obtained in this way have the character of an interfacial wave on the boundary between the regions, and the lower layer acts like a wave-guide along which this can propagate. (See §2.2.2.)

Various three-layer models such as those of Scorer (1954) and Corby and Wallington (1956) have extended these ideas to find particular combinations of layer depths and l values which will lead to large lee waves amplitudes. One more recent example will be given. Pearce and White (1967) have made numerical calculations for the case of an atmosphere consisting of two layers with the same l^2, above and below a thin 'inversion' layer in which $l^2 = l_3^2$ is larger. The possible wave patterns now depend strongly on the

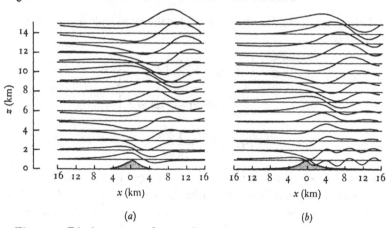

Fig. 2.14. Displacements of streamlines over a ridge, computed for (a) an airstream for which l^2 is nearly independent of height, (b) a flow for which l^2 is substantially larger in the lowest 2 km. (After Sawyer 1960. These figures are Crown copyright and are produced with the permission of the Controller of H.M.S.O.)

waveguide effect of this intermediate stable layer. For a given l_3 and barrier shape, the maximum wave amplitude near the inversion layer has a sharp maximum as a function of its depth. In the range where the amplitude decreases as the depth decreases, the wavelength of the possible waves is increasing. This is consistent with the observed behaviour of atmospheric lee waves, which weaken when the lower layers are heated, thus reducing the inversion thickness through the process of turbulent entrainment from below. (See chapter 9 for a further discussion of this last point.)

It is by no means necessary to restrict the discussion to these special and somewhat artificial variations of l^2 with height. Sawyer (1960) has computed the streamlines for the flow of observed air streams (as well as some idealized flows) over a two-dimensional barrier of the form $z = h_0/(1+(x/b)^2)$, a shape first used by Queney (1941) and incorporated in many subsequent calculations. He was able to deal numerically with continuous variations of l^2, cutting the calculation off at the top by assuming a small constant l^2 above this and using the outward radiation condition. In various circumstances he found lee waves diminishing in amplitude downstream, or staying substantially constant when there are very stable layers near the ground, and two examples are given in fig. 2.14. Note that the line of the crests slopes backwards (against the flow) with increasing height, which is a general property of lee waves. This is consistent with a group velocity and propagation of energy relative to the fluid directed in that same sense, upstream and away from the obstacle (see § 2.2.2). Sawyer confirmed the conclusions arrived at by studying simpler models, and added more information about the behaviour at greater heights. When l^2 increases upwards, waves can maintain a large amplitude up to the stratosphere (the inertial effect of density variation i.e. the factor $(\rho_0/\rho)^{\frac{1}{2}}$ is significant here), though these may decrease in amplitude downstream. If two wavelengths are possible, the shorter is emphasized near the ground and the longer at higher levels.

It should also be noted that problems related to lee waves can arise when there is a layer of convecting fluid below a stably stratified region, rather than a solid obstacle. The large eddies of the convection may then be regarded as moving relative to the stable layer, and generating waves in it. (See §§ 7.3.4 and 9.2.4.)

2.3.3. *Reversals of velocity and critical layers*

It was implicitly assumed above in discussing (2.3.4) that $u \neq 0$, since then the equation becomes singular. When the flow velocity reverses direction with height (so that the mean horizontal velocity $u(z)$ relative to the obstacle is zero at some level), important effects arise which require separate treatment. These will be sketched using the concepts of ray theory and group velocity (§ 2.2.2), following the method used by Bretherton and Garrett (1968) and in previous work referred to by them. They have provided rigorous justification for the ray tracing procedure, and have shown that a medium may be regarded as locally uniform provided the gradient Richardson number is large and the vertical wavelength small compared to the scale of variation of U and N in the region of interest.

When fluid is moving, the frequency of a wave measured relative to a fixed point is no longer given by (2.2.7); that is the frequency relative to the fluid, but it is changed by the advection of wave crests past the point to become

$$\omega = ku \pm Nk(k^2 + m^2)^{-\frac{1}{2}}. \qquad (2.3.6)$$

It is this frequency which is imposed at the source and which must remain constant following a wave group. The rate of transmission of energy relative to the fluid (i.e. the group velocity) is unchanged by the locally uniform motion so the total group velocity relative to the ground again has components $(\partial\omega/\partial k, \partial\omega/\partial m)$ and (2.2.11) are modified only by adding the local wind speed to the horizontal component.

Near a level where

$$\omega_1 \equiv \omega - ku = 0, \qquad (2.3.7)$$

i.e. where the phase velocity ω/k equals the velocity of flow, the wave will not be propagating at all through the fluid; this occurs for stationary lee waves when $u = 0$ since the mechanism of their formation ensures that the phase velocity relative to the ground is zero. As such a critical level is approached, (2.3.6) shows that m must become large. The wavefronts become nearly horizontal and closer together, but in such a way that the horizontal wavenumber k is unaltered following a group. (See fig. 2.9c.) Both components of group velocity become zero, relative to the fluid at the critical level, and so the energy density (and amplitude of the waves) increases to large values here. This is consistent with a general result proved by Bretherton and Garrett (1968): it is no longer the energy density E, but E/ω_i which is conserved following a wave group, where ω_i (defined by (2.3.7)) is called the intrinsic frequency of the wave relative to the local mean flow. More detailed studies of the critical layer have been reported by Booker and Bretherton (1967) and Hazel (1967); these confirm that, when the gradient Richardson number is greater than about unity, very little wave energy gets through the critical level.

The consequences of this concentration of energy (and of the associated absorption of the momentum flux) will be explored further in chapters 4 and 10. It suffices to say now that the momentum lost by the waves can act to accelerate the mean flow, and there can also be instabilities associated with the increased wave amplitudes. The main predictions of this theory of the critical layer have been verified in a simple laboratory experiment (Hazel 1967). A shear flow can be produced in a long closed tube containing stratified fluid by tilting it about a horizontal axis (see chapter 4 where this method is described more fully in another context). A linear stratification produces a linear velocity profile with zero velocity at the central depth. A small obstacle at the bottom of the flow produces lee waves, made visible in fig. 2.15 pl. IV by thin lines of dye placed at regular height intervals. In accordance with the prediction waves reach the centre of the channel but cannot pass through the level where $u = 0$. There is also the suggestion in this experiment of a distortion of the velocity profile corresponding to the injection of momentum at the critical level.

2.4. Weak non-linearities: interactions between waves

The success of linear theory in describing the various phenomena mentioned above gives confidence in the basic assumption that the component waves can be regarded as independent. If this were not so, and energy were transferred rapidly and indiscriminantly between the modes because of non-linear effects which have been omitted, no simple pattern could remain recognizable for long. There is one circumstance, however, in which even a weakly non-linear process can have an important result, and this will be described next. In this extension each individual wave can still be treated by linear theory, but when several waves satisfy certain resonance conditions, energy is interchanged preferentially between them. The other extreme case of strong non-linear interactions, where no such selective effect operates, has more of the characteristics of turbulence, and this will be discussed in later chapters and also briefly in §2.4.3.

2.4.1. *The mechanism of resonant interaction*

All interaction calculations require much tedious algebra to complete the details but the essential physical argument can be given more simply and without at first referring to a particular problem. Consider any system which can be described on linearized theory in terms of a sum of undamped waves $\sin(\mathbf{k} \cdot \mathbf{x} - \omega t)$, and for which the dispersion relation is known, say

$$F(\mathbf{k}, \omega) = 0. \tag{2.4.1}$$

Suppose now that two waves specified by $(\mathbf{k}_1, \omega_1)$ and $(\mathbf{k}_2, \omega_2)$ are of small enough amplitude for each to be adequately described in this way, but that the lowest order cross-product term is retained. That is, the equation describing, for example, the displacement of an interface will now contain an extra term $\sin(\mathbf{k}_1\cdot\mathbf{x}-\omega_1 t)\cdot\sin(\mathbf{k}_2\cdot\mathbf{x}-\omega_2 t)$. This can be regarded as a forcing disturbance acting on the linearized system, and the problem of 'interaction' consists in calculating the response to this weak forcing. As for any vibrating system, the response must depend on the relation between the natural and forcing frequencies.

Now the above product (forcing) term can be expressed as sum and difference waves with wavenumbers and frequencies

$$\left.\begin{aligned}\mathbf{k}_3 &= \mathbf{k}_1 \pm \mathbf{k}_2,\\ \omega_3 &= \omega_1 \pm \omega_2.\end{aligned}\right\} \qquad (2.4.2)$$

Each combination having one of the wavenumbers $\mathbf{k}_3$ will tend to generate secondary waves which are possible free modes of the system, that is, which have frequencies ω_j satisfying $F(\mathbf{k}_3,\omega_j)=0$, and move with the phase speed $\omega_j/|\mathbf{k}_3|$. In general none of the ω_j will equal ω_3 given by (2.4.2), and so the forcing and the secondary waves will not remain in phase and the energy transfer will remain small. If, on the other hand, $\omega_3=\omega_j$ and one of the combination waves moves at the same speed as a free oscillation of the same wavenumber, resonance occurs; the waves stay in phase and a continuous transfer of energy can take place. The growth is limited by the amount of energy available in the primary waves, by the closeness of ω_3 and ω_j and by viscosity in a real fluid. Other factors which influence the time during which $\mathbf{k}_1$ and $\mathbf{k}_2$ overlap (such as the length of the wavetrains) are also relevant, but given long enough, such secondary waves can grow to amplitudes comparable with those of the primary waves, no matter how small the original amplitude.

This process, in which two waves interact to give a third, is called a *quadratic interaction*, and three waves which are related by (2.4.2) and separately satisfy the dispersion relation (2.4.1) constitute a *resonant trio*. If there are no other transfers, the energy in such a trio is just passed back and forth among the components (since the mechanism still operates if another pair of waves is regarded as primary). In some special cases the variations of all three components have been calculated explicitly as functions of time. It is a property of the weak interaction process that the period characteristic of the interchange is much longer than those of the separate wave components, and this period is inversely proportional to the amplitude of the interacting waves (Bretherton 1964). For a lucid general treatment of this problem, see Simmons (1969).

It should be noted that the wavenumbers used above are vector wavenumbers, and so in general waves travelling in different directions can interact. Hasselman (1966) has shown how a wide variety of resonant interaction processes can be conveniently discussed by constructing vector diagrams which are closely related to collision diagrams for elementary particles, but none of these more sophisticated results will be given here.

2.4.2. Interactions of interfacial waves

Particular examples of resonant interactions can now be given, starting with the simple case of interfacial waves in a two layer system studied by Ball (1964). He showed that the conditions for resonance can be satisfied by a triplet consisting of two surface waves $(\mathbf{k}_1, \omega_1)$, $(\mathbf{k}_2, \omega_2)$ travelling in opposite directions, and an interfacial wave $(\mathbf{k}_3, \omega_3)$ having the same direction as one of them. One such triplet is shown diagrammatically by the points $AB_1 C_1$ on the dispersion curves drawn in fig. 2.16. The two branches of the curve are just equations (2.1.7) for the internal mode and (2.1.8) for the surface mode, with particular choices of the density difference and layer depths. For two-dimensional waves this and a similar triplet $AB_2 C_2$ are the only possible ones involving wave A. When a general direction of propagation is allowed, however, wave A can interact with an infinite set of other surface waves (two for each choice of direction) and the result must again be an internal wave.

Herein lies the relevance of this process for internal waves, and the reason why it has been mentioned although surface waves *per se* are not of concern to us here. It is a possible mechanism for generating internal waves from surface waves on the ocean, which can operate over a wide range of conditions of the surface. It is especially significant when compared with purely surface gravity wave interactions, since Phillips (1960) has shown that those can only occur at higher order, between three components to give a fourth (i.e. that no

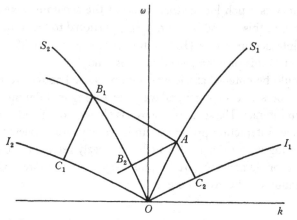

Fig. 2.16. Schematic representation of the conditions for resonant inter-action. The surface wave A belongs to the two interacting wave triplets $A B_1 C_1$ and $A B_2 C_2$. (From Ball 1964.)

resonant *trio* can be found using only one branch of the dispersion curve or surface).

The locus of surface wavenumbers of arbitrary direction, which can interact with a given wavenumber to give an internal wave, is shown in fig. 2.17 for two choices of the frequencies ω_1 and ω_2. As the difference in frequency (and therefore $\omega_3 = \omega_2 - \omega_1$, the frequency of the internal wave) becomes small, the distance between the two branches of this curve is reduced. Thus as shown by the vectors on the right-hand figure, one can get an interaction between two surface waves of nearly the same wavenumber (but different directions) to produce an internal wave of comparable wavenumber in a third direction, having the very low frequency characteristic of internal waves. It is also clear from this vector diagram that short waves travelling in nearly the same direction can interact to produce a much longer (smaller $|\mathbf{k}|$) wave, travelling nearly at right angles to the surface waves. The attenuation of the disturbances with depth should be least important in this latter case, and significant long waves on a shallow thermocline can arise even though the surface wave motions themselves fall off sharply with depth (see Thorpe (1966b) for a more detailed discussion of this problem).

No direct experimental demonstration of this surface–internal wave interaction has been given (experimenters have been deterred

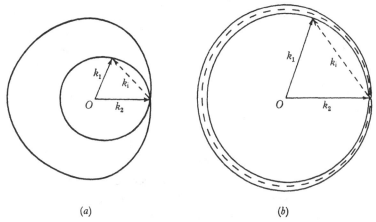

(a)   (b)

Fig. 2.17. Loci of surface vector wavenumbers $\mathbf{k}_1$ which can interact resonantly with $\mathbf{k}_2$ to give an internal wavenumber $\mathbf{k}_i$. The magnitudes of $\mathbf{k}_1$ and $\mathbf{k}_2$ are (a) unequal (b) nearly equal.

by the difficulties of absorbing the surface waves without disturbing the stratification). A comparable interaction has, however, been demonstrated between three wave modes all on the interfacial transition region separating two uniform layers. Davis and Acrivos (1967b) observed that a simple waveform generated at one end of a channel can under certain circumstances become unstable, in the sense that irregularities characteristic of higher modes spontaneously appear on the interface. (See fig. 2.18 pl. III.) They explained this behaviour by showing that resonant triads can exist for which the two waves interacting with the primary wave can begin with arbitrarily small amplitude, and both grow exponentially by extracting energy from the primary wave. (Hasselman (1967) showed that this is an example of a more general result, namely that the sum interaction of the pair described by (2.4.2) will always be unstable in this way.)

This can be illustrated by writing down the equations for the variation of amplitudes in the general form

$$\left.\begin{aligned}
\mathrm{d}a_1/\mathrm{d}t &= \alpha_1 a_2 a_3, \\
\mathrm{d}a_2/\mathrm{d}t &= \alpha_2 a_3 a_1, \\
\mathrm{d}a_3/\mathrm{d}t &= \alpha_3 a_1 a_2.
\end{aligned}\right\} \qquad (2.4.3)$$

Here the α_i are the interaction constants (whose calculation for particular systems will not be undertaken). Suppose that $a_1 = A$ is so much larger than a_2 and a_3 that initially it can be regarded as constant. Then (2.4.3) can be linearized and have solutions

$$a_2, a_3 \sim \exp\{ \pm A \sqrt{(\alpha_2 \alpha_3)} t \}. \qquad (2.4.4)$$

In a real fluid, the growth will be limited also by viscosity, and adding damping terms proportional to amplitude gives

$$\left. \begin{aligned} \mathrm{d}a_2/\mathrm{d}t &= \alpha_2 A a_3 - \beta_2 a_2, \\ \mathrm{d}a_3/\mathrm{d}t &= \alpha_3 A a_2 - \beta_3 a_3. \end{aligned} \right\} \qquad (2.4.5)$$

(For a discussion of the calculation of the coefficients β_2 and β_3 and other effects of viscosity see, for example, Thorpe (1968c).) Thus growth is possible provided $\alpha_2 \alpha_3 > 0$ and

$$A > A_{\mathrm{crit}} \equiv (\beta_2 \beta_3 / \alpha_2 \alpha_3)^{\frac{1}{2}}. \qquad (2.4.6)$$

Davis and Acrivos applied these results to the wave modes on a density profile of the form $\rho \propto \tanh(z/L)$ (some more details will be given in §4.3.3). They showed that the above conditions for growth can be satisfied by a trio consisting of a primary wave, the second mode moving in the same direction (the simplest description of the observed bulges on the interface shown in fig. 2.18) and a third mode component travelling in the opposite direction, which could be produced by reflections at the far end of the experimental tank. The measured relation between the frequencies agreed with that predicted from the dispersion relations using the observed interface thickness. They also showed that the critical amplitude criterion can be translated into a frequency condition, for fixed amplitude of the primary wave. Instability will only occur provided the frequency ω_1 is high enough, and a thick interface is more unstable than a thin one, for a given density difference. All of these statements are in agreement with their own experimental findings, and with some earlier observations reported by Keulegan and Carpenter (1961) which were at that time unexplained.

2.4.3. Interactions with continuous stratification

Thorpe (1966b) has generalized the arguments outlined above to include interactions which occur between internal wave modes with either multilayered or continuous stratification. The only restriction is that not all the waves of a resonant trio can belong to the same mode (for the same reason as purely surface wave quadratic interactions are excluded). Phillips (1968) combined the frequency condition for resonance ($\omega_3 = \omega_1 \pm \omega_2$) with the result (2.2.8) for bodily internal waves. Together these imply that the slopes of the three interacting wavenumbers to the horizontal must form a closed triangle

$$\cos\theta_1 \pm \cos\theta_2 = \cos\theta_3 \qquad (2.4.7)$$

and this geometrical property can be used with the summation condition on the wavenumbers to construct diagrams of the possible interactions with a given internal wave. Interactions between waves and currents can be included by regarding the latter as the limit of a wave as $\omega \to 0$ (Phillips, George and Mied 1968). It has been shown, for example, how periodic variations of density or current speed can lead to the trapping of internal waves in a finite layer of an unbounded fluid, essentially by a resonance mechanism.

Direct observations of a resonant interaction between three internal wave modes in a linearly stratified fluid have been made in a large laboratory channel by Martin, Simmons and Wunsch (1969). They produced the first and third modes using an array of paddles to give appropriate independent displacements at one end of the tank, and absorbed the energy at the other end using a 'beach' in the form of a shallow wedge having a small enough angle not to reflect any of the frequencies of interest (see §2.2.2 and Wunsch 1969). They tuned the paddle frequencies close (but not exactly) to the resonant condition and measured vertical displacements by means of conductivity probe arrays at various depths and distances from the wavemaker.

They found, in agreement with theory, that a fourth mode wave could be detected by numerical analysis of the spectra and cross spectra of the signals from the vertical arrays. Its amplitude grows not with time but with distance down the channel (because this determines how long the waves have been interacting) and its phase velocity is in the opposite direction to that of the original two. This mode is clearly distinguished in the periodograms because it has a much larger amplitude than any other peak corresponding to sums or differences of the original frequencies. There is also evidence of multiple resonances leading to higher modes, which have been more thoroughly documented in later experiments.

McEwan (1971) has reported an elegant experiment in which a resonant interaction was observed between a trio of standing internal waves in a rectangular tank of linearly stratified salt solution. One mode (generally that having one half wavelength in the vertical direction and a few wavelengths in the horizontal) was directly forced, together with its harmonics. Two higher free modes then grew from the background noise, each energized by the interaction of the other with the forced wave. The results are in excellent agreement with a theory developed for the standing wave case, and this configuration seems especially suitable for careful quantitative work. Viscous dissipation round the boundaries and due to the internal shears can be estimated accurately, and the amplitude of forcing necessary to produce growing disturbances to the original mode can readily be calculated (cf. (2.4.6)). McEwan was particularly interested in the processes of degeneration of the wave field and wave breaking at supercritical amplitudes, but a discussion of this aspect of his work will have more relevance in the context of mixing in the interior of large bodies of stratified fluid, and will therefore be deferred until the final chapter.

Before we leave the subject of weak resonant interactions they should be contrasted again with strong interactions, in this context of a continuously stratified fluid. The fact that each wave mode separately satisfies the equations of motion exactly has already been mentioned (following (2.2.9)), and this implies that the amplitudes can become large, while the non-linear terms in the equations of motion still vanish identically. It is now possible for forced modes, generated by an arbitrary pair of internal waves of large amplitude, to attain amplitudes comparable with the primary waves without the help of the resonance mechanism. These can in turn interact with each other and with the original internal wave modes with no restrictions on frequency or wavenumber, and instead of energy being interchanged continuously and slowly between three modes, the result will be an indiscriminant transfer of energy between many modes.

The formal statement of the conditions for each of these mechanisms can be expressed in several equivalent ways (Phillips 1966 a,

p. 178), which physically amount to the following. The resonant mechanism is the more important provided the vorticity associated with the internal waves remains small compared to the buoyancy frequency N, and this restriction also ensures that the interaction must extend over many wave periods for substantial transfer of energy. If on the other hand a typical rate of vertical shear (which is usually the more important contribution to the vorticity) is of the same order as N

$$\partial u'/\partial z \sim N, \qquad (2.4.8)$$

(i.e. the gradient Richardson number (1.4.3) is of order one or less), then the forced modes are comparable in magnitude to the initial modes and they can develop rapidly within a period of the primary wave. The condition for a range of a continuous spectrum near wavenumber magnitude k to have a vigorous ('turbulent') energy exchange can be written in a form comparable to (2.4.8), namely

$$S \equiv \{k^3 E(k)\}^{\frac{1}{2}} \sim N, \qquad (2.4.9)$$

and weak resonant interactions are important when $S \ll N$. This is the appropriate generalization of (2.4.8) since the combination on the left is the only quantity with the dimensions of a shear which can be formed from k and the energy density $E(k)$. (See §5.2.2 for the definition of $E(k)$ and a further discussion of this point.)

CHAPTER 3

FINITE AMPLITUDE MOTIONS IN STABLY STRATIFIED FLUIDS

The previous chapter was based on equations of motion made linear by assuming that the amplitude of wave-like disturbances of the fluid remained infinitesimal. We now consider various large amplitude phenomena which require the inclusion of the non-linear terms for their explanation. First, some of the inviscid wave problems already treated will be extended to finite amplitude, and the essentially non-linear phenomenon of internal solitary waves will be discussed. Then various quasi-steady flows which arise in nature and in civil engineering applications will be treated, using a generalization of free surface hydraulic theory (and thus relating the properties of such flows to the waves which can form on them). Internal hydraulic jumps, the flow of a thin layer down a slope, and the nose at the front of a gravity current come under this heading. Finally we introduce the effects that viscosity and diffusion can have on slow steady motions in a stratified fluid, describing upstream wakes and boundary layers and the process of selective withdrawal.

3.1. Internal waves of finite amplitude

3.1.1. *Interfacial waves*

We refer again to the statement made in §2.1.2, that (2.1.8) is valid for finite amplitude long waves. (Cf. Lamb 1932, p. 278.) This implies that the highest point of any disturbance will move fastest, and so the forward slope of a wave of finite amplitude will tend to steepen, an effect called 'amplitude dispersion'. This is in contrast with the result of the frequency or wavenumber dispersion previously described by (2.1.7) which is valid for general depths. Since long waves propagate faster than short, this will lead to the spreading out of an arbitrary waveform and a decrease of slope. Thus it seems plausible that under some circumstances a balance between

[48]

these opposing tendencies can be achieved, and that it will be possible to have waves of finite amplitude and permanent form.

Some approximate results for internal waves of this kind will be quoted below. These are obtained by an expansion technique first used by Stokes for surface waves (and applied to interfacial waves by Hunt (1961)), and are given in the form of a power series in a small parameter, successive terms of which contain the higher harmonics. The first term represents the linear solution with sinusoidal waveform, and the higher order terms imply a distortion of this shape. One must be careful in each case to define the expansion parameter in such a way that it is indeed small. Initially it will be taken as the wave slope $(a_1 k)$ where a_1 is an amplitude and k the wavenumber, but in some later cases it will be seen that this needs modification, and the fluid depth must be taken into account. Thorpe (1968a, c) has made an extensive study, both theoretical and experimental, of progressive and standing waves at interfaces and in continuous stratification and his papers also contain a good survey of the earlier work. Other features of his experiment will be referred to in other chapters, but here only three special cases will be chosen for illustration and later use. All of these are for progressive waves on an interface, though comparable results were obtained for standing waves.

For two deep layers, the interface displacement η is given to third order by

$$\eta(x, t) = a_1 \cos(kx - \omega t) + \tfrac{1}{2} a_1 (a_1 k) \left(\frac{\rho_2 - \rho_1}{\rho_1 + \rho_2} \right) \cos 2(kx - \omega t)$$

$$+ a_1 \frac{(a_1 k)^2 (3\rho_1^2 - 10\rho_1 \rho_2 + 3\rho_2^2)}{8} \cos 3(kx - \omega t) \qquad (3.1.1)$$

and the dispersion relation is

$$\omega^2 = \frac{gk(\rho_2 - \rho_1)}{\rho_1 + \rho_2} \left\{ 1 + \frac{\rho_1^2 + \rho_2^2}{(\rho_1 + \rho_2)^2} (a_1 k)^2 \right\}. \qquad (3.1.2)$$

If the density difference is small, the distortion of the wave profile is determined by the third order term; this implies that the crests and troughs are symmetrical but flattened relative to a sine curve. The frequency for a given wavenumber increases with increasing amplitude.

When one layer is shallow (kh_1 small) and the other is deep, $(\rho_2 - \rho_1) \ll \rho_2$, and T_1 denotes $\tanh kh_1$, then to second order

$$\eta(x, t) = a_1 \cos(kx - \omega t) - \tfrac{3}{4}a_1(a_1 k)\{(1 - T_1)/T_1^2\} \cos 2(kx - \omega t)$$
$$(3.1.3)$$

and

$$\omega^2 = \frac{gk(\rho_2 - \rho_1)T_1 T_2}{\rho_1 T_2 + \rho_2 T_1}\left\{1 + \frac{(a_1 k)^2}{8T_1}(9 - 22T_1 + 13T_1^2 + 4T_1^3)\right\}. \quad (3.1.4)$$

The form (3.1.3) shows that the wave profile is more sharply peaked on the side pointing into the deeper layer (and this result holds also for more general ratios of layer depths). This asymmetry has been observed in the laboratory and also in the ocean, as shown by the records obtained by La Fond (1966) in a shallow thermocline (reproduced in fig. 3.1). (Note however, that the whole picture is changed if there is a mean shear across the interface.)

For the third example we take that of equal layer depths, with at first an arbitrary density difference. Then with $h_1 = h_2 = h$, $T = \tanh kh$

$$\eta(x, t) = a_1 \cos(kx - \omega t)$$
$$+ \tfrac{1}{4}a_1(a_1 k)\left[\frac{(\rho_2 - \rho_1)}{\rho_1 + \rho_2}\frac{3 - T^2}{T^3}\right]\cos 2(kx - \omega t) \quad (3.1.5)$$

and

$$\omega^2 = gkT\frac{(\rho_2 - \rho_1)}{\rho_1 + \rho_2}\left\{1 + \frac{(a_1 k)^2 Y}{8T^4(\rho_1 + \rho_2)^2}\right\},$$

where $\quad Y = [4(\rho_1 + \rho_2)^2 T^2(2T^2 - 1) + (\rho_2 - \rho_1)^2(3 - T^2)], \quad (3.1.6)$

which in the case of small density differences simplifies to

$$\omega^2 = gkT\frac{(\rho_2 - \rho_1)}{\rho_1 + \rho_2}\left\{1 + \frac{(a_1 k)^2}{2T^2}(2T^2 - 1)\right\}. \quad (3.1.7)$$

When one or both of the layers is shallow, (3.1.4) and (3.1.7) show that there will be a reversal of the frequency dependence on wavelength at a particular depth, which for equal layers occurs at $T^2 = \tfrac{1}{2}$ or a depth to wavelength ratio of 0·140. The experiments of Keulegan and Carpenter (1961) show both that the frequency decreases with increasing amplitude for equal layers when they are

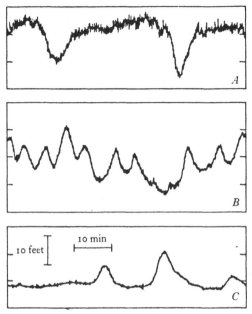

Fig. 3.1. Smoothed records of the shape of internal waves (A) near the sea surface, (B) at an intermediate depth and (C) near the sea floor. (From La Fond 1966. In *Encyclopedia of Oceanography*, ed. R. W. Fairbridge, © 1966 by Litton Educational Publishing, Inc. Reprinted by permission of Van Nostrand Reinhold Company.)

shallow enough, and that the wave profiles are symmetrical and very nearly sinusoidal, in agreement with (3.1.5).

Another consequence of a finite amplitude wave motion at an interface, i.e. the retention of second order terms, is that the orbits of individual fluid particles are changed. A steady horizontal drift in the direction of wave propagation, of order $(a_1 k)^2$, is superimposed on the circular orbits, with a maximum at the interface and an exponential decay on each side (cf. the 'Stokes drift' for surface waves). As Dore (1970) has shown, a small viscosity greatly increases this effect, producing a current proportional to $(a_1 k)^2 Re^{\frac{1}{2}}$, where $Re = (\omega/\nu k^2)$ is a Reynolds number based on the wave motion. The calculated form of the velocity profile in a channel with rigid top and bottom is shown in fig. 3.2. The conditions of no net horizontal flow imposed on this solution is appropriate to the case of a tank of limited length, and measurements of the distortion of dye

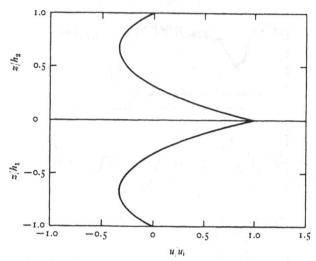

Fig. 3.2. The profile of horizontal mass transport velocity for a progressive wave moving from left to right in a two-layer fluid system bounded by two rigid horizontal planes. (After Dore 1970.)

streaks show the same general features. When the interface is not sharp, however, a backflow is predicted near the centre, and there is the extra complication that the density distribution is distorted near the ends of the tank and the simple drift cannot continue indefinitely. Similar drift motions have been documented by Thorpe (1968c) for the case of continuous stratification.

3.1.2. Cnoidal and solitary waves

It is clear that when one or both of the layer depths is small enough for $\tanh kh \approx kh$ to be small, the above expansions must become invalid. In this limit, the ratio of the coefficient of the second harmonic in (3.1.5) to that of the first is of order

$$\frac{a_1}{k^2 h^3} \frac{\rho_2 - \rho_1}{\rho_1 + \rho_2} = l_1 \qquad (3.1.8)$$

which need not be small just because $a_1 k$ is small. A different procedure is necessary when both layers are shallow which in effect uses l_1 as the expansion parameter, and puts a limitation on its magnitude. Similarly, the ratio of the corresponding coefficients in (3.1.3) is of order

$$a_1/k h_1^2 = l_2, \qquad (3.1.9)$$

and this is the appropiate parameter to use when only one layer is thin. The difference between l_1 and l_2 is of some importance, which will be brought out in the discussion below.

For the case of two thin layers Benjamin (1966) has obtained valid solutions for periodic waves of permanent form (called cnoidal waves because of the symbol 'cn' used to denote the Jacobian elliptic functions which describe their shape). He used a method which takes no account of the fact that the density difference in l_1 may be small, i.e. it does not rely on the Boussinesq approximation. We will write down explicitly only the result for the limiting case as $kh \to 0$, that is, for the internal solitary wave (first considered by Keulegan (1953)). The interface displacement from its state of rest can be written, neglecting terms of order $(a_1/h)^4$, as

$$\eta = a_1 \operatorname{sech}^2 \frac{x}{2b}, \qquad (3.1.10)$$

where the lengthscale b depends on the densities and layer depths. With equal layers and rigid boundaries, for example, the displacement is upwards and

$$b^2 = \frac{1}{3} \frac{h^3}{a_1} \frac{\rho_1 + \rho_2}{\rho_2 - \rho_1} \left\{ 1 + \frac{a_1}{h} \frac{\rho_2 - \rho_1}{\rho_1 + \rho_2} \right\}. \qquad (3.1.11)$$

For unequal layers and small density differences the crest of the wave is towards the deeper layer. If b is identified with the wavelength k^{-1} in (3.1.8) it is clear that the parameter l_1 is of order unity for internal solitary waves. The longer the wave, the smaller must be its amplitude for the required balance between the two non-linear effects to be maintained; higher waves will continue to steepen and even break (see §3.2.2). The phase velocity is

$$c^2 = gh \frac{(\rho_2 - \rho_1)}{(\rho_1 + \rho_2)} \left[1 + \frac{\rho_2 - \rho_1}{\rho_1 + \rho_2} \frac{a_1}{h} \right], \qquad (3.1.12)$$

always greater than that of infinitesimal waves; the solitary wave is in fact the fastest travelling wave of given height and unchanging form.

Analogous results have been obtained by Long (1965) and Benjamin (1966) for solitary waves in a continuous (exponential) density gradient. They showed that in that case it is not possible to use the Boussinesq approximation since this implies an inconsistent neglect of terms which may be of the same order as others which have been retained. A crude indication of what can happen is obtained by examining (3.1.6) (and a similar coefficient in the next term omitted from the expansion (3.1.5)). When T is small, the two terms in the factor Y are generally comparable, and though they are small, an accurate estimate of both of them is essential when discussing the limit corresponding to the solitary wave.

A different kind of internal wave can exist on a thin layer of light fluid lying over a deep layer, or in the equivalent circumstances described below. Benjamin (1967) has produced a comprehensive theory for progressive waves and continuous variations of density confined to shallow regions, but again only the solitary wave solution on uniform layers between rigid boundaries will be written down here. The interface displacement is downwards, and provided a_1/h_1 is fairly small, is of the form

$$\eta = \frac{-a_1 \lambda^2}{x^2 + \lambda^2},$$ (3.1.13)

where
$$\lambda = \frac{4 \rho_1 h_1^2}{3 \rho_2 a_1}.$$

This is clearly consistent with l_2 (defined by (3.1.9)) being of order unity if λ^{-1} is identified with k. The phase velocity is

$$c^2 = \frac{\rho_2 - \rho_1}{\rho_1} g(h_1 + \tfrac{3}{4}a_1);$$ (3.1.14)

note the much stronger dependence on amplitude than in (3.1.12), when the density difference is small. A free surface has the effect of decreasing both the speed and the length of the wave, for a given amplitude.

Solitary waves of this second type are likely to be important on the seasonal thermocline i.e. below a thin light layer at the surface of the ocean. They can also occur on a thin transition region between two uniform layers, and then (by symmetry) take the form of bulges travelling along the interface. They were first observed in the laboratory by Davis and Acrivos (1967a) in a study which brought out the close relationship between these bulges and the second mode waves formed as a result of resonant interactions (see fig. 2.18). It was shown that such a disturbance can propagate back and forth along a tank, being reflected off the ends and persisting until damped by viscosity (fig. 3.3a pl. III). The measured velocities were in good agreement with an expression equivalent to (3.1.14), modified to take account of the actual density distribution. When the amplitude of the disturbance is larger, the character of the motion changes, and a lump of fluid having a circulation like that in a vortex pair propagates bodily along the interface (fig. 3.3b). Davis and Acrivos produced numerical calculations, valid for large amplitude, which contain this feature of closed streamlines (though the equations themselves break down in this range for another reason, which will be discussed in §3.1.3).

3.1.3. Waves and flows in a density gradient

As mentioned above, related expansion techniques have been used to calculate the form of internal waves with continuous stratification (for example by Thorpe (1968c) and Benjamin (1966)) but these problems will not be pursued further here. We turn instead to a restricted class of flows for which exact solutions can be obtained, but with no limitation on amplitude (such as is still required by (3.1.8) and (3.1.9)). Dubreil-Jacotin (1937) and Long (1953a) derived an exact first integral of the equations of motion which is in general non-linear, but which becomes linear (without approximation) for special choices of stratification and velocity. Much of the detailed work on large amplitude waves has been based on this simplification.

Several forms of this equation will be written down below without proof, but the conditions necessary for their validity must be clearly understood. The motion is supposed two-dimensional, incompressible and steady. At some horizontal position upstream of the body (in the negative x-direction) there are no disturbances and the velocity and density distributions are specified functions of z. Finally, there should be no reversal of flow in any vertical section, so that a streamfunction ψ may be defined which varies monotonically with z, and the density $\rho = \rho(\psi)$ (i.e. is constant along streamlines). With these assumptions (which will be discussed further below) it follows from the full inviscid equations of motion that

$$\nabla^2 \psi + \frac{1}{\rho}\frac{d\rho}{d\psi}\left[\tfrac{1}{2}(u^2+w^2)+gz\right] = f(\psi), \qquad (3.1.15)$$

where $\qquad\qquad u = \partial\psi/\partial z, \quad w = -\partial\psi/\partial x$

and $f(\psi)$ is a specified function of ψ.

Yih (1960) gave a transformed version of (3.1.15) in terms of a pseudo-streamfunction $\tilde{\psi}$ defined by

$$\rho^{\frac{1}{2}}u = \partial\tilde{\psi}/\partial z, \quad \rho^{\frac{1}{2}}w = -\partial\tilde{\psi}/\partial x$$

(whose definition is related to the procedure discussed at the end of §1.3, which removes the inertia effects). This may be written

$$\nabla^2\tilde{\psi} + gz\frac{d\rho}{d\tilde{\psi}} = \frac{dH_*}{d\tilde{\psi}}, \qquad\qquad (3.1.16)$$

where $H_* = p + \frac{1}{2}(u^2 + w^2) + g\rho z$ is the 'total head', which is constant along a streamline. From this form it is clear that the equation is linear provided both ρ and H_* are polynomials of degree no higher than quadratic in $\tilde{\psi}$. Yih (1965) has discussed the various possible cases, one of which corresponds to the situation of interest here. If both $d\rho/dz$, the density gradient, and $\frac{1}{2}\rho U^2$, the dynamic pressure, are assumed independent of z in the undisturbed flow, (3.1.16) reduces to an especially simple form. At the same time one can introduce the displacement η_0 of the streamlines from their initial position as the dependent variable, to obtain the equation which is our starting point in the following section on lee waves

$$\nabla^2\eta_0 + \frac{N^2}{U^2}\eta_0 = 0. \qquad\qquad (3.1.17)$$

Note that the Boussinesq assumption has not yet been made, but when it is added, the condition on (3.1.17) is that N^2 and U^2 separately should be constant with height. This equation, with constant coefficients, is the Helmholtz equation, exactly the same form as that obtained for $(\partial^2 w/\partial x^2)$ in the linear theory (2.3.1).

The assumptions leading to (3.1.17) can break down in several ways. Most important in the lee wave application is the restriction that all streamlines must emanate from a position far upstream. It will be shown in §3.1.4 that statically unstable regions with closed streamlines appear in the solutions when the wave amplitude becomes large; though such 'rotors' are physically plausible, they cannot consistently be described using the present model. There is also the problem of 'upstream influence', the possibility that the introduction of a finite obstacle into a flow on which lee waves can form will result in 'columnar' disturbances extending upstream (i.e. those having $k = 0$; see §2.3.1). From the first of equations (2.2.11), it is clear that energy could then propagate upstream provided the vertical wavenumber $m < N/U$; if such motions were present they would change the approaching flow and thus violate the assumed conditions far upstream.

The columnar disturbances are not needed to satisfy the boundary conditions in the lee wave problem, but they can nevertheless arise in connection with non-linear transient effects. Benjamin (1970) gave an integral argument concerning the inviscid transient problem which suggested that there will always in principle be an upstream influence in a confined flow which is subcritical in the sense used in §2.3.1. McIntyre (1972) has modified this conclusion by showing that, while there will always be a columnar disturbance, this need not appear ahead of the body. For the first lee wave mode there is an upstream influence (to second order in the wave slope) only when $NH/U > 2\pi$, i.e. under conditions where a second mode can also appear. Though the question has a basic philosophical importance, it is often of less practical significance. Even when upstream influence is predicted it is weak, and it is still useful in practice to investigate the effects of an obstacle for a given flow upstream, disregarding the changes in the flow caused by the introduction of the obstacle. In an unconfined flow the disturbances fall off with distance, and the assumption of no upstream influence is formally valid at large times after the introduction of the obstacle (though of course disturbances could arise in other ways not considered here, for example through the mechanism of resonant interaction described in §2.4.3).

An associated difficulty raised by Benjamin (1966) has not yet been fully resolved. He has questioned whether the special boundary conditions imposed to bring the equations to linear form are in fact representative of the full non-linear steady problem (which requires also the specification of a downstream boundary condition). Certainly the 'critical layer' phenomena of §2.3.3, as well as the formation of the rotors mentioned above and of hydraulic jumps (§3.2.2) are excluded from the formulation based on the assumptions leading to (3.1.17).

3.1.4. *Finite amplitude lee waves*

Because of the identical form of (2.3.1) and (3.1.17) the solution to some non-linear problems can be obtained directly from the linearized solutions described in §2.3. The boundary condition at the obstacle is still essentially non-linear, however, since the lowest streamline must follow its surface, which may have an arbitrary height. Long (1955) showed (for the case of a closed channel) how an *inverse* method can be used to calculate related flowpatterns and obstacle shapes. He first calculated streamlines for a convenient shape of infinitesimal barrier, using a superposition of linear solutions, and then increased the amplitude of the motion to a finite value. The resulting solution still satisfies (3.1.17) but it no longer satisfies the boundary conditions on the obstacle (which on the linearized theory were applied at $z = 0$). The final step is to replace the original obstacle by another having the shape of the bottom streamline calculated in this way, on which the conditions are automatically satisfied.

Long (1955) compared his calculated streamlines for various obstacle shapes and heights with flow patterns observed in a series of laboratory experiments (which remain the best and most comprehensive in this field). All his photographs are worth detailed study, but only two are reproduced here. In fig. 3.4a, b pl. v are compared the observations and calculations in a case where only one lee wave mode can occur, and fig. 3.4c, d correspond to the range with lower velocity ($Ri_0^{\frac{1}{2}} > 2\pi$) where a second mode appears. In the second case rotors are observed, but apart from the fact that the observed wave motions are weaker than predicted, the agreement between theory and experiment is surprisingly good, and the flow outside these regions of reversed flow is apparently little affected by their presence. Some alteration of the flow ahead of the obstacle is also seen in fig. 3.4d, a first indication of the 'blocking' which becomes important in very slow flows or for high barriers. (This phenomenon is discussed further in §§3.2.1 and 3.3.2. Note that it is a different limiting case from that described earlier in connection with 'upstream influence', when flows of finite velocity over small barriers were of interest.)

Yih (1960) used a different inverse approach, in which instead of specifying a barrier shape he investigated (using a linearized form of (3.1.16)) the effect of introducing various singularities into a stratified flow in a channel. A vertical line of cross-stream vorticity, for example, can produce a region of closed streamlines in its neighbourhood, as well as wave-like disturbances downstream. By choosing the boundary of this closed region as the shape of the obstacle, one again obtains a possible exact solution for the flow over a rather blunt body. Source and sink distributions (with total strength zero) can also be used, a device closely analogous to that used by Rankine to generate streamlined bodies in a uniform inviscid flow. Pao (1969) has extended Yih's method by using distributions of vortex pairs and doublets to generate barriers and satisfy upstream boundary conditions in a semi-infinite as well as a confined flow. (Again these calculations were continued well past the point where the equations no longer give a valid description of the flow.)

One more result for channel flows should be mentioned. Claus (1964) used a comparable equation to (3.1.17) for a compressible flow, and calculated exact lee wave patterns for particular density and velocity distributions using Long's method. He then compared this with a similar calculation for an incompressible fluid having the same velocity profile but a density stratification the same as the potential density distribution in the compressible flow. The wavelengths in the case compared were identical, but the maximum amplitude in the compressible case was nearly twice that of the incompressible flow, reflecting the fact that it is density and not potential density which determines the energy density.

When indirect methods are used, the shape of a barrier producing waves cannot be specified *a priori*. It is much more convenient to be able to fix the shape, and then investigate systematically what happens to the lee waves as the stratification and flow velocity are changed. Drazin and Moore (1967) were the first to obtain such a solution, with boundary conditions prescribed on a thin vertical strip projecting above the floor of a channel. Important as this solution was theoretically, it is an unrealistic case physically (because the flow cannot remain attached to such a barrier – see below). A series of papers by Miles (1968a, b)† and Miles and Huppert (1969) has shown how a modified diffraction theory, familiar in

† Miles (1969) has also given an excellent review of the whole subject.

other contexts, can be used to solve the problem for any obstacle shape in a closed channel and for an infinite half-plane. Only the latter will be considered here; for this case, we recall, it can be shown that there will be no 'upstream influence' extending to $x = -\infty$.

The governing equation (3.1.17), obtained by making the assumptions $d\rho/dz = $ const and $\rho U^2 = $ const, can be written in the non-dimensional form

$$\nabla_1^2 \eta + \kappa^2 \eta = 0, \qquad (3.1.18)$$

where both $\eta = \eta_0/h$ and the dimensionless operator ∇_1^2 are scaled with the maximum height h of the barrier, and $\kappa^2 = (Nh/U)^2$ has the form of an overall Richardson number. The lee waves depend on the *obstacle* height through the parameter κ^2, which is different from the Ri_0 used to characterize the whole flow in §2.3.1. Ri_0 is always large since a lengthscale $H \gg h$ comparable with the scale height of the whole region of interest must be used. The range of interest of the constant κ is roughly $0 < \kappa < 1.5$, as the barrier height for a given flow is increased from zero to that at which the flow becomes unstable, in the sense that overturning and backflow first appears. At this point, when $\kappa = \kappa_c$, a streamline becomes vertical somewhere in the flow and the model is strictly invalid at larger κ, though outside the regions of overturning the solutions again seem to describe observed phenomena reasonably well. Another kind of instability, which draws energy from the shear flow due to the waves, will be discussed in chapter 4.

Streamlines calculated by Huppert for the case of a semi-circular obstacle are shown in fig. 3.5, for three values of $\kappa : \kappa = \kappa_c = 1.27$ and two values on either side of this. The increase of amplitude of the waves and the tendency to overturn as κ is increased are clearly seen. Values of κ_c have also been calculated for a wide variety of obstacle shapes, using a representation in terms of appropriate distributions of dipole sources (cf. the indirect methods). Given the functional form, two parameters are enough to specify the shape, namely κ and $\epsilon = h/b$ where b is a measure of the width. Huppert and Miles (1969) explored the whole range of semi-elliptical obstacles, for which κ_c varies smoothly from 1.73 for $\epsilon = \infty$ (corresponding to the vertical strip) through $\kappa_c = 1.27$ at $\epsilon = 1$ (the semi-

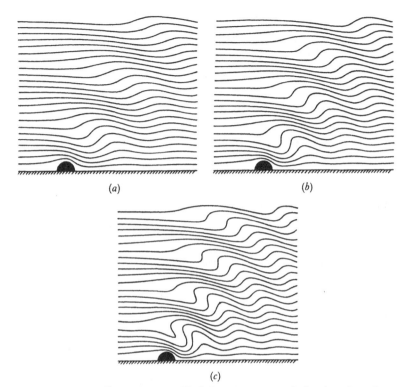

Fig. 3.5. Streamlines of a stratified flow over a semi-circular obstacle, calculated by Huppert (1968) using (3.1.18) (a) $\kappa = $ 1.0, (b) $\kappa = \kappa_c = $ 1.27, (c) $\kappa = $ 1.5.

circle), to 0·67 at $\epsilon = $ 0. In the last paper of the series they investigated various limiting cases and compared them with the exact solutions. When $\kappa \to $ 0 with ϵ fixed (i.e. weak waves are formed behind a body of *given shape*) a simple dipole field results, with amplitude proportional to the moment of the dipole, and therefore to the cross-sectional area of the obstacle hb (as remarked in §2.3.1). The limiting solutions as $\epsilon \to $ 0 with κ/ϵ fixed (i.e. for a *fixed length* of obstacle) show that the linear theory is accurate only for $\kappa \ll $ 1. The largest value of κ_c ever found explicitly for a wide flat body is $\kappa_c = $ 0·85, for the shape $z = h[1 + (x/b)^2]^{-1}$ used in so many linear lee wave computations (see §2.3.2).

The direct calculations of Miles and Huppert also lead to numerical estimates of the drag coefficient (for the limited class of

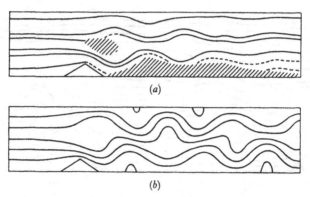

(a)

(b)

Fig. 3.6. Flow over a triangular obstacle with $\kappa = 2.7$. (From Davis 1969.) (a) observed flow, with hatched areas denoting mixing regions; (b) calculated flow, showing rotors.

flows described by (3.1.18); reference should again be made here to the work of Benjamin (1970) who has shown that the drag exerted by an obstacle on a stream is intimately connected with the question of upstream influence). A knowledge of the flow pattern immediately allows a computation of the change in the flux of horizontal momentum, and hence of the wave drag D. The drag coefficient C_D is defined by comparing D to the force due to the dynamic pressure $p_1 = \frac{1}{2}\rho U^2$ acting over the height h of the body, so

$$C_D = D/p_1 h. \tag{3.1.19}$$

The drag coefficient for tall ellipses is a monotonically increasing function of κ at any ϵ, which varies little with ϵ, whereas for squat ellipses it depends more strongly on ϵ as well as on κ. For all elliptical bodies, however, the maximum C_D for unseparated flows with $\kappa < \kappa_c$ varies little, between 2·3 at $\epsilon = \infty$ to 3.0 at $\epsilon = 0$. A comparison of the exact C_D with the drag calculated using the several asymptotic approximations of Miles and Huppert (1969) referred to above showed that there is a substantial range of κ/ϵ where only the full solution gives an accurate estimate of C_D.

Davis (1969) has used a related method to obtain numerical solutions for the flow over a triangular obstacle in a channel of finite depth. He has also compared his solutions with laboratory experiments which are especially interesting in that they emphasize the differences between the assumed and actual flows for the same

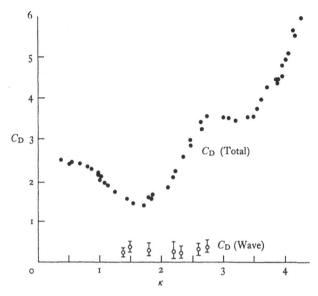

Fig. 3.7. Comparison between measured total drag and wave drag co-efficients for a stratified flow over a thin barrier. (After Davis 1969.)

nominal upstream conditions. Using triangular and also thin barriers, he recorded flowpatterns (fig. 3.6) which are a combination of a turbulent wake due to the separation of the flow from the sharp-edged bodies, lee waves produced by the barrier and the wake, and also the turbulent rotors which form as a result of the waves but are not adequately dealt with by the theories based on (3.1.18). He made an estimate of the pure wave drag from the observed amplitude of the waves, and compared this with direct measurements of drag, made by towing a floating obstacle along the channel and measuring the force on it. As shown in fig. 3.7, the total drag is considerably greater than the wave drag, and this must be due to phenomena which are not described by lee wave theory. Outside the 'stable' range of $\kappa < \kappa_c$, the wave solutions provide no guide at all to the behaviour of the flow or to the drag. (The properties of the turbulent wake as such will be discussed further in §5.3.1.)

3.2. Internal hydraulics and related problems

The flow of a layer of heavy fluid under lighter fluid has many features in common with the hydraulics of a stream with a free surface. Some of these analogous properties will now be summarized, as a prelude to the discussion of the new effects which become possible when the density differences are small and the fluids can mix freely. In nature, most flows of this kind are turbulent, but they can for many purposes be accurately described using perfect fluid theory, with sometimes the addition of terms to take into account friction on the bottom and at the interface, and the overall energy losses.

3.2.1. *Steady frictionless flow of a thin layer*

Consider first a two layer system in two dimensions, with a layer of depth $h_1(x)$, density ρ_1 flowing with a velocity U in the x-direction under a deep layer at rest. Suppose that an obstacle is introduced, with height $h = h(x)$ above the horizontal bottom, and length $L \gg h$ so that the flow is slowly varying and can be regarded as uniform at any section. Applying Bernoulli's equation just below the interface gives

$$\frac{p_1}{\rho_1} + \tfrac{1}{2}U^2 + g(h + h_1) = \text{const}, \qquad (3.2.1)$$

where p_1 is the pressure at the interface, and just above:

$$\frac{p_1}{\rho_2} + g(h + h_1) = \text{const} \qquad (3.2.2)$$

(any change in velocity in the upper layer will be small because its depth is large). The continuity equation in the lower layer is

$$Uh_1 = Q \quad \text{say}, \qquad (3.2.3)$$

and eliminating p_1 and U from the above gives

$$\frac{1}{2}\frac{Q^2}{g'h_1{}^2(x)} + h_1(x) = C - h(x), \qquad (3.2.4)$$

where C is a constant. This is of exactly the same form as the ordinary hydraulic relation with $g' = g(\rho_1 - \rho_2)/\rho_1$ replacing g, and is sketched in fig. 3.8.

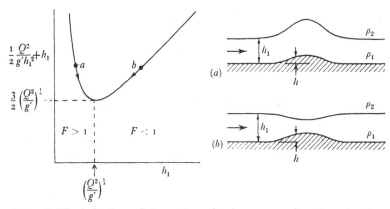

Fig. 3.8. The behaviour of a heavy layer flowing over an obstacle and under a deep lighter layer at rest, as described by the 'hydraulic' relation (3.2.4). If the flow is supercritical upstream, (situation (a) $F > 1$) the interface bows upwards over the obstacle; if it is subcritical (as at (b) with $F < 1$) the interface drops.

Given the form of the obstacle $h(x)$ the volume flux per unit width Q and the constant C, the depth of the layer h_1 can be found from (3.2.4). Its behaviour depends critically on the upstream value of the internal Froude number defined by (cf. (1.4.5))

$$F = \left(\frac{Q^2}{g'h_1^3}\right)^{\frac{1}{2}} = \frac{U}{(g'h_1)^{\frac{1}{2}}},$$

since the left-hand side of (3.2.4) has a minimum at $F = 1$. If $F > 1$ far upstream, h_1 will increase as $h(x)$ increases and the interface bows upwards over the obstacle (as shown in fig. 3.8a). If $F < 1$ on the other hand, the level drops ($h + h_1$ decreases) as the flow draws on its potential energy to gain enough kinetic energy to surmount the obstacle (fig. 3.8b).

The 'critical' internal Froude number $F = 1$ corresponds to the state in which the flow velocity just equals the velocity of *infinitesimal* long waves on the layer, called the critical velocity U_c. (See (2.1.8).) When the flow is supercritical ($F > 1$), small disturbances cannot propagate upstream against the flow and any obstacle will have a purely local effect. When the flow is subcritical ($F < 1$), short waves can remain at rest relative to the obstacle, and so a phenomenon analogous to lee waves on the interface can be observed downstream of shorter obstacles (Long 1954). (This criterion for wave formation can be compared with that found in §2.3.1 for a continuous gradient, namely $F = Ri_0^{-\frac{1}{2}} < \pi^{-1}$, remembering of course the different definition of the internal Froude number.)

This subcritical case also sheds some light on the difference between the phenomena of upstream influence and blocking. In general, when stationary waves appear downstream, longer waves can propagate upstream, but if h is small these will be damped by friction. Though there is undoubtedly an upstream effect, its net result is small. If h is increased beyond the point where the right-hand side of (3.2.4) falls below $\frac{3}{2}(Q^2/g')^{\frac{1}{3}}$, however, no solution is possible, essentially because the fluid no longer has enough energy to get over the barrier. In this case true blocking occurs, and the depth upstream must increase until a new steady state is established with $F = 1$ over the obstacle, which now acts as a control section for the whole flow. Upstream the flow is subcritical, with depth h_0 say, and above the obstacle (where the depth is $\frac{2}{3}h_0 = h_c$, the critical depth) it changes smoothly to a supercritical flow downstream.

3.2.2. *Internal hydraulic jumps*

Changes in the flow can also take place more abruptly. For example, when the blocking described above is produced by suddenly increasing the height of an obstacle, the front of the layer of increased depth may be very sharp, and will propagate upstream as a travelling hydraulic jump (also called a bore or surge). If the supercritical flow below the barrier is obstructed again further downstream, it can pass again into a subcritical flow through a stationary hydraulic jump (fig. 3.9). This arises as a result of finite amplitude long waves, which can travel faster than U_c and therefore move upstream, transmitting information faster than is possible on linear theory. (Such waves, whose properties depend only on the local mass flux and the continuity equation have been called 'kinematic waves' by Lighthill and Whitham 1955.) If their amplitude becomes so large that frequency dispersion can no longer balance the steepening effect (see §3.1.2), then the upstream face of the wave eventually breaks, with the accompanying turbulence which characterizes jumps and surges.

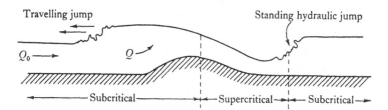

Fig. 3.9. Flow of a layer over a high obstacle: production of internal hydraulic jumps.

The conditions at a jump can be investigated using the conservation of momentum and mass, i.e. by applying (3.2.1) and (3.2.3) to the states before and after the jump where the depths are h_1 and h_1'. It follows on eliminating the velocities that

$$h_c^3 = \tfrac{1}{2}h_1 h_1'(h_1 + h_1'); \qquad (3.2.5)$$

thus a stationary jump can only occur between two 'conjugate' states, one shallower and the other deeper than the critical depth. The final depth h_1' may be expressed in terms of h_1 and F_1 upstream as

$$h_1' = \tfrac{1}{2}h_1[(1 + 8F_1^2)^{\frac{1}{2}} - 1]. \qquad (3.2.6)$$

By comparing the fluxes of kinetic energy before and after a jump, it can be shown that a jump can only take place from a supercritical to a subcritical flow, and that an amount of energy equal to

$$\frac{(h_1' - h_1)^3 g' Q}{4 h_1 h_1'} \quad \text{per unit span}$$

is dissipated in the process. (The opposite case would imply a creation of energy at the jump.) When F_1 is only just supercritical (a weak jump), some of the energy can be radiated away in the form of internal waves, but at higher F_1 not enough energy can be dissipated in the waves alone and they must break (Benjamin 1966, 1967). In strong jumps little energy goes into waves.

Phenomena which can be interpreted in terms of each of these processes have often been observed in the atmosphere (Ellison 1961). A sudden pressure change, accompanied by a change in wind speed, can signal the passage of a strong jump. Regular oscillations in a barometer record (with the number of crests increasing in time as the energy travels away from the jump) correspond to an undulating jump. Sometimes condensation and precipitation effects complicate and intensify the atmospheric jump phenomenon (producing a 'squall line'); the same is true if a moist supercritical flow passes over a mountain, where the thickening and lifting can give rise to heavy orographic rainfall. Other related processes which are important in the atmosphere, but which do not have close parallels in ordinary hydraulic theory, remain to be discussed in later chapters, namely the effect of a continuously stratified environment above the flowing layer (§9.2) and the turbulent mixing produced by the kinetic energy released in a jump (§6.2.3).

The above ideas have been generalized to apply to several layers having arbitrary depths and velocities, notably by Long (1954). He concentrated on the case of comparable depths and equal velocities, simulating this experimentally by moving an obstacle through a tank containing immiscible layers, initially at rest. He produced convincing demonstrations of the various regimes of flow over the obstacle, and of the jump phenomena (fig. 3.10 pl. v), and applied the ideas (Long 1953 b) to explain the violent winds and turbulence observed in the flow over the Sierra Nevada range (see fig. 3.11 pl. vi). More recent developments in this field, which will also be important in the study of changing flows in shallow estuaries, have been summarized by Yih (1965, p. 130), and reference has already been made to the observations by Thorpe et al. (1972) of large amplitude internal seiches in a lake, which take the form of surges travelling along the thermocline and reflecting from the ends.

3.2.3. Flow down a slope

In considering changes in the flow along a horizontal bed which take place over a relatively short distance, it was reasonable to neglect friction as we did above; but when the flow of a heavy layer down a long slope is of interest, friction must certainly be added if the velocity is not to increase without limit. Here, following the usual hydraulic assumption, we take the drag to be proportional to the square of the velocity and to include the effects of friction at the bottom and at the interface. (The form of the latter is treated more fully in §6.2.) The momentum equation parallel to a slope at a small angle θ to the horizontal is then

$$\frac{d}{dx}(U^2 h_1) = g' h_1 \theta - g' h_1 \frac{dh_1}{dx} - C_D U^2, \qquad (3.2.7)$$

where the terms on the right represent respectively the component of gravity accelerating the layer, the pressure force due to a possible change in depth, and the frictional force (C_D now being used to denote a combined drag coefficient). Eliminating the velocity using (3.2.3) gives

$$\frac{dh_1}{dx}(1 - F^2) = (\theta - C_D F^2). \qquad (3.2.8)$$

When $F = 1$ (and $h_1 = h_c$, the critical depth), $\theta = C_D$, the critical slope; a smooth change from subcritical to supercritical flow or vice versa can only occur when the slope is critical. More generally, when the flow is uniform and $dh_1/dx = 0$, $F^2 = \theta/C_D$, a constant which depends only on the slope and the friction, not at all on the rate of supply of fluid or its density. The corresponding *normal depth* of the flow h_n is from the equation on p. 65

$$h_n^3 = C_D Q^2/\theta g'. \qquad (3.2.9)$$

When a flow is started with depth $h_0 \neq h_n$ its behaviour depends on the relative magnitudes of h_0, h_n and h_c. If $h_0 > h_n > h_c$, dh_1/dx is positive and tends towards θ for large x; this flow can remain subcritical everywhere and spread out smoothly to merge with a layer of heavy fluid overlying a horizontal plane at the bottom of the slope (see fig. 3.12). Supercritical flows, on the other hand, can only change to subcritical through a jump, which may occur either on the slope or further out over the horizontal plane. An especially convincing application of this result was made by Ball (1957). He explained the sudden changes from light to strong winds at a coastal station in Antarctica in terms of the movement of a jump, which is formed as the katabatic wind (the downslope flow formed by cooling of the air over the continent) flows over the sea. Sometimes it is not clear if one should describe observations in these terms, or as a flow over an obstacle (fig. 3.9); the important feature in either case is that the flow is supercritical at the slope. Although a line of clouds, strong rotary motions and sometimes waves downstream are associated

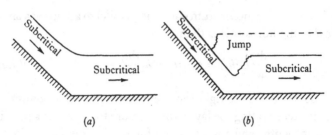

Fig. 3.12. Flow of a heavy layer down a slope and onto a horizontal plane (a) a subcritical flow can spread out smoothly, (b) a supercritical flow must become subcritical through a hydraulic jump, whose position can change as the flow rate fluctuates.

with such hydraulic jumps, they arise in a rather different way from ordinary 'lee waves' and the rotors discussed previously, which can only be formed when the flow over a mountain is subcritical.

3.2.4. *The 'lock exchange' problem*

We turn now to phenomena in which the intrusion of a front or 'nose' into a fluid of different density is the main concern, rather than the steady flow of the layer behind it. Important examples are the motion of a front of cold air under warmer air in the atmosphere (on a small enough scale for rotation to be neglected, say a sea-breeze front rather than one associated with a large depression), or the intrusion of salt water under fresh when a lock gate is opened at the mouth of a river. Following Benjamin (1968) we will concentrate here on flows along a horizontal plane. It has been shown by Middleton (1966) that for small slopes the motion of the nose can be described using the depth and density difference at the nose itself, with the slope and the depth of the layer behind playing only a secondary role. (See the next section and especially §6.3.4 for a discussion of noses on steep slopes.)

Consider first a two-dimensional frictionless flow in a closed channel of depth d, referred to axes fixed relative to the nose (see fig. 3.13 for the notation). As in §3.2.1, continuity and Bernoulli's equation applied just above the interface give two equations involving the final velocity U_2 and depth h of the flowing layer. In the absence of friction a third relation follows from the fact that the

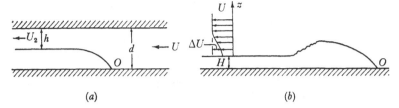

Fig. 3.13. Two-dimensional flow of a heavy layer. (a) Frictionless flow in a channel, (b) a real flow under a deep lighter layer.

'flow force' (the total pressure force plus the momentum flux per unit span) is constant, which gives

$$\tfrac{1}{2}\rho_1(U^2d + g'd^2) = \rho_1(U_2^2 h + \tfrac{1}{2}g'h^2). \qquad (3.2.10)$$

For consistency of the three relations, Benjamin (1968) showed that

$$U_2^2 = 2g'(d-h) = \frac{g'(d^2 - h^2)d}{(2d-h)h}, \qquad (3.2.11)$$

whose only non-trivial solution is $h = \tfrac{1}{2}d$. Thus the only steady, energy conserving flow (this is implied by the use of the Bernoulli condition) is one in which the advancing layer fills half the channel. In this symmetrical case, the Froude number of the approaching flow is subcritical

$$U/(g'd)^{\frac{1}{2}} = \tfrac{1}{2} \qquad (3.2.12)$$

and that of the receding flow supercritical

$$U_2/(g'h)^{\frac{1}{2}} = \sqrt{2}. \qquad (3.2.13)$$

Benjamin has also calculated accurately the shape of the interface; this agrees with an earlier result of von Kármán (1940), who found that the slope of the nose at the stagnation point is 60° to the horizontal.

Many measurements have been made of the flow generated by removing a vertical barrier separating two layers of equal depth and slightly different densities, in channels of various sizes and cross-sections (see Barr 1967). At low flow rates, the Reynolds number must be used explicitly in the interpretation, but for flows on a large enough scale for the direct effects of viscosity to be small, the experiments confirm the near constancy of the nose velocity U over a considerable length of travel. In completely enclosed wide conduits, the measured velocity is about

$$U = 0.44(g'd)^{\frac{1}{2}}, \qquad (3.2.14)$$

not very different from the inviscid value predicted by (3.2.12). In a channel with a free surface, the flow becomes unsymmetrical, the corresponding multiplying constants being 0.47 for the underflow and 0.59 for the overflow.

Lofquist (1960) has made direct measurements of the stress between a single layer flowing along a horizontal plane and a stationary fluid above, and has shown that this is much smaller than the bottom stress. The latter must therefore account for most of the discrepancy between (3.2.12) and (3.2.14), though the interfacial stress can still be important in describing stationary noses (see the next section). Observations of the interface (when one of the layers is dyed) show that it is covered by an irregular pattern of cusped short waves (see fig. 4.13). These have been interpreted either as disturbances due to turbulence generated in each of the layers by the shear at the boundaries, or as a manifestation of a small scale instability of the shear flow at the interface. They will be discussed further in §4.1.4 in the context of interfacial stability, and also in §9.1 when describing the process of mixing at an interface.

3.2.5. *Gravity currents and noses*

When an advancing nose is shallower than half the total depth (as it often is in practice) Benjamin's argument shows that there must be a loss of energy near the front. He has demonstrated further that this energy cannot be carried away in the form of a smooth wave train but must be accompanied by 'breaking'. This is in accord with laboratory observations of gravity currents in deep water (Keulegan 1958) which show that they have 'head' just over twice the depth H of the layer, behind which is a turbulent region and an abrupt drop to the uniform layer behind (see fig. 3.13*b*). The drag associated with the velocity change in the deep layer due to this turbulent wake must be taken into account when considering the force balance on the layer. In no other way can the net force $\frac{1}{2}(\rho_1 - \rho_2)gH^2$

acting in the direction of motion be balanced and a steady state attained.

If one neglects mixing (as we are doing throughout this 'hydraulic' approach), and supposes that the pressure in the 'wake' is hydrostatic, then the velocity of advance U of a nose advancing under a deep fluid can be calculated as follows. The dynamic pressure $\frac{1}{2}\rho_1 U^2$ at the stagnation point must equal the difference in hydrostatic pressure measured at the boundary far upstream and downstream i.e. $g(\rho_1 - \rho_2)H$, so

$$U = 2^{\frac{1}{2}}(g'H)^{\frac{1}{2}}. \qquad (3.2.15)$$

Benjamin (1968) has made the point that these pressures should be compared by taking a circuit passing between the stagnation point and the layer far downstream through the deeper overlying fluid. An earlier argument due to von Kármán (1940), (though giving the same expression for the velocity) is erroneous, since it applied Bernoulli's equation along the interface, an invalid procedure in a dissipating, rotational flow. We note too that the reasoning used by Prandtl (1952, p. 369) is sound only for the initial transient phase of the motion. He equated the dynamic pressures due to the two fluids at the front, neglecting any effects of hydrostatic pressure due to the heavy fluid ejected upwards, and deduced a front velocity half that of the layer behind. As soon as the extra fluid arriving at the nose has fallen back on the underlying layer, a steady state will be achieved in which the nose and layer velocities are nearly equal.

The development of a mixed region of increasing length complicates the argument somewhat, and so does the observed lifting of the stagnation point above the bottom, an effect of bottom friction. Taking the latter into account empirically reduces the multiplying constant in (3.2.15), to give closer agreement with the experimental value of 1.1 suggested by the experiments reported by Keulegan (1958) and Wood (1966). In making this comparison with experiment one must also be careful to allow for the effect of a finite channel depth. When expressed in terms of the layer thickness $H = d - h$, the nose velocity changes monotonically (and almost linearly with H/d) between $2^{-\frac{1}{2}}(g'H)^{\frac{1}{2}}$ for $H = \frac{1}{2}d$, as implied by (3.2.12), and the value (3.2.15) as $H/d \to 0$. Even at $H/d = 0.1$, a common condition in laboratory experiments, the correction can be significant.

The above arguments of course take no account of cross-stream non-uniformities and three dimensional motions, though these have been clearly documented by Simpson (1969) in the laboratory and the atmosphere. (See fig. 3.14, pl. VI.) Something more should be said here too about observations of noses on inclined beds. As the slope is increased the ratio of the front to the layer velocity steadily decreases, and it is down to about 0.7 on a slope of $\frac{1}{20}$. Through the whole of this range of slopes, however, the velocity of the nose can be described using the local density difference and the height D of the head by the relation

$$U = 0.75(g'D)^{\frac{1}{2}}. \tag{3.2.16}$$

A rather more complicated example of the hydraulic approach to 'noses' is provided by studies of the intrusion of a salt wedge far upstream against an opposing fresh water flow. (See Ippen 1966 chapter 11.) Here we consider only a stationary wedge over a flat bottom and a fixed total depth d. The extra factor is that the interfacial stress must now be taken into account explicitly in determining the shape of the interface, and so a momentum equation of the form (3.2.7) is applied to each layer. Assuming again a quadratic dependence of stress on velocity, and that the velocity in the wedge is so small that the bottom stress under it can be neglected compared with that at the interface, one obtains (cf. (3.2.8))

$$(F_1^2 - 1)\frac{dh_1}{dx} = C_D F_1^2 \left(\frac{d}{d-h_1}\right), \tag{3.2.17}$$

where $F_1 = U_1/(g'h_1)^{\frac{1}{2}}$ is the internal Froude number of the upper layer. This can clearly be written as a relation involving only h_1 and the distance x (measured from the nose say), using $U_1 h_1 = $ const.

The final step necessary to produce a closed solution is the recognition that a natural control section for the flow of the upper layer occurs where the cross-section increases abruptly either in depth or width (see Wood 1968), for instance at the mouth of an estuary as a river enters the sea. In general, when both layers are moving, the condition here is

$$F_1^2 + F_2^2 = 1. \tag{3.2.18}$$

For a single flowing layer $F = 1$, and no external disturbances can propagate upstream past this point. The integration of (3.2.17) with

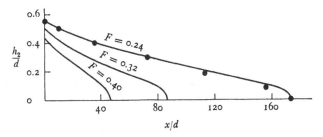

Fig. 3.15. Observed shapes of arrested saline wedges for three fresh water flows (i.e. different internal Froude numbers), with the theoretical points fitted to one of them by Harleman (1961).

boundary conditions applied at the control gives an explicit relation between the initial value of $F = F_0$ say, the depth h_1 and the distance x along the wedge; this is compared with observations due to Keulegan (1957) in fig. 3.15. The total length of the wedge is given by

$$\frac{L}{d} = \frac{1}{4C_D}[\tfrac{1}{5}F_0^{-2} - 2 + 3F_0^{\frac{2}{3}} - \tfrac{6}{5}F_0^{\frac{4}{3}}]. \qquad (3.2.19)$$

If the mean value of C_D can be regarded as constant, then a knowledge of L for one F_0 allows the evaluation of C_D, which can be used to predict the length of the intrusion for other flow conditions and to calculate the shape of the nose. This method gives results in good agreement with Keulegan's laboratory experiments, though its extension to the full scale phenomenon still involves a good deal of empiricism.

The gradual change in length of intrusion predicted by (3.2.19) should be contrasted with a later result (§6.2.3). When the motion of the bottom layer is taken into account explicitly, there can be a rather rapid transition between a regime of slow flow against the upper current, and a more rapid flow in the same direction. This arises as a result of a change in the rate of mixing (and therefore of the stress) between a turbulent bottom current and the flow outside it, and it is more appropriately discussed in the context of mixing.

3.3. Slow motions in a stratified fluid

3.3.1. *The problem of selective withdrawal*

In many applications (an important example is the design of cooling water intakes for power stations) it is necessary to be able to predict the maximum rate of withdrawal of fluid with desired properties which can be attained before fluid from a different level also begins to flow. The simplest approach to this problem is to use the hydraulic method, and one problem of this kind has already been solved implicitly in §3.2.1. The flow of a layer of heavy fluid through a control section (or over a barrier), where $F = 1$, represents the maximum flowrate for which the lighter fluid is at rest. If fluid is removed at a line sink located on the bottom downstream of the control at any rate less than this, then only the lower fluid will flow out, but if F (based on the mass flux and h_c) rises above a critical value of order unity, then the upper fluid will be drawn in too. Wood (1968, 1970) has shown how this idea can be extended to multi-layer systems flowing through horizontal contractions which act as a control, and has calculated the fluxes in each layer as the total withdrawal rate is increased.

The theory becomes considerably more difficult when a sink discharges from a large body of fluid where there is no clear point of control. Craya (1949) and Huber (1960) considered various cases of sinks in the end wall of a channel, at a different level from the interface between two layers, and proposed methods for calculating the form of the interface as well as the critical condition. Such calculations are at best approximate, and it is often necessary to find the conditions for 'drawdown' experimentally. A fundamental difficulty is that such experiments can never be really steady, because the upstream boundary conditions are altered as the flow continues (see §3.2.2). Nevertheless, practically useful estimates can be made and fig. 3.16 summarizes some of these results for a two-layer system. It is clear on dimensional grounds that, if the orifice width or diameter D is small, the critical condition must always be expressible as $F = F_{\text{crit}}$ where the critical Froude number is defined in terms of g', some flux and the vertical separation h of the sink and interface. For point sinks

$$F_{\text{crit}} = Q_3/g'^{\frac{1}{2}}h^{\frac{5}{2}} \qquad (3.3.1)$$

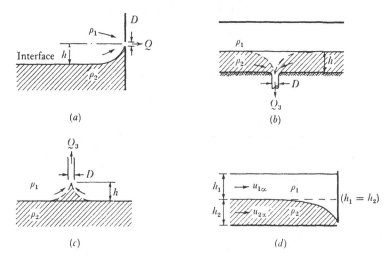

Fig. 3.16. Selective withdrawal of fluid from a two-layer system: critical Froude numbers (as defined in the text) for various boundary conditions. (a) Orifice or slot in end wall, $F_{crit} = 2.6$ (orifice), 1.5 (slot). (b) Orifice in bottom, $F_{crit} = 1.6$. (c) Large tube above interface $F_{crit} = 4.5 (D/h)^{\frac{1}{2}}$. (d) Slot in corner (Huber, theoretical), $F_{crit} = 1.66$.

and for line sinks $\qquad F_{crit} = Q_2/g'^{\frac{1}{2}}h^{\frac{3}{2}}$ $\qquad$ (3.3.2)

where Q_3 is a volume flowrate and Q_2 the flux per unit width. This simplification is not always clear from the original papers, where the form in which the results are given puts undue emphasis on D. An explicit dependence on D/h appears in the results of Rouse (1956) shown in fig. 3.16c because the orifice diameters ranged up to $D = 10h$.

Similar results hold for continuously stratified fluid, though theoretically they are approached rather differently. Yih (1965, p. 82) used the linear equation (3.1.16) to find a series solution for the flow into a line sink at the end of a channel, with the velocity specified upstream (uniform U in the Boussinesq approximation). He showed that this solution only remains valid provided $F = U/NH$ is sufficiently large, H being the channel depth. For slow flows, when $F < \pi^{-1}$, wavelike solutions also become relevant upstream (cf. §2.3.1) and it is no longer possible to satisfy the boundary conditions at $x = -\infty$, that is, to draw fluid in uniformly from all heights in the channel. The experiments of Debler (1959) have confirmed that

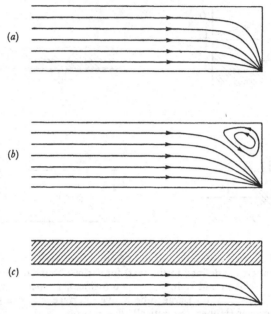

Fig. 3.17. Selective withdrawal from a continuously stratified fluid, (a) complete withdrawal, (b) the formation of a corner eddy, (c) a stagnant upper region.

these solutions give realistic flowpatterns for $F > \pi^{-1}$. As F approaches π^{-1} from above a closed corner eddy appears in the solutions (as shown schematically in fig. 3.17). As remarked earlier in connection with rotors this is not properly described by (3.1.16), but it does (perhaps fortuitously) correspond to what was observed in this range.

For $F < \pi^{-1}$ the fluid is divided into a flowing region and a stagnant layer extending all the way upstream. Debler's results suggest that a similar criterion to (3.3.2) can be used to describe the amount of fluid withdrawn. Defining

$$F_{\mathrm{crit}} = \frac{Q_1}{d_1^2} N^{-1} = \frac{Q_1}{g'^{\frac{1}{2}} d_1^{\frac{3}{2}}} \tag{3.3.3}$$

where Q_1 is the flux per unit width, d_1 is the depth of the flowing layer far from the source and g' is based on the density difference across d_1, the experiments correspond to a value of $F_{\mathrm{crit}} = 0.25$. It turns out to be a good general rule for all kinds of slow flows that this internal Froude number, based on the properties of the flowing layer (or each of the layers, if there is more than one), lies between $\frac{1}{4}$ and $\frac{1}{3}$, the lower values probably being due to viscous effects which have so far been neglected here. Yih (1965, p. 90) has shown how the formation of a stagnant region can be postponed to smaller F by inserting an obstacle near the sink (whose form was calculated by an indirect method, using an array of singularities and choosing a closed streamline). At even lower F blocking must inevitably occur, when a layer of fluid at the bottom of the channel cannot flow over the barrier.

3.3.2. Blocking ahead of an obstacle

It is clear both from the lee wave experiments described in §3.1.4 and the sink flows just discussed that it is not profitable to treat slow flows in a density gradient, for which $F \ll 1$, using a steady model and fixed upstream boundary conditions. Now it is appropriate to recall the result quoted in §2.2.2; in the limit $F \to 0$, a two-dimensional body in steady horizontal motion will carry along with it a plug of fluid bounded by tangents above and below it and extending to infinity both up and down stream. What is observed experimentally (see fig. 3.18 pl. VII) just after such a body is set into slow motion is not so simple, however. In addition to a plug of finite length directly ahead of the body, an array of alternating jets develops at other horizontal levels. These jets become more numerous as F is reduced, and they all propagate upstream in time, altering the approaching flow. All of these results can be reconciled provided the flow is treated as *time-dependent*, and the approach to a steady state is examined as an initial value problem.

Using a model of this kind Trustrum (1964) obtained solutions which contain features resembling Long's and Debler's observations (and also some analogous results for rotating fluids, which will not be described here). She assumed a steady inviscid flow with constant U and linear density gradient, into which at $t = 0$, $x = 0$, a disturbance (in the form of a source–sink distribution) is inserted. The governing equations in this essentially linear theory were derived using the Boussinesq assumption, and also one of the Oseen type in which the perturbation velocity is ignored in the convective terms. With the convention (to be used also in §3.3.3) that the mean flow is in the negative x-direction, her solutions as $t \to \infty$ are periodic in both x and z for $x < 0$ (i.e. for appropriate boundary conditions they describe lee waves downstream), whereas the upstream solutions are periodic in z and independent of x, and therefore describe one-dimensional flows extending to $x = +\infty$ (the columnar modes). The magnitude of both parts of the flow is proportional to the strength of the source–sink distribution, as in Yih's solutions, but the new terms (those independent of x) were previously excluded by the initial assumption of uniform upstream velocity. When the method is applied to sink flows, realistic looking closed streamlines are obtained. The approximations are strictly valid only far from the sink, however, and even there the predicted perturbations upstream are finite and not small, so the basic assumption leading to the linearized theory becomes questionable again.

Another approach to the time-dependent problem was made by Bretherton (1967) who examined the development of the flow in terms of waves propagating away from the body. (His analysis too was carried out for inertial waves in a rotating fluid, i.e. for a 'Taylor column', but the translation of the results to the case of plane gravity waves in a stratified fluid is immediate.†) His physical interpretation is as follows. If a cylindrical body is rapidly accelerated from rest to a small horizontal velocity, gravity waves will be generated, which spread out in all directions. Far from the body, the larger scale waves arrive first, since for any angle of propagation the group velocity depends inversely on k (2.2.11), and a broad velocity profile is at first produced (fig. 3.19a). In the region ahead of the body the waves have zero frequency but finite group velocity relative to the body. As more energy is put into the fluid, the length of the steady flow region increases at a finite rate comparable with the group velocity of waves whose half wavelength is equal to the diameter of the obstacle.

Regions of larger velocity gradients can build up later as the slower shorter waves arrive, and fig. 3.19b shows the velocity profile

† The analogy has been important in the development of the subject, though it will not be pursued in detail here. For a full discussion see Veronis (1967), which is referred to in another context on p. 243.

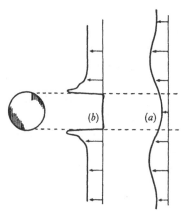

Fig. 3.19. The development of a blocked flow far from a cylinder moved slowly in a horizontal direction through a stratified fluid. (*a*) Broad velocity profile due to the first long waves. (*b*) Plug flow produced much later as the slower short waves arrive. (From Bretherton 1967.)

at a later time; the central plug flow is now nearly at rest relative to the body. This profile was calculated by superposing the velocities due to all the waves which had arrived at the time of interest (which implies the restriction of infinitesimal motion and hence linear equations). Very near the surface of the body a singularity tends to develop because of the cumulative effect of very small scale waves which have not had time to propagate away. Bretherton went on to discuss the effect of a small viscosity on his inviscid solutions. He argued that this could be treated separately from the problem of generation and propagation, and that each wave component will decay exponentially at a rate which increases as the wavelength decreases. The singularities will be removed in a real fluid because the shortest waves are rapidly damped, but the longest waves are little affected. Ultimately a steady state is predicted in which the 'blocked' regions are of finite length.

Whatever detailed model is used to describe these slow flows it is clear that an inviscid solution must become physically unrealistic in two regions. Very near an obstacle a viscous boundary layer must dominate, and at large distances viscosity must eventually cause all the disturbances to die away. Problems in which viscous (and also diffusive) terms have been taken explicitly into account are discussed in the following sections.

3.3.3. *Upstream wakes and boundary layers*

Martin and Long (1968) have given a careful derivation of the relevant equations (to which reference should be made for the details), but the essential points can be made simply from first principles. Consider a steady motion in which fluid particles preserve their identity and properties (i.e. there is no diffusion), and in which viscous and buoyancy forces are dominant. One can then neglect the time-dependent and inertial terms on the left of (1.3.3) to obtain the component Navier–Stokes equations

$$\left.\begin{aligned}
0 &= -\frac{\partial p'}{\partial x} + \nu\nabla^2 u, \\
0 &= -\frac{\partial p'}{\partial z} - g\frac{\rho'}{\rho_0} + \nu\nabla^2 w,
\end{aligned}\right\} \tag{3.3.4}$$

where p' and ρ' are the perturbation pressure and density (cf. (1.3.4)). Now eliminating the pressure by cross-differentiation, and at the same time using a boundary layer type of approximation to neglect the higher x derivatives, gives

$$\frac{g}{\rho_0}\frac{\partial \rho'}{\partial x} = -\nu\frac{\partial^4 \psi}{\partial z^4}, \tag{3.3.5}$$

where ψ is the streamfunction defined as in §3.1.3 by $u = \partial\psi/\partial z$, $w = -\partial\psi/\partial x$. This last assumption is justified by the fact that the flows of interest (as in the inviscid case), are elongated horizontally, and have much smaller variations in the x- than in the z-direction.

Again the discussion is simplified if one assumes that far upstream there is a uniform flow, say in the negative x-direction, with a linear density gradient (cf. §3.1.3), i.e.

$$\psi = -U_0 z, \quad g\rho'/\rho_0 = -N^2 z.$$

Since without diffusion ρ' must be constant along streamlines, combining these two gives a relation between ψ and ρ' which is valid even when the flow is disturbed by the introduction of a thin obstacle:

$$\frac{g\rho'}{\rho_0} = \frac{N^2}{U_0}\psi. \tag{3.3.6}$$

Substituting this in (3.3.5) gives a single equation for ψ

$$\frac{\partial^4 \psi}{\partial z^4} + \beta \frac{\partial \psi}{\partial x} = 0, \qquad (3.3.7)$$

where $\beta = N^2/U_0\nu$ has the dimensions of (length)$^{-3}$. If this is made non-dimensional using a lengthscale L (say the length of the body or the distance along it) so that $x = Lx_1$, $z = Lz_1$, then

$$\frac{\partial^4 \psi_1}{\partial z_1^4} + \beta_1 \frac{\partial \psi_1}{\partial x_1} = 0. \qquad (3.3.8)$$

Here $\beta_1 = N^2L^3/U_0\nu$ is the ratio of buoyancy to viscous forces, which can be written in the form of a Richardson number Ri_0 times a Reynolds number Re, where $Ri_0 = N^2L^2/U_0^2$, $Re = U_0L/\nu$. Martin (1966) has shown further that the terms neglected in arriving at (3.3.7) are of order $(Re/Ri)^{\frac{1}{2}} = Re\,\beta_1^{-\frac{1}{2}}$ if $Re \gtrsim 1$; this latter ratio is therefore a measure of the importance of higher order inertia terms relative to those retained. We also note here for later reference his result that the neglect of diffusion is justified provided

$$\kappa\,Ri^{\frac{1}{2}}/\nu\,Re^{\frac{1}{2}} \ll 1,$$

where κ is the molecular diffusivity of the property causing the density variations.

The parabolic nature of (3.3.8) implies that, just as in the ordinary one-dimensional diffusion equation in which time is analogous to the variable x_1, boundary conditions imposed upstream of a given flow section (cf. those imposed at future times) cannot influence the flow there. This approximation therefore means that in the region behind a body there is no way in which its presence can be felt by the fluid. Ahead of the body there will, however, be an *upstream wake*, whose form can be calculated from (3.3.7) using the boundary conditions imposed on the body and upstream (in the positive x_1-direction). This asymmetry in the behaviour can also be deduced using a physical argument based on vorticity; only upstream of a body is the vorticity generated by buoyancy in the opposite sense to that generated by viscosity, so that a balance between the two becomes possible.†

† It is of course the restriction to very slow motions which has eliminated the possibility of the more familiar kind of wake, whose properties in the limit of *large* velocity will be treated in chapter 5.

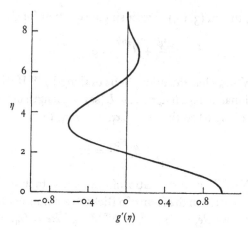

Fig. 3.20. The horizontal velocity profile in an upstream wake.
(From Martin and Long 1968.)

Two special solutions of (3.3.7) with different boundary conditions are of interest here. Far upstream of the body the form of the solution does not depend on the details of the flow around it, but only on the drag D or the momentum defect flux

$$J = \frac{D}{\rho} = \int_{-\infty}^{\infty} p' dz \qquad (3.3.9)$$

which is constant at all sections upstream. Long (1959) showed that, by choosing a certain combination of powers of x and z as a new independent variable, (3.3.7) with the condition (3.3.9) can be reduced to an ordinary differential equation, i.e. he found a similarity solution which can be written as

$$\psi = \frac{J}{\nu} (\beta x)^{-\frac{1}{2}} g(\eta), \qquad (3.3.10)$$

where
$$\eta = z \left(\frac{x}{\beta}\right)^{-\frac{1}{4}}. \qquad (3.3.11)$$

The substitution of these expressions into (3.3.7) gives the equation for the non-dimensional function $g(\eta)$ (dashes denoting derivatives),

$$4g^{iv} - \eta g' - 2g = 0, \qquad (3.3.12)$$

which was solved numerically by Long to obtain (for example) the horizontal velocity profile reproduced in fig. 3.20. Qualitatively this agrees with the alternating jet structure which is observed, but

Martin and Long (1968) showed that it is difficult to go far enough upstream to make the direct effect of a body negligible and obtain a quantitative experimental check.

Pao (1968 a) also produced comparable solutions for the upstream wake, and carried out experiments (see fig. 3.21 pl. VII) covering a range of values of Re/Ri_0 between 0.2 and 10. At the higher values of this ratio (as inertia effects became increasingly important) he showed that the wake becomes weaker but that successive velocity maxima spread out further vertically. These features are predicted by the theory of Janowitz (1968), who, instead of neglecting inertial terms entirely, linearized the equations of motion using the Oseen approximation (compare §3.3.2). He also showed that decaying lee waves can exist downstream, which attenuate more slowly as the speed of the body increases.

Graebel (1969) has extended the theory based on (3.3.7) to include bodies with a finite vertical thickness $2b$, taking b as the lengthscale in β_1 instead of L. He showed that the blocked region upstream has a length of order β_1, and that here viscosity must act to raise a fluid particle against gravity so that it can pass the body, while behind gravity is the principal controlling factor. The resulting difference in pressure distribution, nearly constant immediately upstream in the stagnant region and varying hydrostatically downstream as in the original undisturbed fluid, dominates the drag, which can be expressed to this approximation as

$$D = \rho J = \tfrac{2}{3} g b^3 \left| \frac{d\rho}{dz} \right|, \qquad (3.3.13)$$

independent of both viscosity and speed (though of course the existence of viscosity and some motion is essential to establish the pressure distribution).

A second similarity solution of (3.3.7) has been found to describe the motion immediately over a thin plate (i.e. the upstream boundary layer). Taking the plate to be semi-infinite $0 \leqslant x \leqslant \infty$ (and the origin therefore at the back of the plate), the solution corresponding to boundary conditions of no slip at the surface and small disturbances at large z is

$$\psi = U_0 \left(\frac{x}{\beta}\right)^{\frac{1}{4}} f(\eta), \qquad (3.3.14)$$

with η again defined by (3.3.11). This represents a boundary layer which grows upstream from the back of the plate, and it is also applicable to finite plates because of the property that conditions upstream cannot affect the flow at any given section. Substitution of (3.3.14) into (3.3.7) gives an ordinary differential equation for $f(\eta)$ comparable to (3.3.12) whose numerical solution gives the horizontal velocity profile.

Several of the predictions of this theory have been verified in a careful series of experiments carried out by Martin (1966), using streak photography of aluminium particles suspended in a large channel of stratified salt water. He measured the heights of the first two stagnation surfaces (i.e. the positions where the flow was moving with the free stream velocity, in a cordinate system moving with the plate), and obtained excellent quantitative agreement with (3.3.14) in the region of its validity. He also clearly demonstrated the upstream growth of the boundary layer from the back of the plate. (See fig. 3.22 pl. VII.) The two conditions mentioned following (3.3.8) do, however, place severe (and opposing) restrictions on the range of applicability of (3.3.14). Inertia becomes important at large velocities and very close to the rear of the plate. More serious is the neglect of diffusion, which becomes important near the plate at *low* velocities, because of the wrong boundary condition imposed on the salinity field there. A diffusive boundary layer grows from the front of the plate, and this can become thick enough at the rear edge to have a visible effect on the flow. For these reasons, the experiments are limited in practice to velocities of about 1 mm/s with very slowly diffusing substances, and it is impossible to get a range of conditions where both these effects are negligible if a temperature gradient in water is used to produce the density variation.

3.3.4. *Viscous diffusive flows*

When diffusion is introduced into the problem, it is no longer permissible to use the assumption that ρ' is a function of ψ alone, which led to (3.3.6) and to the elimination of ρ' from (3.3.5). Instead, one must add to (3.3.4) (or the more general form of these which includes inertial effects) a diffusion equation

$$u\frac{\partial \rho'}{\partial x} + w\frac{\partial \rho'}{\partial z} = \kappa\frac{\partial^2 \rho'}{\partial z^2}, \qquad (3.3.15)$$

where κ is the diffusivity of the substance producing the density gradient. Equations (3.3.5) and (3.3.15) must now be treated as a coupled pair, from which ρ' can be eliminated to give a single sixth-order equation for ψ. With the extra assumptions that the density gradient is linear and that perturbations to this gradient can be neglected (and again making the boundary layer assumption and taking the velocity to be small), (3.3.15) becomes

$$N^2 \frac{\partial \psi}{\partial x} = \frac{g\kappa}{\rho_0} \frac{\partial^2 \rho'}{\partial z^2}. \qquad (3.3.16)$$

Eliminating ρ' from (3.3.5) and (3.3.16) gives

$$\frac{\partial^6 \psi}{\partial z^6} + B \frac{\partial^2 \psi}{\partial x^2} = 0, \qquad (3.3.17)$$

where $B = N^2/\kappa\nu$.

In all problems where diffusion replaces inertia as a significant parameter (and buoyancy and viscosity remain important), the combination B, having dimensions of $(\text{length})^{-4}$, always emerges, and replaces the β which appeared in (3.3.7). The non-dimensional form of this parameter, $B_1 = L^4 B$, where L is a characteristic horizontal lengthscale of the problem, is a Rayleigh number (see chapter 7). Koh (1966) showed how higher order approximations (in which (3.3.17) corresponds to the zeroth order term) can be developed using an expansion technique. The appropriate expansion parameter is equivalent to $Pe \cdot B_1^{-\frac{1}{4}}$ (where $Pe = Q/\kappa$ is a Péclet number based on the volume flux per unit width Q). This combination is therefore a measure of the importance of inertia relative to the terms retained and its smallness ensures the adequacy of (3.3.17), in which B (or B_1) alone determines the behaviour. (Compare with the different roles of $Re \cdot \beta_1^{-\frac{1}{2}}$ and β_1 discussed in the preceding section after (3.3.8).)

In just the same way as before, similarity solutions of (3.3.17) can be found for wakes and boundary layers by imposing appropriate boundary conditions.† It should be noted immediately, however, that the governing equations are no longer parabolic, and so the

† All the solutions considered here are for infinitely large regions; the extra problems introduced by vertical boundaries are discussed by Foster and Saffman (1970).

solutions for the wake which include diffusion are symmetric, valid downstream as well as upstream of the body. Long (1962) found the stream function for the wake far from a body, which can be expressed in terms of the drag ρJ, the parameter B and the geometry as

$$\psi = \frac{J}{\nu}(Bx)^{-\frac{1}{3}}g_1(\gamma) \qquad (3.3.18)$$

with
$$\gamma = B^{\frac{1}{3}}z/x^{\frac{2}{3}}. \qquad (3.3.19)$$

(Compare with the viscous, non-diffusive case (3.3.10), and (3.3.11) in which there was a different dependence on x.) The function $g_1(\gamma)$, obtained by numerical integration of the resulting ordinary differential equation, is very similar to that for $g(\eta)$ (reproduced in fig. 3.20) and will not be drawn here.

The diffusive boundary layer, which grows from the *front* of a thin plate (now taken as the origin), is described by

$$\psi = -U_0 x^{\frac{1}{3}}B^{-\frac{1}{3}}f_1(\gamma) \qquad (3.3.20)$$

with γ defined by (3.3.19). Again the form of $f_1(\gamma)$ is not very different from that of $f(\eta)$ in (3.3.14). The ratio of the coefficients in (3.3.20) and (3.3.14), at a fixed value of x, can be used to assess the importance of diffusion; this agrees with the criterion stated in the previous section.

A third problem in which diffusion matters as well as viscosity and buoyancy is that of the very slow withdrawal of a continuously stratified fluid from a two-dimensional slit placed at the origin (Koh 1966). In this case the stream function must be everywhere proportional to the volume flux per unit width Q, and to the same order of approximation as before

$$\psi = -Qf_0(\gamma), \qquad (3.3.21)$$

where f_0 is another non-dimensional function of the same similarity variable $\gamma = B^{\frac{1}{3}}z/x^{\frac{2}{3}}$. The fluid removed will come from a confined region near $z = 0$ (like an internal boundary layer), and the horizontal velocity has a sharp peak at this level. The shape of the withdrawal layer (see fig. 3.23) can be deduced from the form of γ, using the numerical solution for f_0 only to find the numerical con-

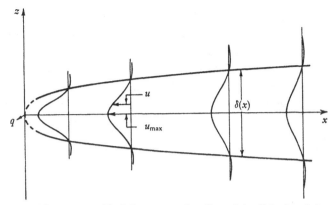

Fig. 3.23. Viscous stratified flow towards a line sink. Calculated form of the withdrawal layer. (From Koh 1966.)

stant. Thus the edge of this region corresponds to the value of γ where $f_0(\gamma) = 0$ i.e. to $B^{\frac{1}{6}}z/x^{\frac{1}{3}} = \gamma_\delta = $ constant.

Koh (1966) evaluated the constant by assuming that this theory remains valid right to the origin, and predicted that

$$\delta(x) = 7.14x^{\frac{1}{3}}B^{-\frac{1}{6}}. \tag{3.3.22}$$

Later work (Imberger 1972) has shown, however, that the observed thicknesses are consistently greater than this, and in agreement with a theory which takes inertial forces properly into account. Near a strong sink the inviscid theory discussed in §3.3.1 will apply to a layer of uniform depth in which inertial effects matter, but at large distances viscosity and diffusion eventually dominate, and the layer must spread out further vertically (by an amount which is given approximately by (3.3.22)). In some practical situations, for example in reservoirs, $Pe \cdot B_1^{-\frac{1}{6}}$ may never be small enough for inertia to be neglected even at the far end of the region of interest.

Vertical profiles of horizontal velocity were also predicted by Koh, using the numerical solutions for f_0 to calculate $u = \partial \psi / \partial z$ from (3.3.21). These were compared with measurements in a large channel of water stratified with either temperature or salinity, by recording photographically the distortion of vertical streaks of dye as fluid was withdrawn steadily at one end. The measured shapes were in good agreement with theory at all positions along the tank, though the magnitudes differed from the simple prediction based on the buoyancy–diffusion theory. This was attributed to the neglect of inertial effects, but we should emphasize again the fundamental problem of time dependence, which will be just as serious for viscous as for inviscid flows (see §3.3.1). Experiments of this kind cannot achieve a steady state, though in the absence of a simple alternative they are interpreted using relations like (3.3.22) which are based on this assumption.

CHAPTER 4

INSTABILITY AND THE
PRODUCTION OF TURBULENCE

In this chapter we consider various mechanisms whereby laminar flows of a stably stratified fluid can break down and become turbulent. The first task is to summarize some results of hydrodynamic stability theory as it applies in this context, that is, the investigation of the conditions under which small disturbances to the motion can grow. The logical development of the preceding chapters will be followed by restricting the discussion to dynamic instabilities due to shearing motions of a statically stable initial stratification. 'Convective' instability associated with an increase of density with height will be left until chapter 7, and two-component systems, in which different diffusivities play a vital role, will be treated separately in chapter 8.

The problem of instability of layers across which both density and velocity are rapidly varying functions of height is given special attention here, since such shear layers are very common in the atmosphere and ocean, and the vertical transports of properties such as heat and salt depend strongly on what happens near them. The topics covered now will, however, go beyond what is usually meant by the term 'instability' in the strict sense. It seems useful at the same time to outline other ways in which energy can be fed into limited regions of a more extensive flow, at a rate sufficient to cause a local breakdown. Some of the phenomena have already been mentioned in the context of steady flows, and the wider geophysical implications of the results will be discussed in chapter 10.

4.1. The stability of a free shear layer

4.1.1. *The various types of instability*

We cannot do justice here to the extensive literature on hydro-dynamic stability, or even to the small part of it which deals with stratified shear flows. Before considering special cases, however, it seems desirable to outline the methods used and to give a physical picture of the various possible mechanisms which can lead to instability. Details are to be found for example in the book by Betchov and Criminale (1967), and the review by Drazin and Howard (1966) (the latter concentrates specifically on the inviscid stratified case).

Consider first a homogeneous, inviscid two-dimensional flow in the x-direction, with velocity $u(z)$. If this has a wave-like disturbance with horizontal wavenumber k and phase velocity c imposed on it, then it follows from the equations of motion that the amplitude of a small vertical velocity perturbation w' can be described by

$$\frac{d^2 w'}{dz^2} = \left[(u-c)^{-1}\frac{d^2 u}{dz^2} + k^2 \right] w'. \qquad (4.1.1)$$

This equation, obtained by Rayleigh, has been extended by others to include the effects of viscosity and density gradients, through the addition of terms involving the Reynolds and Richardson numbers respectively. In the latter case, this adds a term $(u-c)^{-2}(du/dz)^2 Ri$ to the square brackets on the right (where Ri is the gradient Richardson number defined by (1.4.3), and the Boussinesq approximation has been used). The equation poses in each case an eigenvalue problem, which allows one to determine for given boundary conditions a criterion for neutrally stable waves. This takes the form of a relation between k, the phase speed and the appropriate physical parameter (e.g. Re or Ri). It is also possible to calculate growthrates of the unstable disturbances as a function of these parameters, and numerical methods have been developed which now allow quite general flows to be investigated.

Let us turn now to the problem of particular interest, a density interface across which there is a velocity difference. Benjamin (1963) has distinguished three types of instability which can occur in this situation, and though his ideas were applied initially to the motion of a fluid near a flexible solid boundary they are readily adapted to the two-fluid interface. What he has called Class A instability takes the form of waves which are of the same kind as Tollmien–Schlichting (T–S) waves, the mechanism of breakdown of flows near fixed solid boundaries. The presence of viscosity is essential for the growth of these waves. The disturbance originates at the critical level where $u = c$ (using the notation of (4.1.1)), but in order for this to grow, momentum must be transferred from the wall to the critical layer (by Reynolds stresses). A small but non-zero viscosity makes it possible to initiate this process in flows for which the inviscid solution is stable; the loss of energy by the mean flow exceeds the gain by the oscillatory motion. (A larger viscous effect can again lead to damping, and this of course is the origin of the critical Reynolds number criterion for transition to turbulence.)

Class B disturbances are related on the other hand to the free surface waves which can exist in the flexible medium when there is no flow over it. In the two-fluid context, they are like waves on the surface of the sea, and are the kind considered by Miles (1957) in his theory of wind-generated waves. (See also Lighthill (1962) for a physical interpretation of the mechanism.) The shear of the 'wind' profile is an essential factor for the growth of such waves: provided the curvature is negative at the critical height z_c, energy and momentum can be extracted from the mean flow at that level and transferred to the surface below. This mechanism is effective only for wavelengths very much longer than z_c, and dissipation acts now always to inhibit the growth. Care is necessary when trying to apply these ideas to 'internal' interfaces, since in the limit of small density differences which is of special concern to us here there is a range of velocities for which unstable waves are of Class A (i.e. viscosity becomes a destabilizing influence).

The Class C, or Kelvin–Helmholtz (K–H) instability, occurs when waves of the above two types (in the fluid and on the 'flexible boundary' or interface) coincide in both speed and wavelength. Historically this was the first kind to be described, and it will be discussed explicitly in §4.1.2. It leads to a violent breakdown at an interface and in a region on each side of it, whose character is practically independent of viscosity. It can be interpreted alternatively as a result of the action of the pressure disturbance in phase with the interface elevation in overcoming the 'stiffness' of the surface (due to the restoring force of gravity in the case of interest here), and in terms of the instability of a vortex sheet (or more diffuse vortex layer).

The results quoted later suggest that the form of instability which is actually observed first on an interior shear layer in a stably stratified fluid is of the Kelvin–Helmholtz type, and that the breakdown can therefore be treated as an inviscid problem. Benjamin (1963) has shown that when the density difference is small, a Class A wave is the first which can theoretically become unstable, at a velocity difference across the interface of $(1/\sqrt{2})$ times that necessary for the K–H mechanism to operate; but the growthrate of the former is so much slower that it is overwhelmed by the latter in any situation where the shear continues to increase.

4.1.2. *The Kelvin–Helmholtz mechanism*

The results of inviscid stability analyses for various density and shear profiles will be considered in turn, beginning with the simplest case of a vortex sheet between two deep uniform layers moving with different velocities. This can be treated using the method described in §2.1 (and see Lamb (1932), p. 373 for the history of this problem) by defining the velocity potentials so as to include the mean flow as well as the disturbance. Using the same notation as before,

$$\phi_0 = -U_0 x + \phi_0', \quad \phi_1 = -U_1 x + \phi_1', \qquad (4.1.2)$$

where U_0 and U_1 are the layer velocities in the positive x-direction.

A theorem due to Yih (1955) has shown that the first disturbances to go unstable as Ri is decreased are two-dimensional, so it can be assumed without loss of generality for the present purpose that the displacements are again of the form (2.1.2). Applying the conditions that the displacement and the pressure must be continuous across the interface, and eliminating a, A_0 and A_1 now leads to

$$\rho_0(\omega - kU_0)^2 + \rho_1(\omega - kU_1)^2 = gk(\rho_1 - \rho_0). \qquad (4.1.3)$$

The phase velocity $c = \omega/k$ is

$$c = \frac{\omega}{k} = \frac{\rho_0 U_0 + \rho_1 U_1}{\rho_0 + \rho_1} \pm \left\{ \frac{g}{k}\frac{\rho_1 - \rho_0}{\rho_1 + \rho_0} - \frac{\rho_0 \rho_1}{(\rho_0 + \rho_1)^2}(U_1 - U_0)^2 \right\}^{\frac{1}{2}} \qquad (4.1.4)$$

which reduces to (2.1.5) when $U_0 = U_1 = 0$. The first term is a weighted mean velocity of the layers, and the waves move relative to this with a phase speed related to the shear across the interface. The stability of the interfacial waves depends on the character of the second term in (4.1.4). The square root is imaginary (so that the disturbances are exponentially growing and stationary relative to the mean velocity, rather than oscillatory) and the flow unstable if

$$(\Delta U)^2 = (U_1 - U_0)^2 > \frac{g}{k}\frac{{\rho_1}^2 - {\rho_0}^2}{\rho_0\rho_1}. \qquad (4.1.5)$$

For a given velocity difference, the motion will be unstable and waves will grow if k is sufficiently large.† In the limiting case of zero density difference, *all* disturbances are unstable. This example illustrates a more general result: when the phase velocity of waves is expressed as a complex function $c = c_r + ic_i$, the magnitude of the imaginary part of the frequency kc_i is a measure of the growthrate, and the criterion for *marginal* stability is $c_i = 0$.

For comparison with later results we note that (4.1.5) can, in the case of a small density difference, be written as

$$\frac{(\Delta U)^2}{g'\lambda} > \frac{1}{\pi}, \qquad (4.1.6)$$

i.e. in the form of a critical internal Froude number (or inverse Richardson number) based on the wavelength of the disturbance. This indeed is the only form which can arise when buoyancy and inertia forces alone are acting and there is no other relevant length scale in the problem.

The result (4.1.3) can easily be generalized to give the condition that there should be stationary, neutrally stable waves ($c = 0$) on an interface between layers of finite thickness h_0 and h_1:

$$\rho_0 U_0^2 \coth kh_0 + \rho_1 U_1^2 \coth kh_1 = \frac{g}{k}(\rho_1 - \rho_0). \qquad (4.1.7)$$

In the limit of very long waves, this reduces to the form (3.2.17), quoted in the context of hydraulic theory. The limiting relation is not a stability criterion of the same kind as (4.1.5): even if it is satisfied,

† As in chapter 2, surface tension effects are neglected here; but surface tension would limit this instability by providing an extra restoring force which is important for small wavelengths.

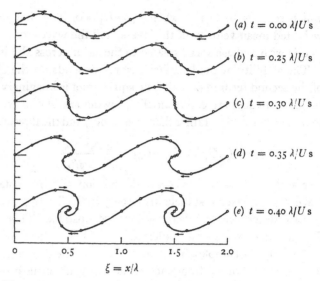

Fig. 4.1. The rolling up of a vortex sheet which has been given a small
sinusoidal displacement. (From Rosenhead 1931.)

the flow may already be unstable to other disturbances of shorter
wavelength. (It is related to the propagation rather than the growth
of waves, and implies that a two-layer flow cannot be changed
gradually through a state where (3.2.17) applies without leading to
the violent disruption of the flow by an internal hydraulic jump.)

It must be emphasized again that the above results are based on
linear theory, so that one can strictly obtain information only about
the initial stages of growth, and cannot assume that the exponential
growth will continue indefinitely (until the waves 'break' for
example). In the limiting case of a vortex sheet with no density
difference across it, however, Rosenhead (1931) made a calculation
taking non-linear terms into account which should be valid at later
times. The development of the shape of an originally sinusoidal
interface is shown in fig. 4.1, which indicates that it winds up into a
spiral form due to the interaction of the various parts of the vortex
sheet and the distortion of the waveform by the mean flow. It will be
seen later that patterns very like this are observed when a density
interface becomes unstable.

4.1.3. *Interfaces of finite thickness*

Taylor (1931 *a*) and Goldstein (1931) extended these calculations to more realistic shear and density distributions, using (4.1.1) with the term proportional to Ri added:

$$\frac{\mathrm{d}^2 w'}{\mathrm{d}z^2} = \left[(u-c)^{-1}\frac{\mathrm{d}^2 u}{\mathrm{d}z^2} + (u-c)^{-2}\left(\frac{\mathrm{d}u}{\mathrm{d}z}\right)^2 Ri + k^2 \right] w'. \quad (4.1.8)$$

Again the effect of viscosity was not considered explicitly, only indirectly through its effect on the velocity distribution (whose form, as a function of z, will also be called the velocity profile). They discussed for simplicity a series of layers, in each of which Ri is zero or constant, and matched the conditions across the interfaces. The examples which are most relevant here are illustrated in fig. 4.2.

Case (*a*) is a layer of intermediate density and thickness h between two deep uniform layers, with the velocity changing linearly between constant values over this same depth. If an overall Richardson number is defined using the total density and velocity changes and the lengthscale h, $Ri_0 = g(\Delta\rho/\rho)h/(\Delta U)^2$, then the boundaries of the unstable region are as shown in fig. 4.3, which is a plot of Ri_0 against non-dimensional wavenumber $\alpha = \frac{1}{2}kh$. In contrast with the result (4.1.5) for the vortex sheet, the distribution of the vorticity over a finite depth h has stabilized the higher wavenumbers, and only disturbances in a narrow band of intermediate scales are unstable. The unstable range is centred on the state where the waves on both interfaces move with the same velocity, and it becomes narrower at higher values of Ri_0.

The second case (*b*) has the same linear velocity change but a continuous exponential density variation through the intermediate layer (which is also effectively linear in the Boussinesq approximation). Goldstein's approximate calculation has been clarified and extended by Miles and Howard (1964), whose stability diagram is drawn in fig. 4.4, using the same variables as in the previous case. At low Ri_0 (which is now also the gradient Ri through the interface) there is again a range of intermediate wavenumbers which are unstable. The significant new point is that above $Ri = \frac{1}{4}$ small

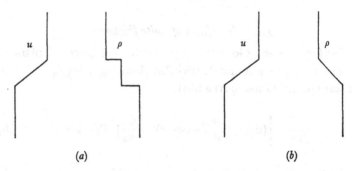

Fig. 4.2. Velocity and density profiles used in the
first interfacial stability calculations.

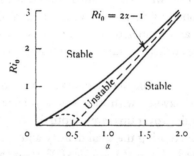

Fig. 4.3. The stability characteristics of a shear layer corresponding to
fig. 4.2 a. Waves which can grow on the density discontinuities contribute
significantly to the instability.

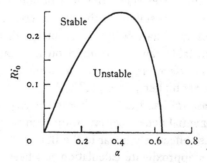

Fig. 4.4. The stability characteristics of a shear layer corresponding to
fig. 4.2 b. Note the difference in scales, and the very much smaller region of
instability compared to fig. 4.3. (The stability boundary in fig. 4.4 has
been added to 4.3 for comparison, as the dotted line.)

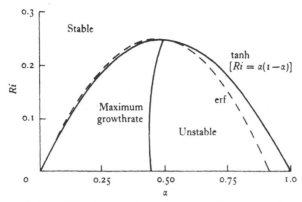

Fig. 4.5. The stability boundaries and wavenumbers for maximum growth-rate for interfaces with tanh and error function profiles of both velocity and density. (After Hazel 1972.)

disturbances of *all* wavenumbers are stable.† The 'most unstable' wave, defined as the first to go unstable as Ri is reduced below $\frac{1}{4}$, is given by $kh = 0.83$, i.e. it has a wavelength

$$\lambda = 2\pi/k = 7.5h, \qquad (4.1.9)$$

and it travels with the mean velocity of the layers. Miles and Howard also calculated growthrates in the unstable range.

Using mostly numerical methods, calculations have now been carried out for a variety of smoothly varying profiles of both density and velocity. Drazin (1958) assumed an exponential density variation and a tanh velocity profile, and showed that the critical Ri, above which all small disturbances are stable, is again $\frac{1}{4}$, and occurs at $kh = \sqrt{2}$ (where Ri and h are now based on the gradients at the centre and the total velocity change). The use of similar tanh profiles for both density and velocity gives a parabolic boundary of the unstable region, with $Ri = \frac{1}{2}kh(1 - \frac{1}{2}kh)$, and the most unstable wave-number is at $kh = 1$. Hazel (1972) has shown that this last result is not sensitive to the exact form of the distributions (it is little changed when error function profiles are used), and his computations of the stability boundaries and maximum growthrates for lower values of Ri are shown in fig. 4.5. The most unstable wave-

† This is a context in which the use of a 'gradient Froude number' $F_g = Ri^{-\frac{1}{2}}$ would be an advantage, so that large values of F_g correspond to instability; current usage decrees otherwise, and we continue to use Ri here.

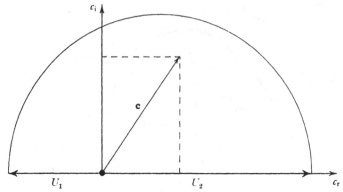

Fig. 4.6. Howard's 'semi-circle theorem', summarizing the conditions on the complex wave velocity which must hold if disturbances are to grow on an interface.

length can vary with the profiles chosen, but its range (from the value given in (4.1.9) to $\lambda = 2\pi h = 6.3h$ for the smoothly varying profiles) is small for distributions which represent the kind of interface which is of most interest here.

Various general criteria for instability have also been obtained from the stability equation without reference to particular profiles. For a detailed account of these see Miles (1961, 1963) and the review by Drazin and Howard (1966, p. 60), but three results will be mentioned here. Rayleigh's result for homogeneous inviscid fluids, which states that there must be a point of inflexion in the velocity profiles for instability to occur, has been generalized. The form with stratification is not so useful (nor so simple), since it involves the unknown wave velocity c, and it is not known if an inflexion point is necessary. Secondly, Howard (1961) showed that the complex wave velocity for any unstable disturbance must lie inside the semi-circle in the upper half of the complex plane which has the total range of velocities (i.e. the velocity of the upper and lower layers far from the interface) as diameter (see fig. 4.6). This sums up a number of earlier results which put limits on the wave velocity and growth-rates, and it can considerably simplify numerical calculations. The simplest and most readily applied condition is that of Miles (1961), which states that the *sufficient* condition for an inviscid continuously stratified flow to be *stable* to small disturbances is that $Ri > \frac{1}{4}$ everywhere in the flow.

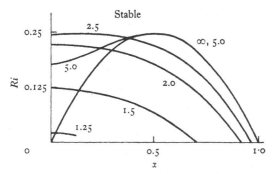

Fig. 4.7. The effect of horizontal boundaries on the stability of a density interface to a shear (after Hazel 1972). The numbers against the curves are the ratios of the total depth to a characteristic thickness of an interface with tanh profiles of density and velocity.

We should note particularly the wording of the last result; it does *not* imply that the flow must become unstable if Ri falls below $\frac{1}{4}$ somewhere, as one might be tempted to say on looking at the particular cases shown in figs. 4.4 and 4.5. Counter examples have been found; for instance, with a jet-like velocity profile $u \propto \mathrm{sech}^2 z$ and an exponential density profile, the flow can become unstable if $Ri_{\min} < 0.214$, somewhat less than the Miles limit. Hazel (1972) has shown too that rigid boundaries can have a stabilizing influence. The change in the stability boundaries for a flow with tanh profiles is shown in fig. 4.7, as symmetrically placed horizontal walls above and below are moved in from infinity. (The numbers against the curves represent the ratio of the depth of the region to the thickness of the transition layer.) The first, and very surprising, effect is to make long waves less stable, but at smaller separations all wavenumbers are stabilized. The lesson to be learnt here is that the whole profile matters in determining the 'critical Richardson number'.

The stability also varies considerably when the profiles in the transition region are assumed to have the same shape but different thickness. For example Hazel considered the case $u = \tanh z$, $\rho = \tanh(rz)/r$, and showed that the stability depends strongly on the factor r. The variation of gradient Richardson number with height is shown in fig. 4.8 for various values of r; for $r < \sqrt{2}$ (and so certainly for all cases $r < 1$ where the density interface is wider than the shear) the minimum Ri occurs at the centre, and there is insta-

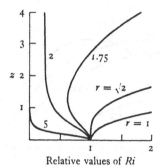

Relative values of Ri

Fig. 4.8. The variation of gradient Richardson number with height as a function of the ratio r (the ratio of velocity to density interface thickness).

bility when $Ri < \frac{1}{4}$. For $\sqrt{2} < r < 2$ the minimum value is smaller and displaced from the centre, and disturbances at these points will grow fastest. At even higher values of r where the density profile is very thin compared to the shear no critical value of Ri is relevant and the behaviour is more nearly like that considered in §4.1.2.

Many interesting problems remain to be investigated in this field. It would be useful, for example, to have results for flows with boundaries unsymmetrically placed with respect to the density profile (see §4.2.3).

4.1.4. *Observations of the breakdown of parallel stratified flows*

The predictions of the above theory have been tested in several recent laboratory experiments. Scotti and Corcos (1969) made measurements in a carefully designed wind tunnel, in which they set up a two-layer flow stratified with temperature (taking precautions to avoid heating and cooling effects at the walls). They worked always at large enough velocities for the flow through the entry section to be supercritical in the sense implied by (3.2.18), so that all disturbances were swept downstream. A small shear across the interface was generated because the lighter air stream was accelerated more than the heavier by the imposed pressure gradient through the contraction, and this shear then remained constant downstream. The interface thickened downstream due to diffusion of heat and vorticity, but the profiles of Ri remained similar: the distribution must be like that shown in fig. 4.8 for low r, with a minimum at the centre (the approximate r in air is about 0.8, since

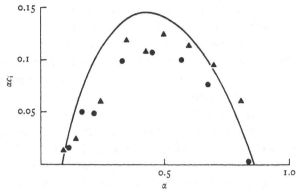

Fig. 4.9. Comparison of measurements of waves on a sheared interface with the predictions of inviscid theory. Minimum $Ri = 0.07$; the symbols denote two experimental runs, and the line represents the computed growthrates. (From Hazel 1972.)

the temperature profile spreads faster than the velocity). A slight adjustment of the depth of the tunnel allowed Ri to be kept nearly constant throughout the length of the working section and for a long time.

Very small waves were generated using a thin wire oscillated with controlled frequency at the centre of the interface, and their growth-rate (in time) measured by observing the change in amplitude down the tunnel. When $Ri_{min} < \frac{1}{4}$ the waves grew at a rate which de-pended on Ri_{min} and wavenumber, and when $Ri > \frac{1}{4}$ they were attenuated, the decay rate being independent of Ri. The measured growthrates at $Ri_{min} = 0.07$, for example, are compared in fig. 4.9 with Hazel's (1972) calculation for the same profiles, based on inviscid theory. The agreement is good, and the waves travelled with the mean flow in accordance with the theoretical prediction.

The experiment of Thorpe (1968b, 1971) has already been referred to in passing in §§2.3.3 and 3.2.4. He produced a shear flow in a stratified fluid contained in a closed tube of rectangular section, by tilting the tube away from the horizontal. The velocity profiles generated as gravity accelerates the flow are closely linked to the density distribution, and can be calculated from it using inviscid theory or with viscous corrections added. Until the surges formed at the ends of the channel reach the centre, the flow is parallel, and this is therefore a very useful experimental configuration. Observations of stability can be compared with theory which has been extended to allow for the acceleration. Experiments have been carried out also with linear salinity gradients (in this case no instability was observed although Ri fell as low as 0.014) and immiscible fluids (Thorpe 1969b) but only the case of two miscible layers will be discussed in any detail here.

Two layers, a layer of fresh water above a salt solution, were put into the tube while its long axis was nearly vertical. It was then brought carefully to a horizontal position and the interface allowed to spread by diffusion for a known time (so that the density distribution through it could be calculated). The tube was tilted again rapidly through a small angle and left there, and after a few seconds a regular array of waves suddenly appeared on the interface. These waves were stationary when the layer depths were equal, and grew; in about a second they rolled up (as shown in the photographs in fig. 4.10 pl. VIII) into a form closely resembling that of fig. 4.1. The wavelengths, and the velocity difference at which the waves appeared, were in good accord with theory applicable to an accelerating flow. Thorpe also showed that, though viscosity affected the velocity profile in some of his runs, fair agreement with the measured growthrates was obtained by treating these profiles using the inviscid theory, i.e. no damping effect of viscosity was apparent down to Reynolds numbers (based on the maximum shear and the momentum thickness of the shear region) of about 100. This conclusion is supported by the more recent stability calculations of Maslowe and Thompson (1971).

At later times, the interfacial rolls break up completely to produce a thickened interface (Thorpe 1971). Visual observations suggest that this consists of a relatively well-mixed interior bounded by sharp gradients at the edges (as sketched in fig. 4.11, and shown in the sequence of shadowgraph pictures in fig. 4.12 pl. VIII). Detailed measurements of concentration profiles (Thorpe, personal communication) indicate, however, that the final mean density gradient is more nearly linear through this thickened interface, with many small scale steps and interfaces superimposed on it, rather than two large steps at the edges as was first supposed. (For further discussion related to this interfacial structure, see §§4.3.1, 5.3.2 and 10.3.1.)

When the original layers are deep and their velocity is held

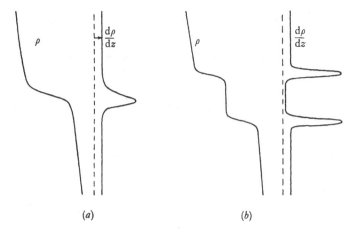

Fig. 4.11. Sketch of the assumed density and density gradient profiles, (a) before and (b) after mixing produced by the K–H instability.

constant after the instability is observed (by returning the tube to the horizontal), an energy argument (cf. §10.2.3) shows that the growth of the interface will be limited and the layers remain distinct. If the flow continues to be accelerated, a new instability can arise on the thickened interface (see §5.3.2). When the layers are not deep compared to the thickened interface, the rolls fill the whole of the channel, and turbulence generated at the solid boundaries will also help to mix both layers thoroughly together.

In subsequent sections we will need to distinguish carefully between instabilities (and mixing) driven by processes occurring at the interface itself, and those due to turbulence produced by some other mechanism elsewhere in the flow. By keeping this difference in mind now, one can explain another observation reported by Thorpe. If, during the setting up of the experiments just described, the axis of the channel was tilted *suddenly* from vertical to horizontal, the equivalent of the 'lock exchange' flow (§3.2.4) was set up, with two 'noses' advancing along the tube in opposite directions. At no time were large overturning instabilities observed on the interface in this case, but instead smaller, less regular travelling disturbances appear, having the cusp-like form shown in fig. 4.13 pl. IX. These are very like the forced waves produced by turbulent stirring in the layers (see chapter 9), and the difference in behaviour can be interpreted as follows.

In the first case, with accelerating laminar layers, the density and velocity gradients are sharp and both confined to an interface of thickness h. The interfacial (gradient) Ri is smaller than the overall Ri_0 based on the layer depth H and related to it by

$$Ri \approx \frac{h}{H} Ri_0. \qquad (4.1.10)$$

Not only is $h/H < 1$ in this situation, but Ri_0 may be made small by waiting long enough, and so Ri will eventually fall below the critical value. If, on the other hand, the velocity profile is determined independently of the sharpness of the density step (as is likely in the steady turbulent counter flows well behind the 'noses'), it will have a lengthscale more nearly comparable with H, so that

$$Ri \approx \frac{H}{h} Ri_0. \qquad (4.1.11)$$

It has been shown (3.2.13) that in this latter case $Ri_0 = \frac{1}{2}$ in a horizontal channel, so whatever the value of $H/h > 1$, no instability of the K–H type is to be expected (but remember fig. 4.8).

A laboratory observation which at first sight contradicts the above statement has been made by Simpson (1969); he showed that K–H 'billows' can form immediately behind the front of a gravity current (§3.2.5). In that case, however, the velocity gradient in the non-turbulent accelerating flow over the nose must again be determined by the density gradient, and a small gradient Ri can be obtained.

Several striking atmospheric phenomena can be understood in terms of the instability of a steady parallel shear across a region of strong temperature gradient. Some kinds of billow clouds certainly arise in this way; Ludlam (1967) has documented some good examples, one of which is reproduced in fig. 4.14 pl. x. Recently high power radar techniques have been used in conjunction with vertical soundings of wind and temperature to document the onset of K–H instabilities even when there is no cloud. Atlas et al. (1970), for example, have obtained echoes from a thin layer of high temperature gradient but small Ri, and have followed billows as they grow, overturn and finally break down to leave a patch of 'clear air turbulence' (fig. 4.15 pl. ix). The original single sharp radar echo is split into two, and this structure has been explained in terms of larger gradients at the edge of the turbulent layer, just as in the laboratory experiments (figs. 4.11 and 4.12).

The formation of billows is often associated with internal waves, which can produce quasi-steady shears on density interfaces. For many purposes these are locally equivalent to the steady shear described above, but they will be discussed separately in §4.3.3.

4.2. The combined effects of viscosity and stratification

Much less is known about the stability of flows in which both viscosity and buoyancy as well as inertia forces are important (but see Drazin (1962) for a general survey of the whole range of possible problems). One difficulty is that sometimes viscosity has been invoked explicitly when an alternative explanation can be given with viscosity playing only a secondary role. Several problems of this kind are included below, as well as one in which viscous effects enter more directly.

4.2.1. *Viscous effects at an interface*

Thorpe's experiments (§4.1.4) have made it clear that the stability of a density interface is well described by inviscid theory, even when the velocity profile is influenced by viscosity. The only likely exception to this rule in an unbounded fluid is a flow at very low Reynolds number, where viscous damping will then reduce the growthrate of disturbances. (See Betchov and Criminale 1967, p. 79.) Much earlier, however, Keulegan (1949) carried out an extensive and much quoted series of experiments, using a pool of sugar solution with a laminar or turbulent flow of fresh water above it. He invoked viscosity to interpret his results, but they should now be re-examined in the light of the more recent work.

Keulegan's essential result can be obtained using dimensional reasoning. Given a velocity U (the mean velocity of the upper layer), a density difference or $g' = g\Delta\rho/\rho$ and the kinematic viscosity ν (of the lower fluid), only one non-dimensional parameter,

$$K = U^3/\nu g', \qquad (4.2.1)$$

can be formed. With a laminar flow, the onset of instability (as judged by the appearance of waves on the interface) was shown to depend mainly on K, and occurred at a critical value of about $K = 500$.

Now let us substitute for ν in (4.2.1) using the viscous lengthscale $d = (\nu x/U)^{\frac{1}{2}}$ (the thickness of the boundary layer formed in time x/U as the flow travels a distance x from the entrance of the channel) to obtain

$$K = \frac{U^2}{g'd}\left(\frac{xU}{\nu}\right)^{\frac{1}{2}}. \qquad (4.2.2)$$

This criterion for instability can therefore be thought of as a Richardson number condition based on the depth of the shear layer, with an additional weak dependence on a Reynolds number Re. With this experimental arrangement the velocity profile is determined completely by the diffusion of vorticity and is not linked to the density profile (as it is in the tilted channel). Since the density distribution was not measured, a quantitative comparison of the two ideas is not possible in retrospect; but the range of mean velocities and hence Re used was small and a completely inviscid explanation of the instability now seems more likely than the original one.

In the turbulent case, Keulegan observed some disturbance of the interface at all velocities (cf. figs. 4.13), and defined 'instability' as the point where the rate of mixing suddenly increased (due to the formation of larger scale K–H waves). This occurred at a lower constant value of $K = 180$, presumably because of the sharpening of the gradients through the interface at a given mean velocity. This is again consistent with an inviscid interpretation, but it is at first sight contrary to the explanation given in §4.1.4 of the contrast between the photographs of figs. 4.10 and 4.13. It must be remembered, however, that the velocity profiles in the two laminar experiments are produced in quite different ways, and are likely to be sharper when the motion is driven by buoyancy differences across a sharp interface.

4.2.2. Thermally stratified plane Poiseuille flow

Definite theoretical results have been obtained for a class of plane viscous flows, the most important example being a parabolic velocity profile between solid walls maintained at different temperatures (so that there is a conductive heat flux through the walls and the undisturbed temperature gradient is linear). In this case both viscosity and also the diffusivity of the property determining the density (e.g. heat) are genuinely important in the stability problem. The limiting case of a destabilizing density gradient and zero velocity will be discussed separately in chapter 7. Here we present some of the results for the combined problem, emphasizing the case where the density distribution is heavy at the bottom.

This flow can be defined in terms of three non-dimensional parameters formed from the maximum velocity U_*, the depth d, the density difference $\Delta\rho$ between the bottom and top walls, and the molecular properties ν and κ. Following the original development of Gage and Reid (1968) (and setting aside for the moment the extension to other profiles due to Gage (1971)) we use

$$Re = \frac{U_* d}{2\nu}, \quad Ra = -\frac{g(\Delta\rho/\rho)\,d^3}{\kappa\nu}, \quad Ri = \frac{g(\Delta\rho/\rho)\,d}{16U_*^2} \quad (4.2.3)$$

i.e. a Reynolds number, a Rayleigh number (see §3.3.4 and chapter 7), and an overall Richardson number which corresponds exactly to the gradient Richardson number near the wall. (For a fixed value of $Pr = \nu/\kappa$, taken by Gage and Reid to be unity in their numerical calculation, Ra and Ri are alternative parameters, not independent ones.) When the density difference is destabilizing (Ra positive), there are two kinds of instability. The first is purely 'thermal' in origin; it sets in at a value of Ra which is independent of the shear and takes the form of rolls with axes in the direction of mean flow. The other leads to the Tollmien–Schlichting type waves characteristic of a shear flow instability over a solid boundary. There is an abrupt transition between the two at about $Ri \approx -10^{-6}$, the thermal instability dominating at larger values of $|Ri|$; the smallness of this number emphasizes the dominant role of the density gradient.

When Ri is positive (and the stratification stabilizing) Gage and Reid again found numerically the form of the stability curves, plotting wavenumber against Re for various fixed values of Ri. With $Ri = 0$ there is an unstable loop in the k, Re plane, the minimum of which gives the critical value of Re. As Ri is increased, the

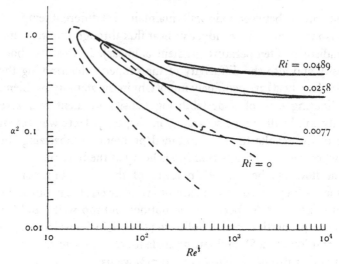

Fig. 4.16. The neutral stability curves (Reynolds number Re as a function of non-dimensional wavenumber α) for various values of Richardson number in stably stratified plane Poiseuille flow. (After Gage and Reid 1968.)

area of the loop decreases (see fig. 4.16) and the critical value Re_{crit} increases. The theory thus predicts that above $Ri = 0.0554$ the flow will be stable to small disturbances no matter how large is Re. It also shows that with any smaller Ri the flow is stable if Re is sufficiently small, and at large Re the instability is confined to a particular narrow range of wavenumbers. Extending the arguments to velocity profiles with inflexion points, Gage (1971) has shown that this conclusion is not changed qualitatively, though the numerical values are different. For a particular profile he showed that complete stabilization is still achieved, at a Richardson number (evaluated at the critical point) of about 0.107.

The much smaller value of the critical Richardson number obtained here (compared to that for a free shear layer) suggests that a boundary always has a stabilizing influence, in spite of the viscous mechanism for instability which it introduces. (The stabilizing effect of symmetrical boundaries on an inviscid interface has already been shown in fig. 4.7.) A similar deduction was made much earlier by Schlichting (1935), though his theory was less satisfactory since it neglected diffusion. He applied his results to the discussion of the boundary layer under the heated roof of a wind tunnel, which

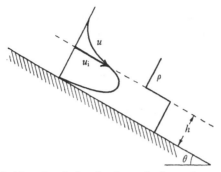

Fig. 4.17. Definition sketch for the flow of a heavy layer down a slope.

was observed to remain laminar up to high Reynolds numbers provided $Ri > \frac{1}{20}$.

When comparing these results with those treated later we must keep clearly in mind the two features which distinguish the problem discussed in this section. There must be a constraining effect, as provided here by the solid boundary (but perhaps a region of much stronger stability could behave like a wall for this purpose). Secondly, the stable density gradient must continue right down to this 'wall', which is only possible when there is a flux of buoyancy through it. (See §4.3.1.)

4.2.3. *Flows along a sloping boundary*

Another case where the role of viscosity is ambiguous is the flow of a gravity current along a sloping boundary. This has been discussed in §3.2.3 (using hydraulic theory) and will be mentioned again in §6.2 (where mixing is of prime concern), but here we investigate the stability of the laminar flow. In addition to being driven by the component of gravity acting down the slope, this will be subject to a stabilizing component across the flow and therefore is related to the other problems treated in this section. We first summarize the properties of the *steady* flow of a uniform layer of heavy viscous fluid flowing under a lighter fluid (sketched in fig. 4.17), whose stability is to be investigated.

Integration of the viscous equation of motion through the depth h of the lower layer leads to

$$u = \frac{g' \sin \theta}{2\nu} [zh - z^2] + u_1 \frac{z}{h}, \qquad (4.2.4)$$

where $u(z)$ is the velocity parallel to the slope and u_1 is the velocity of the interface. In terms of the mean velocity $\bar{u}$ through the layer this may be written

$$\frac{u_1}{\bar{u}} = 2 - \frac{1}{6}\frac{g'\sin\theta\cdot h^2}{\bar{u}\nu} = 2 - \tfrac{1}{6}H. \qquad (4.2.5)$$

The dimensionless parameter H (the inverse of the J defined by Ippen and Harleman (1952)) can be written variously as

$$H = Re\sin\theta/F^2 = Re\cdot Ri_0\tan\theta, \qquad (4.2.6)$$

where

$$Re = \frac{\bar{u}h}{\nu}, \quad F = \bar{u}/(g'h)^{\frac{1}{2}} \quad \text{and} \quad Ri_0 = g'h\cos\theta/\bar{u}^2$$

(the overall Richardson number Ri_0 being defined to conform with §6.2, with the slope θ included in its definition). The numerical value of H can range from 3 for a free surface flow, to 12 when there is a fixed boundary at $z = h$. Ippen and Harleman found experimentally that $H = 7.3$ for the case of interest here, corresponding to $u_1/\bar{u}$ of about 0.59. This agrees with a theoretical result of Lock (1951) who showed that the overlying fluid is dragged forward by the layer to this extent. The important point is that H has a fixed value, so although both Re and F (or Ri_0) enter the problem, in fact only one of them can be chosen independently when the slope is given, the other being related to it through (4.2.6).

The different parameters used in the literature to describe the relative importance of buoyancy and viscosity can all be regarded as combinations of Re and F, and which is most appropriate depends on the boundary conditions. If, as here, some lengthscale is basic and the velocity is produced by buoyancy, then the Grashof number

$$Gr = \frac{g'h^3}{\nu^2} = Re^2/F^2 \qquad (4.2.7)$$

should be chosen, since this retains h but removes $\bar{u}$. When the velocity difference across an interface is imposed and no external lengthscales are relevant then the parameter $K = \bar{u}^3/\nu g' = Re\cdot F^2$ used by Keulegan is appropriate (see §4.2.1). Thus it seems logical to use Gr to describe laminar flows driven along a slope by gravity, always bearing in mind that the use of a relation like (4.2.6) will convert it to an equivalent value of either Re or F alone.

PLATE I

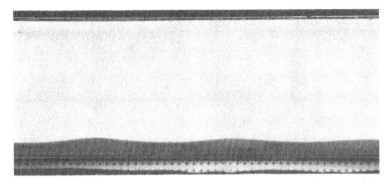

Fig. 2.1. A wave on the interface between two homogeneous fluid layers with different densities and depths. The lower layer is dyed.

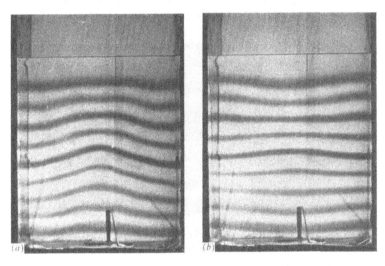

Fig. 2.6. Laboratory experiments on standing internal waves in a continuously stratified fluid (a) mode (2, 1), (b) mode (2, 3). The dyed layers marking surfaces of constant density were inserted during the filling of the tank. (From Thorpe 1968a.)

PLATE II

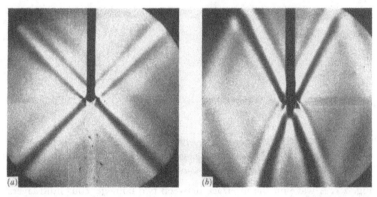

Fig. 2.7. Schlieren pictures of the density variations produced by internal waves propagating along rays and originating at a horizontally oscillated body (a) $\omega/N = 0.615$, (b) $\omega/N = 0.900$. (From Mowbray and Rarity 1967.)

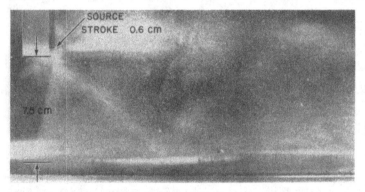

Fig. 2.11. Streak pictures of particle motions associated with internal waves propagating away from a vertically oscillated source ($\omega/N = 0.60$).

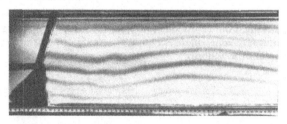

Fig. 2.12. Waves produced by oscillating a flap at one end of a tank of salt solution with linear density gradient. Disturbances described by the 'mode' and 'ray' theories can both be seen. (From Thorpe 1968c.)

PLATE III

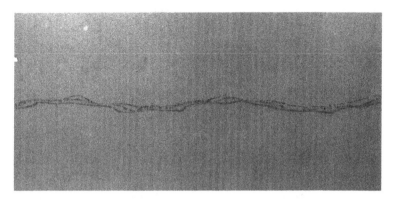

Fig. 2.18. Instability of an interfacial wave caused by the growth of higher modes by resonant interaction. (From Davis and Acrivos 1967 b.)

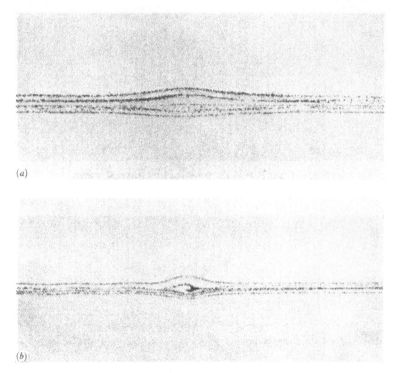

Fig. 3.3. Solitary waves on an interfacial transition region between two layers (a) small amplitude, (b) large amplitude, showing closed streamlines. (From Davis and Acrivos 1967 a.)

PLATE IV

Fig. 2.3. Surface slicks produced by internal waves
(British Columbia Air Photographic Service).

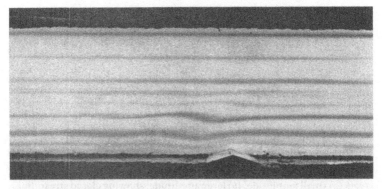

Fig. 2.15. Absorption of energy at a critical level in a shear flow: lee waves
cannot penetrate beyond the centre-line of the channel where $u = 0$.
(From Hazel 1967.)

PLATE V

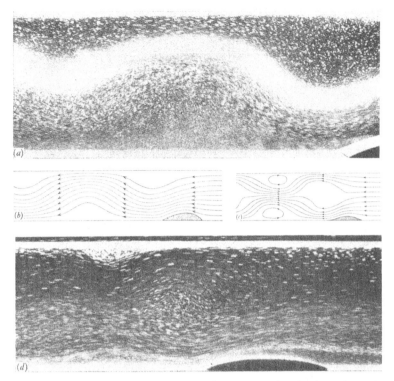

Fig. 3.4. Stratified flow (from right to left) over a barrier in a channel.
(a), (b) Comparison between observed/calculated flow patterns, with one lee mode present and parameters (defined following (3.1.18)) $\kappa = 0.60/0.65$ and $\epsilon = 0.27/0.38$. (From Long 1955.) (d), (c) Comparison between observed/calculated flow patterns for a case where there are two lee modes, and rotors are formed; $\kappa = 0.65/0.70$, $\epsilon = 0.17/0.15$. (From Long 1955.)

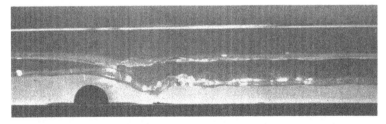

Fig. 3.10. Flow of a three-layer fluid system over a ridge, produced by moving a two-dimensional obstacle from right to left. There is a weak hydraulic jump on the upper interface near the obstacle, and a stronger jump downstream on the lower interface; the latter can be compared to the Sierra Wave phenomenon shown in fig. 3.11 pl. VI. (From Long 1953b.)

PLATE VI

Fig. 3.11. A hydraulic jump in a supercritical airflow over the Sierra
Nevada range, made visible by the formation of cloud, and by dust
raised from the ground in the turbulent flow behind the jump. (Photo-
graph by Robert Symons, published in *Communications on Pure and
Applied Maths*, **20**, no. 2 (review by M. J. Lighthill), © John Wiley &
Sons, Inc., 1967.)

Fig. 3.14. The front of a gravity current in the atmosphere: a Sudanese
haboob, in which thick dust swept up by a cold outflow from a thunder-
storm contributes to the density difference which drives the flow. Note
the overhang, and the clefts in the front which are probably associated
with convective motions arising from the unstable density distribution
near the ground. (Photograph: Flight International.)

PLATE VII

Fig. 3.18. Streak picture of a very slow flow (from right to left) over a high barrier, showing the complete blocking upstream of the obstacle, and the formation of multiple jets. (From Long 1955.)

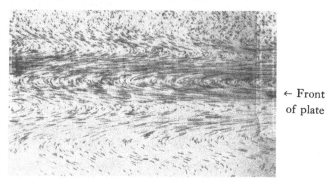

← Front
of plate

Fig. 3.21. Streak picture of an upstream wake, produced by a plate moving from right to left; camera stationary, centred about 10 cm upstream. (From Pao (1968), Boeing Scientific Labs. Document D1–82–0488.)

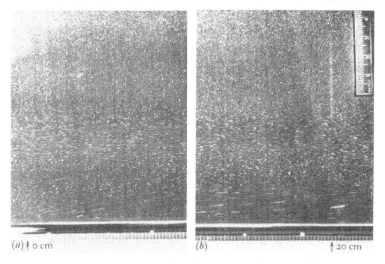

Fig. 3.22. The upstream boundary layer, at two positions along a plate moving from left to right past a stationary camera. The alternating jet structure, and the growth of the boundary layer with distance from the back of the plate, are clearly shown. (From Martin 1966.)

PLATE VIII

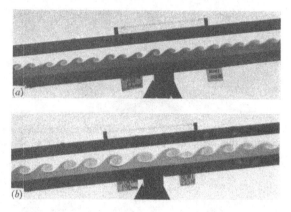

Fig. 4.10. The breakdown of a density interface in a shear flow produced by tilting the containing tube. The interface in (b) has been allowed to diffuse (and thicken) about three times as long as that in (a). (From Thorpe 1971.)

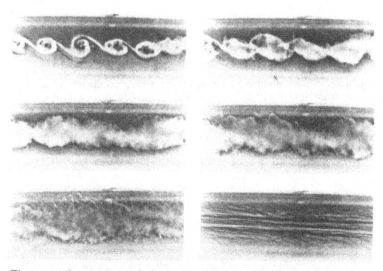

Fig. 4.12. A sequence of shadowgraph pictures of a density interface subjected to a steady shear. Reading from left to right, top to bottom, an instability of the K–H form grows, turbulence is produced, the interface thickens and finally the turbulence is suppressed. (From Thorpe 1971.)

PLATE IX

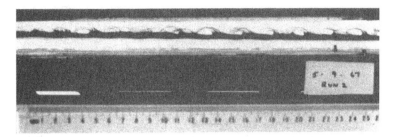

Fig. 4.13. 'Cusped' breaking waves on the interface between the layers which follow two noses advancing in opposite directions along a closed tube (the lock exchange flow). (From Thorpe 1968 b.)

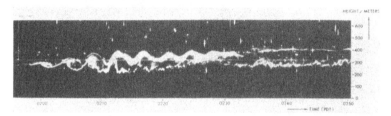

Fig. 4.15. Time–height record of growing and overturning waves obtained using radar in clear air. (From Atlas et al. 1970.)

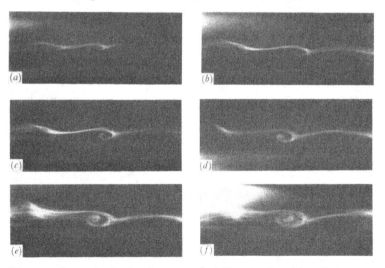

Fig. 4.20. Stages in the development of a billow produced by a long wave travelling along an interface in the thermocline. (From Woods 1968 b.)

PLATE X

Fig. 4.14. Billow clouds, formed by the K–H shear instability mechanism. (Photograph: P. M. Saunders.)

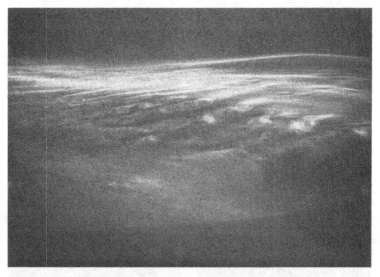

Fig. 4.21. A patch of billows produced by the K–H mechanism on the crest of a 10 m long internal wave on a thermocline sheet. (From Woods 1968b.)

PLATE XI

Fig. 4.23. Clouds formed by lee waves over a mountain. The lowest cloud is associated with a turbulent rotor; above it is a smooth lenticular cloud formed on a wave crest which is not overturning. (Photograph: Betsy Woodward.)

Fig. 7.2. Convective clouds in an unstable layer, aligned in 'streets' along the direction of shear. (Compare with fig. 4.14 pl. x, which shows clouds formed by a shear instability and aligned across the flow. The form of 'billow' clouds can vary widely according to the relative importance of shear and convection.) (Photograph: R. S. Scorer.)

PLATE XII

Fig. 4.22. Shear instability occurring at a node of a standing interfacial gravity wave. (From Thorpe 1968a.)

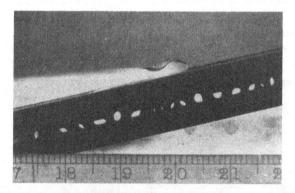

Fig. 4.24. The breaking of an interfacial wave as it approaches a slope. (From Thorpe 1966a.)

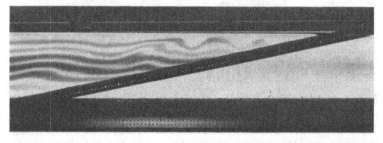

Fig. 4.25. The breaking of an internal wave in a continuous density gradient as it approaches a slope. (From Thorpe 1966a.)

PLATE XIII

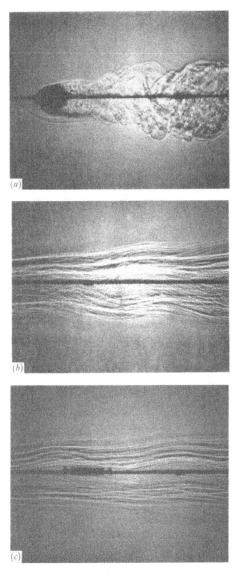

Fig. 5.10. Shadowgraph pictures due to Pao (1968b), of a turbulent wake behind a circular cylinder in a stratified fluid (a) near the body: the wake resembles that in a homogeneous fluid, (b) 50 diameters behind the body: the large scale motions have been damped, (c) 100 diameters behind the body: the fine structure has also decayed, leaving horizontal striations of concentration. (From Pao (1968), Boeing Scientific Labs. Document D1–82–0959.)

PLATE XIV

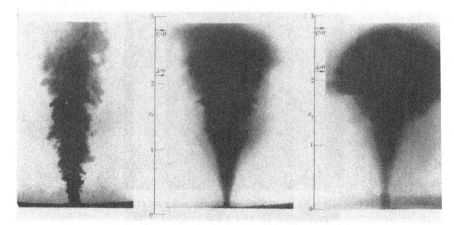

Fig. 6.2. Steady turbulent plumes of dyed buoyant fluid (a) in a uniform environment, with a short exposure which shows the large eddy structure, (b) in a stably stratified environment: a time exposure during an early stage of release, (c) in a stably stratified fluid: a time exposure at a later stage when a layer is spreading out sideways at the top. (From Morton, Taylor and Turner 1956.)

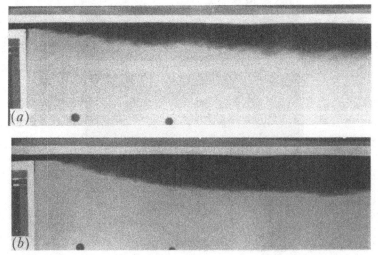

Fig. 6.9. (a) An internal hydraulic jump of the maximum entraining type, with entrainment along its whole length. (b) An internal hydraulic jump controlled by a broad-crested weir downstream: the flow upstream of the jump is identical to that in (a). Entrainment is occurring only at the far upstream end of the jump, and the remainder consists of a roller zone. (From Wilkinson and Wood 1971.)

PLATE XV

Fig. 6.13. Streak picture of the flow in and around an isolated thermal of dyed salt solution, showing approximate streamlines relative to axes at rest. (Photograph by P. M. Saunders.)

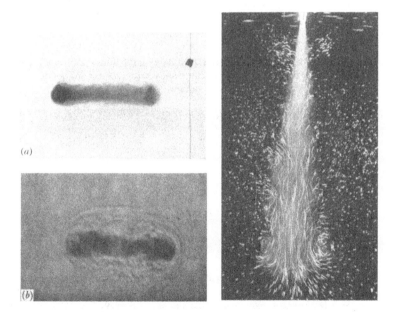

(a)

(b)

Fig. 6.15. (*Left.*) Buoyant vortex rings. (*a*) A turbulent, buoyant core. (*b*) A shadowgraph picture, showing the buoyant core and the surrounding volume of fluid which is carried along with it.

Fig. 6.16. (*Right.*) Streak picture of a 'starting plume' of salt solution in fresh water, showing the cap (resembling a thermal) followed by a steady turbulent plume.

PLATE XVI

Fig. 7.3. Convection rolls forming in a layer of silicone oil with a free surface. The concentric ring pattern shows the strong influence of the circular boundary. (From Koschmieder 1967.)

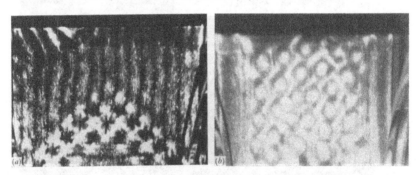

Fig. 7.4. The formation of cells in a square geometry, with solid boundaries above and below. In (a) the mean temperature is fixed, and roll cells form, in (b) it is changing slowly, and hexagons are observed. (From Krishnamurti 1968.)

PLATE XVII

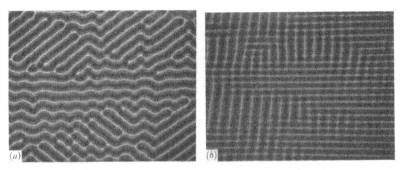

Fig. 7.5. Instability of convection rolls formed on a fluid layer held between two glass plates (a) the zig-zag form, (b) a cross-roll instability. (From Busse and Whitehead 1971.)

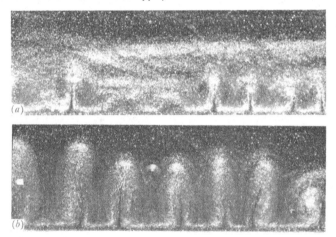

Fig. 7.13. Two stages in the development of a high Rayleigh number flow in silicone oil, marked with aluminium particles, following a sudden increase in temperature of the lower boundary. (From Elder 1968.)

Fig. 7.14. 'Thermals' rising from a heated horizontal boundary under a layer of water. (From Sparrow, Husar and Goldstein 1970.)

PLATE XVIII

Fig. 7.15. The establishment of a stable stratification by a continuous plume of salt solution in a tank filled initially with fresh water. The layers were formed by the injection of dye at fixed time intervals; more dye was being added when this photograph was taken. (From Baines and Turner 1969.)

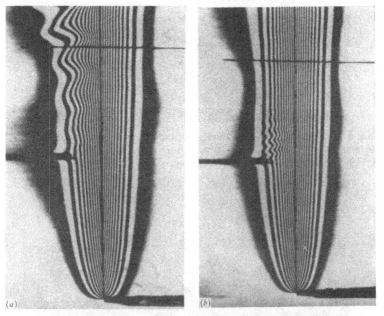

Fig. 7.19. Interferograms of oscillations in the boundary layer near a heated vertical plate. In (a) the disturbance is amplified and in (b) it is damped. (From Polymeropoulos and Gebhart 1967.)

PLATE XIX

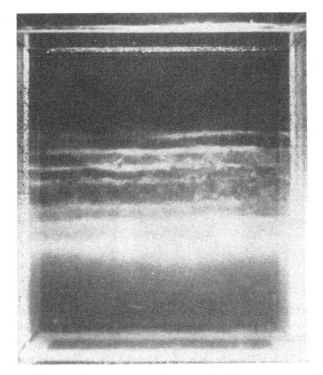

Fig. 8.5. Showing the layers formed in a laboratory tank of smoothly stratified salt solution, by heating from below. The fluid is marked with fluorescein dye (originally put into the bottom layer) and by suspended aluminium powder.

Fig. 8.11. The formation of layers in a two-component system due to the introduction of a sloping boundary. The original distribution of solutes, (linear gradients of NaCl with maximum concentration at the top and zero at the bottom, and of sugar with maximum at the bottom and zero at the top) was stable when contained by vertical walls.

PLATE XX

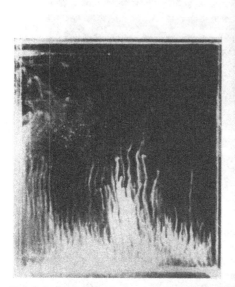

Fig. 8.8. (*Left.*) Vertical cross-section of salt fingers, marked by fluorescein dye added to the upward moving fingers, and lit with a thin sheet of light perpendicular to the viewing direction.

Fig. 8.9. (*Right.*) The formation of convecting layers from a smooth stable gradient of salt, driven by a flux of sugar originating in the dark dyed layer at the top. (From Stern and Turner, *Deep-Sea Res.* **16**, 497–511. © Pergamon Press Ltd, 1969.)

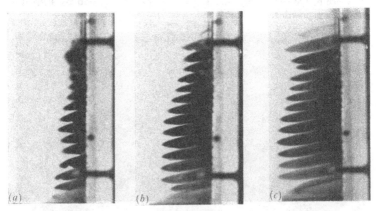

Fig. 8.10. The development of layers in a stratified salt solution subject to heating through a vertical side wall. The photographs were taken (*a*) 19.5 min, (*b*) 24 min and (*c*) 28.25 min after heating began. (From Thorpe, Hutt and Soulsby 1969.)

PLATE XXI

Fig. 8.17. Shadowgraph picture of an interface about 2½ cm deep containing sugar–salt fingers, with convecting layers above and below. (From Shirtcliffe and Turner 1970.)

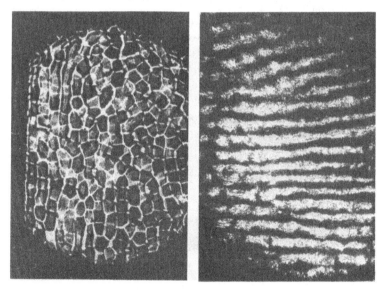

Fig. 8.18. (*Left.*) Plan view of sugar–salt fingers, obtained by a shadowgraph method. The fingers tend to be square, but with gradually changing orientation, except where they are aligned by boundaries (as at left of picture). (From Shirtcliffe and Turner 1970.)

Fig. 8.19. (*Right.*) Plan view of a sugar–salt 'finger' interface, across which there is a shear. The fingers are changed into sheets aligned in the direction of the shear. (Photograph: P. F. Linden.)

PLATE XXII

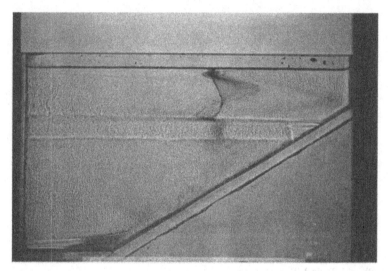

Fig. 8.21. Shadowgraph picture of a two-layer experiment (sugar above salt solutions) in a container with a sloping boundary. The motions are indicated by the distortion of a dye streak. Note the flow right to the bottom, which has produced a reversal of the relative gradients, as indicated by the absence of fingers.

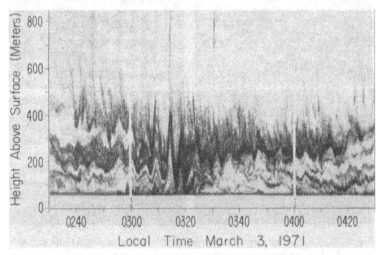

Fig. 10.8. An example of the multiple atmospheric layers observed during the night time radiation inversion, using an acoustic sounding technique. There are many discrete layers, with a widely varying intensity of temperature fluctuations and hence echo strength, all oscillating vertically in phase. (Record obtained by the Wave Propagation Laboratory, NOAA, U.S.A.)

PLATE XXIII

Fig. 9.2. Photographs of a stable interface between salt and fresh water, with stirring on one side of the interface. Dye has been added to (a) the stirred (lower) layer, and (b) the stationary layer. (From Turner 1968b.)

PLATE XXIV

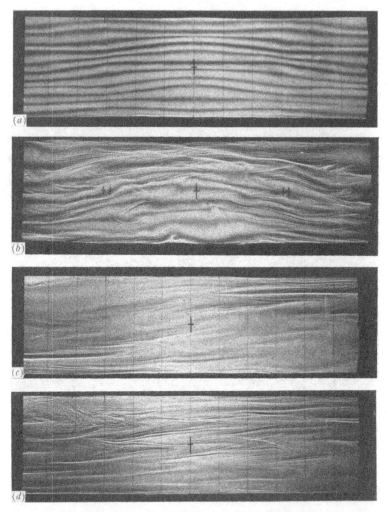

Fig. 10.5. Shadowgraph photographs due to McEwan (1971, and un-published), which illustrate the various features of internal wave breakdown described in the text. The depth of the fluid is 32.6 cm in all cases. (a) Self-limited distortion of the 2/1 forced mode, due to higher modes produced by resonant interaction. (b) Violent breakdown of the 2/1 mode at larger forcing amplitude, again caused by resonant interaction. (c) The growth of instabilities with strong forcing of the 1/1 mode, a case for which there is no resonant interaction. (d) The same, taken six cycles after (c).

Few definite theoretical results are available for the stability of these flows, though we can with some confidence make a generalization from the corresponding free surface cases (see Yih 1965 and Lin 1969). The stability boundary will be expressible as a relation between wavenumber k, slope θ and only one of Re or F (because of the relation between these, established above), and there will always be a minimum Re (or F) above which the flow becomes unstable. The experiments of Ippen and Harleman (1952) suggest that the two-layer flow always becomes unstable when $F = 1$, in the sense that the first wave-like disturbances appear then, with wavelength $\lambda \approx 3h$. This result was independent of slope (and therefore of Re) up to about $\theta = 9°$, and it implies that the breakdown is one which involves interfacial waves on the top of the layer, i.e. it is determined by the inviscid hydraulic criterion discussed in §3.2 and §4.1.2. (Compare with (4.1.6).) Sustained turbulence, on the other hand, was only observed when Re rose above about 1000, and so this must be associated still with the growth of T–S waves at the solid boundary.

It is instructive to compare this layer flow with the related case of a laminar boundary layer under a heated sloping plane in air, which has a continuous stable density profile right to the wall. (This, and its stability, will be described for a vertical wall and for inclined flows *above* the heated plate in §7.4, in the context of convective motion.) It suffices to say now that for a fixed $Pr = \nu/\kappa$, the flow can again be specified by a single parameter, usually taken as $Gr \sin \theta$ where the Grashof number Gr is based on the distance along the plate, but again equivalent to Re through a relation like (4.2.6). Tritton (1963) showed experimentally that for a range of large slopes between 40° and 90° from the horizontal, fluctuations first appear when Gr reaches a value which is nearly independent of the slope and which corresponds to a local value of Re (based on the volume flux) of about 250. He also carefully distinguished this first instability from the fully turbulent state, and both sets of results are shown in fig. 4.18, where critical values of $(Gr)^{\frac{1}{3}}$ are plotted against the angle θ to the horizontal. While the first instability is insensitive to slope, the attainment of the fully turbulent flow is definitely retarded by the stabilizing effect of the density gradient as the plane is tilted towards the horizontal. This is therefore another example

.

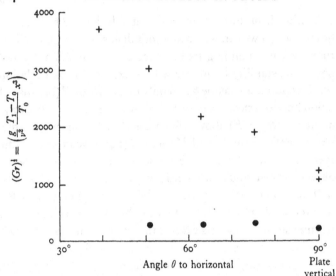

Fig. 4.18. Flow under a heated sloping plate: critical Grashof numbers corresponding to: ● first fluctuations and + fully turbulent flow at various slopes. (After Tritton 1963.)

of the kind of flow considered in §4.2.2, in which the gradient right up to the wall is important.

4.2.4. Transition to turbulence

Before leaving the subject of hydrodynamic stability as such, it is important to be clear about the relevance of the criteria it produces to the question of whether or not a given flow will be turbulent. The answer is already implied by the results of earlier sections, but some further discussion will emphasize the point. Linear stability theory only gives information about the initial growth of small disturbances, and can say nothing about their development at larger amplitude. Drazin (1970) has extended the calculations of the K–H instability into the slightly supercritical range, but no such theory can provide a reliable guide to the detailed behaviour under highly supercritical conditions. The most that can be said then is that the non-linear superposition of many unstable modes produces a truly turbulent motion.

Special care is needed when several physical parameters are

relevant. For layer flows at large slopes, (4.2.6) shows that the criterion $F = 1$ governing the appearance of the first wave-like instability will be satisfied at quite small values of Re. At such values, however, the effect of viscosity on the growth of disturbances can still be large, and a fully turbulent flow is not observed until Re is much greater, comparable with that required for turbulent pipe flow. When there is a density gradient sustained right to the wall, it is clear from Tritton's results (and the theory of §4.2.2) that the Reynolds number for transition to turbulence depends on Ri; but at small enough Re it is this parameter, not the buoyancy parameters Ri or F, which determine the point of transition.

It will be more generally true that the transition to turbulence will be a function of both Re and F, and that only when Re is sufficiently large can its effects be ignored entirely. Based on the experience with flows near boundaries one might suppose that decreasing either Re or F should always increase the stability, but very little is known about the critical values for other boundary conditions, for instance an intruding fluid layer bounded above and below by density interfaces. (Caution is suggested by the results of Thorpe (1969a), who showed theoretically that a certain stratified flow is unstable whereas the corresponding homogeneous (inviscid) shear flow is stable, and Hinwood (1967), who reported experiments which indicated transition in a pipe flow at lower Re with stratification than without.) In chapter 10 it is shown that the important limitation on turbulence at internal shear layers in a large body of stratified fluid (such as the ocean) may often be one of Reynolds number, but with the relevant vertical lengthscale set by the stable stratification.

Finally, it is worth remarking again that the basic property of a stably stratified fluid is its ability to sustain internal wave motions. These need not be unstable, and can certainly exist without there being any breakdown to turbulence. On the other hand, when the motion can be described as turbulent, in the sense that mixing and an enhanced rate of diffusion are observed, it is rarely possible to neglect wave motions entirely. Though these two aspects of a turbulent stratified flow will be considered separately in some of the following sections, the interaction between them must always be kept in mind, and this will be taken up again explicitly in the final chapter.

4.3. Mechanisms for the generation of turbulence

We now turn to the wider question of turbulent flows which are beyond any condition of marginal stability, and identify and compare the processes whereby turbulence and mixing can be produced in a stably stratified fluid. The source of energy for the maintenance of the turbulence must always be examined carefully: it seems fundamental to distinguish clearly between turbulence generated directly at a boundary, and that arising in the interior in various ways. A systematic classification can be based on this principle, and for convenience of later reference, the range of possibilities will first be described in general terms in §4.3.1. The detailed discussion of some of these cases must be deferred to later chapters, but we will follow up here several special problems which arise from topics which have already been introduced.

4.3.1. Classification of the various mechanisms

At the two extremes, it is easy to say whether the energy which is producing turbulence comes from an 'external' or 'internal' process. The mixing of a nearly homogeneous fluid flowing through a pipe or channel is determined entirely by boundary generated turbulence, and the mixing is carried out by eddies extending right across the flow (as shown diagrammatically in fig. 4.19a). At a density interface well away from boundaries (fig. 4.19b), across which a steady shear is established, the breakdown and the production of turbulence are certainly internal processes (§4.1.4), and the turbulent motions are confined to the interfacial region.

A special case of internal mixing will be identified in §5.1.4; it will be shown that the shear and density distributions can be maintained in a kind of marginally stable state, the turbulence being self-regulated by the mixing it produces. The outer edge of a turbulent gravity current flowing down a steep slope can be described in these terms (§6.2.3), since the scale of the turbulence is so strongly limited by the vertical density gradient that this region is effectively isolated from the boundary (fig. 4.19c).

The interpretation of other situations requires more care. Paradoxically, the very stable case of a thin, heavy layer flowing along

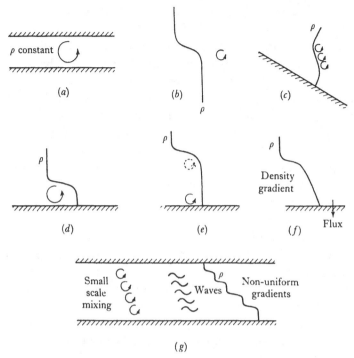

Fig. 4.19. Classification of mixing processes. (a) External mixing of homogeneous flow in a channel. (b) Internal mixing across a density interface remote from boundaries. (c) Self-regulated ('internal') mixing at the outer edge of a gravity current on a steep slope. (d) Gravity current on small slope: mixing across interface is driven by turbulence generated at boundary. (e) Small scale turbulence decays with distance from source. (f) Buoyancy flux also makes externally-generated turbulence ineffective at interface. (g) In strongly stratified channel flows the density gradient must be non-uniform, and both wave motion and turbulence will be important.

a horizontal boundary under lighter fluid (fig. 4.19d) has many features in common with the unstratified case (a). The mixing and stress across the interface are now very small (see §6.2) so that such a layer will be practically homogeneous and stirred by turbulence generated at the boundary. The mixing of fluid across the interface can then be regarded as a consequence of this turbulence, i.e. as a boundary-driven mixing process. This idea can be applied to layers which are stirred in other ways (see §9.1).

There are two circumstances in which turbulence produced at a solid boundary will become less effective in mixing across a density interface some distance away. If, instead of being generated on the scale of the whole layer (as implied in (d)), the turbulence has a much smaller scale than this depth (fig. 4.19e), viscous dissipation will cause the turbulent energy to decay with distance from the source region. When the boundary-interface distance becomes large, this contribution to the energy will have a negligible effect on the interface compared with local processes, and we approach case (b) again. Secondly, when there is a buoyancy flux through the boundary in the sense which will produce a stable density gradient in the otherwise homogeneous layer (fig. 4.19f), all the turbulent energy made available at the boundary may be used in working against gravity in this region, leaving none for mixing across the sharper interface beyond.

On the other hand the boundary stress can remain an important parameter in channel flows having a substantial density gradient (and a large overall Richardson number), although one might at first sight expect these to be 'internally' limited. As discussed more fully in chapter 10, this can be so because such flows must be intermittently turbulent, part of the stress being carried by turbulence and part by waves (fig. 4.19g). In one sense the mixing is an interior process (because the turbulent motions will be limited in vertical extent, as in (c)). However, the whole of the vertical flux of horizontal momentum comes ultimately from the boundary, and this will continue to have an indirect effect on the interior turbulence, operating through the wave contribution to the momentum transport.

4.3.2. Flows near boundaries

It is obvious from fig. 4.19 that one cannot rely on the superficial appearance of a flow to decide how the energy is supplied, and whether the flow will be turbulent. With the above ideas in mind, it is instructive to reconsider some aspects of flows near solid boundaries which bear on the generation of turbulence.

For all large scale natural flows of this kind (for example the wind flowing over the ground), the maintenance of turbulence can be taken for granted if the Richardson number criterion permits it. Such flows are extreme examples of those whose stability was discussed in §4.2, and provided the Reynolds number is large, this latter parameter no longer enters the problem explicitly. (This assumption is the starting point of §5.1.) It is always necessary, however, to consider whether or not there is a buoyancy flux through the boundary, and the precise way in which the stress is transmitted to the fluid.

When the boundary flux is small, and the surface is 'aerodynamically rough', the stress is applied through the overlapping of the turbulent wakes of individual roughness elements, in a region of essentially homogeneous flow. The same is often true of much larger irregularities on a boundary: as pointed out by Davis (see §3.1.4), separation can occur behind rugged mountains and tall buildings, resulting in a direct, local injection of turbulent energy into the flow downstream.

The theories of lee waves described in §§2.3 and 3.1 assume on the contrary that the flow remains attached to the boundary, and that a stable density gradient extends right to the ground. The latter condition, which implies that heat is being extracted at the ground, provides some physical justification for the assumption that separation does not occur when the circumstances are right. Thus the sudden onset of lee waves over some ranges of hills in the evening can be attributed to a change between separated and unseparated flow. During the day convection off a heated mountain will encourage separation, whereas cooling after sunset can produce a katabatic wind (§6.2) in the lee which attaches the flow, and allows the wave motions to be set up. Much of the momentum put into the flow at the boundary can in this case be carried into the interior, only later producing turbulence in the various ways described in following sections.

Other turbulence-producing phenomena which are more directly associated with the flow over obstacles are described in chapter 3. When a stratified airflow blows over a mountain with supercritical velocity (so no lee waves can be formed), an internal hydraulic jump can occur over a horizontal plane downstream. A large amount of energy is lost during this process, and it can appear in several forms. In strong jumps, most of it is in the form of turbulent kinetic energy, some of which is used for mixing (§6.2.3). In weaker jumps, some energy is radiated away in the form of waves (on an interface or into the body of the fluid) which are not now strictly lee waves. A spectacular example of this effect is the cloud formed behind the Sierra Nevada range (see fig. 3.11 pl. VI). A related phenomenon is the head wave behind a nose or small front (§3.2.5). Here, too, a consideration of the momentum balance shows that energy must be dissipated, and it is probably better to regard the generation of turbulence in this region as a property of the whole flow rather than the result of a local shear instability.

4.3.3. *Shear instabilities produced by interfacial waves*

In this and the following sections it will be assumed that internal waves of various forms have by some unspecified means been produced in the interior of a stratified fluid. We examine the possible mechanisms for the breakdown of such waves, first at an interface and later with continuous stratification. Internal waves certainly cannot 'break' in the same way as surface waves. For the surface wave mechanism to be relevant, the downward accelerations must be comparable with the acceleration due to gravity g (not g'), and before such a condition can be approached for internal waves, they will already have become unstable for other reasons.

In §2.1.1 it was shown that the passage of an internal wave along a sharp boundary between stratified layers produces a vortex sheet at the interface. When the wave period is long compared to the time needed for any disturbances to grow, this flow can be regarded locally as quasi-steady, and an instability of the K–H type is to be expected, as described in §4.1.2. When the interfacial region is more spread out, the vorticity produced by the internal wave will also be distributed, and the criterion for instability must depend on the gradient Richardson number (§4.1.3). Phillips (1966a, pp. 168, 187) showed how one can relate Ri to the properties of the wave which gives rise to the shear; we will summarize his argument and then discuss some of the limitations on its use.

We are interested in the shear through a region where the buoyancy frequency N^2 is varying, and so (2.2.6) is taken as the starting point. The amplitude of the vertical and horizontal velocities associated with a wave are related through the continuity equation

$$\frac{\partial \hat{w}}{\partial z} = -ik\hat{u}(z). \qquad (4.3.1)$$

On substitution in (2.2.6) this gives

$$\frac{\partial \hat{u}}{\partial z} = -\,i \left\{ \frac{N^2(z)}{\omega^2} - 1 \right\} k\hat{w}$$
$$= -\left\{ \frac{N^2(z)}{\omega^2} - 1 \right\} \omega k a,$$

$$(4.3.2)$$

where a is the amplitude of the vertical displacement. The maximum value of the shear occurs where $N(z)$ is a maximum (say $N = N_m$) and (for travelling waves) near the crests and troughs. There is a local minimum of the Richardson number here, given by

$$Ri = \frac{N_m^2}{(\partial \hat{u}/\partial z)^2} = \left(\frac{N_m}{\omega} - \frac{\omega}{N_m} \right)^{-2} (ka)^{-2}. \qquad (4.3.3)$$

When $\omega \ll N_m$ and the wavelength is large compared with the interface thickness, the shear can be regarded as steady, suggesting the use of the stability criterion $Ri_{crit} = \frac{1}{4}$. (This value was shown in §4.1.3 to be appropriate for interfaces produced by diffusive spreading far from boundaries.) Applying this condition to (4.3.3), the shear flow associated with a wave of wavenumber k and amplitude a should be unstable if

$$ka > 2\omega/N_m \qquad (4.3.4)$$

which can be quite a small slope. Notice that for a given density difference *thin* interfaces and hence *large* N_m make the interface more unstable to this kind of disturbance, because the shear is also proportional to the density gradient but enters as the square in the denominator of (4.3.3) (cf. the tilted tube experiments of §4.1.4).

Though the above model describes the most important qualitative feature of interfacial stability, the relation (4.3.4) is probably not a satisfactory single criterion for several reasons. The time factor is important too: Ri must be lower than the limiting value and stay subcritical long enough for significant growth to take place before the wave passes by. It is not clear either that it is sufficient to consider only the region near the local (central) minimum of Ri. The horizontal velocity associated with a first mode internal wave will have other inflexion points in the deeper layers on each side where the density gradient (and hence Ri) is small. It may in fact be essential to consider the profiles as a whole before one can say when and where a breakdown first occurs. More work is also needed before we can claim to understand the other limit of short steep waves with frequency comparable with N_m.

Beautiful observations of this kind of instability have been made in the ocean by Woods (1968b). Using skin diving techniques, he has shown that the seasonal thermocline near Malta is characteristically made up of a series of 'layers' in which the temperature gradient is weak, separated by thin interfaces or 'sheets' where the gradients are much larger (see chapter 10 for further discussion of this structure). When an interface is marked with dye and a train of long waves passes along it, shorter waves of the K–H type can be observed to grow at the crests or troughs. The photographs reproduced in fig. 4.20 pl. IX and fig. 4.21 pl. X show that eventually there is a breakdown to give a patch of turbulence which thickens the interface locally. The observations are in reasonable agreement with the theoretical predictions based on the interfacial structure both for the wavelength (cf. (4.1.9)) and the growthrate. There is sometimes a preferential growth at the crest or trough alone, because of the asymmetry introduced by a steady mean shear.

The energy dissipated by turbulence generated in this way comes from the internal waves, and so this provides a mechanism for limiting their amplitude (as breaking does for surface waves). Phillips (1966a, p. 188) has gone further, and based a calculation of the form of a 'saturated' interfacial wave spectrum on this hypothesis. This will not be pursued here, but we note that, even without taking into account any change in amplitude of the basic waves, the instability must be self-limiting, because as the interface spreads out, the energy supplied by the shear is no longer sufficient to mix fluid against the stabilizing effect of the stratification (cf. §10.2.3).

Shear flow instabilities have also been observed (in the laboratory) associated with standing waves, for which the maximum shear is now at the positions of maximum slope (§2.1.3). Carstens (1964) produced the lowest order internal seiche in a long closed tank containing two immiscible fluids, by oscillating the tank horizontally. He observed a K–H type of instability close to the conditions predicted by his theory (which included surface tension), and also showed that the theoretically possible class A disturbance (see §4.1.1) did not have time to grow to an observable amplitude. Thorpe

(1968a) produced much shorter, steeper, internal waves using oscillating plungers at the ends of an experimental tank, and observed a breakdown of the interface between two layers of miscible fluids. One of his photographs is shown in fig. 4.22 pl. XII; the overturning at the node rather than the crest is very clear.

4.3.4. The interaction between wave modes

Another kind of interfacial instability has already been mentioned in §2.4.2, which arises from a resonant action between interfacial modes. At first sight its effect on mixing at the interface may appear similar to that produced by the shear mechanism (see fig. 2.18), but the conditions for breakdown are quite different. Note again the result first obtained by Keulegan and Carpenter (1961), who observed such disturbances only when the interface thickness became *larger* than some critical value (for fixed k and a), the exact opposite of the behaviour predicted by (4.3.4). Interaction between a resonant triplet of waves on an interfacial layer leads to instability in the following way. Energy is fed from the basic wave to the other modes, which can grow exponentially from a small initial amplitude and become manifest as a growing disturbance to the original wave. The most prominent feature of the disturbance shown in fig. 2.18 pl. III is clearly of the second mode (and see also fig. 3.3). The ultimate breakdown to turbulence occurs in the form of an overturning motion, which results from the distortion of the second mode by the shear associated with the original mode.

The criterion for growth of a disturbance depends on viscosity, since the rate of transfer of energy between modes must be at least great enough to overcome viscous dissipation. For comparison with (4.3.4), we record the form of the critical slope for the first mode with wavenumber k_1 on an interface with density profile $\rho \propto \tanh z/L$; according to Davis and Acrivos (1967b) this is

$$(ak)_{\text{crit}} = 2\nu(g\delta)^{-\frac{1}{2}} L^{-\frac{3}{2}} F(m). \tag{4.3.5}$$

Here $\delta = \frac{1}{2}\ln(\rho_{\max}/\rho_{\min})(=\frac{1}{2}\rho'/\rho_{\max}$ for small density differences), L is a measure of the interface thickness and F is an increasing function of the smaller wavenumber m of the two which interact with k_1 ($F(m) \approx 1.7$ when $m = 2$). This form is clearly consistent with the observations that, of all the possible waves formed by triplet interactions, the low wavenumber disturbances are the ones observed, and appear at smaller amplitudes when L is larger.

Two calculations which involve both the resonance mechanism and a mean shear flow should also be mentioned here. Kelly (1968) has considered spatially periodic disturbances superimposed on a particular stratified antisymmetric shear layer, and showed that resonance can lead to a growth in time of two-dimensional waves having twice the wavelength and half the frequency of the periodic component. Their growthrate is faster than that of the most unstable disturbance on the mean flow, provided the amplitude of the imposed wave is large enough; such waves nevertheless draw their energy from the basic flow.

Craik (1968) has predicted a strong 'resonant' instability at a sharp density interface which can be initiated by an interfacial wave. The mean flow considered is unidirectional, with constant velocity gradients which must be sufficiently large but unequal in the two layers. For a basic wave travelling in the direction of the mean flow, the two growing waves which complete the resonant triad can have the same component wavenumber in that direction but propagate obliquely (between 60 and 90° to it). At resonance, all three waves have the same critical layer (see §4.1.1) and the bulk of the energy transfer takes place near that level. The growing waves draw their energy from the mean shear, leaving the original wave disturbance essentially unchanged, and small viscosity plays a dominant de-stabilizing role in the non-linear transfer process (though the un-stable waves are of Class B, according to Benjamin's classification).

4.3.5. *Internal instabilities with continuous stratification*

For completeness we should also mention here the various mechanisms which can lead to the generation of turbulent patches in the interior of a stratified fluid, away from solid boundaries or sharp interfaces. These too are associated with waves propagating through the fluid, and several of them will be described more fully in § 10.3.

At sufficiently large amplitudes, the streamlines associated with lee waves can become vertical (§3.1.4), and circulating regions with closed streamlines ('rotors') appear in the flow. The density distribution is overturned and so becomes unstable, and a patch of turbulence is produced by the conversion of potential energy into kinetic. This behaviour is shown clearly in Long's experiments (fig. 3.4 pl. v), and spectacular examples of this 'kinematic' type of instability are observed in the atmosphere when condensation makes the rotors visible (see fig. 4.23 pl. xi). There is in lee waves the additional possibility of a 'centrifugal' instability occurring in parts of the flow where the curvature is large, before the fluid actually overturns (Scorer & Wilson 1963).

Local, transient overturning can also be produced both by the internal resonance mechanism described in §2.4.3, and by the random superposition of waves arising from many sources, which cannot so easily be associated with particular topographical features. Energy can be concentrated through the 'critical layer' mechanism described in §2.2.3; and we should also mention now several other ways in which the energy of progressive internal gravity waves can be concentrated in a smaller mass of fluid, so increasing the amplitude and the likelihood that they will 'break'. There is the internal equivalent of surface waves approaching a beach, and transferring their energy into a layer of decreasing depth. This effect has been observed in the ocean (see Defant 1961), but it will be illustrated here using the laboratory results of Thorpe (1966a) who studied both the two-layer and continuously stratified cases. Interfacial waves at first steepen at the front as the lower layer becomes shallow, but unlike breakers on a free surface, the 'crests' break backwards (as shown in fig. 4.24 pl. xii); the flow up the slope behaves like the 'nose' at the front of a gravity current (§3.2.5).

The phenomena observed in the continuous case are more complicated (see fig. 4.25 pl. xii, and also compare with the results of Wunsch (1969) referred to in §2.2.2). Near the horizontal boundary, which is the free surface in this case, there is a reduction in wavelength as well as a steepening, accompanied by regions of overturning. The flow near the bottom again runs up the slope, but without forming a raised head, and the breaking process now looks more like surface breakers. After a short time, turbulent mixing near the slope begins to change the density distribution. Mixed fluid flows back into the interior in the form of layers, which are especially prominent and regular when the boundary slope β is critical (i.e. when $\beta = \frac{1}{2}\pi - \theta$ in the notation of §2.2.2, so that the particle motions are parallel to the slope – see also Hart (1971c)). Mixing near a boundary, caused by this and other mechanisms, introduces another possibility: an increase in amplitude leading to breakdown will occur when internal waves propagate horizontally into a region of decreasing density gradient which can no longer sustain them. Thus waves on the thermocline can be expected to grow as they approach a well-mixed region near a shoreline.

Another geometry which can lead to the growth of wave amplitude is a channel of constant depth which converges horizontally, so increasing the energy per unit span for a wave travelling towards the narrower end. Provided the scale is large, so that the viscous damping can be kept small, and the convergence is slow enough to avoid reflection, this seems a promising, but largely unexploited, experimental technique for studying large amplitude waves far from the direct influence of a wavemaker. In just the same way, a tank with decreasing width in the vertical could be a useful device for studying the stability of the vertically propagating waves described by the ray theory of §2.2.2. (It is not known, for example, how large the amplitude can be before some kind of instability occurs, probably in the shear layers bounding the strong in-phase motions.) This latter system is analogous to an atmosphere whose density decreases strongly with height (§3.1.4); the conservation of wave energy per unit mass will lead to an increase of amplitude in both cases. Lee waves of modest amplitude near the ground, and stable to all of the mechanisms discussed here, could achieve a sufficiently large amplitude to become unstable merely by propagating upwards to a level where the density is very much reduced.

CHAPTER 5

TURBULENT SHEAR FLOWS IN A STRATIFIED FLUID

The theme of this chapter will be a more detailed discussion of various kinds of turbulent shear flows, which are (at least at the beginning of the period of interest) well past the state of marginal stability. Some knowledge of the properties of a turbulent shear flow in a homogeneous fluid must be assumed (see, for example, Townsend 1956), and we consider here the additional effects introduced by the presence of density gradients. Turbulent flows in which gravity plays an essential role in driving the mean motion (e.g. turbulent gravity currents) will be treated separately in chapter 6.

These flows will be discussed against the background of the classification introduced in §4.3.1, emphasizing the turbulent features, but also referring to the wave aspects when necessary. We consider first a shear flow near a horizontal boundary, and the effect of a vertical density gradient on the velocity and density profiles, and on the rates of transport. Next we discuss the few theoretical results which are available to describe 'boundary' and 'interior' turbulence in stratified shear flows. Finally, we present and discuss laboratory and larger scale observations which can be used to test these ideas.

5.1. Velocity and density profiles near a horizontal boundary

The most important example of a shear flow near a boundary is the wind near the ground, in a shallow enough layer for the effects of the earth's rotation to be ignored. (The meaning of this limitation will be made clearer in §9.2.4.) Many of the results described below have been developed in the meteorological context (with high Reynolds number flows in mind), and a good summary is given by Priestley

(1959). We begin with a brief description of flows of uniform density before discussing the changes introduced by stratification.

5.1.1. *The logarithmic boundary layer*

Consider a semi-infinite fluid of density ρ, bounded by a horizontal plane which is being moved horizontally at such a rate that it applies a shear stress τ to the fluid near it. In the absence of a pressure gradient, τ must remain constant with height z above the plane, and at large times this produces a steady (in the mean) distribution of velocity $u(z)$ characteristic of a 'constant stress layer'. This idealized model also gives a good description of the flow close to the wall of a pipe or channel, and in the lowest few tens of metres of the atmospheric boundary layer (where the stress has changed little from its value at the boundary).

The velocity gradient responsible for maintaining this stress can be deduced by dimensional arguments. Except very close to the wall (where viscosity may be relevant), it is assumed to be a function of z, τ and ρ alone, so that

$$\frac{du}{dz} = \left(\frac{\tau}{\rho}\right)^{\frac{1}{2}} \bigg/ kz \equiv u_*/kz. \qquad (5.1.1)$$

Here u_* is called the 'friction velocity', and k is a universal constant (von Kármán's constant, with an experimentally determined value of approximately $k = 0.41$). Integrating (5.1.1) gives

$$u = \frac{u_*}{k}(\ln z + c) \qquad (5.1.2)$$

the well known logarithmic profile. The constant of integration c depends on conditions right at the surface, and changes in its value imply the addition of a uniform velocity to the whole flow with no change in its internal structure (see fig. 5.1).

When the boundary is aerodynamically smooth the stress is transmitted by viscosity; there is a viscous sublayer whose thickness is of order $\delta \sim \nu/u_*$ and in which the velocity varies linearly, like $u_*^2 z/\nu$. Matching this to (5.1.2) gives (with axes fixed in the boundary)

$$u = \frac{u_*}{k}\left(\ln\frac{u_* z}{\nu} + c_1\right), \qquad (5.1.3)$$

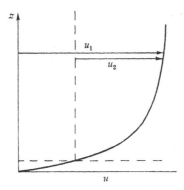

Fig. 5.1. Sketch of the velocity profile in turbulent flow past a solid plane, showing the effect of changing the boundary condition.

where c_1 is now a universal constant with an experimentally determined value of about 2.4 (see Prandtl 1952, p. 126). At an aerodynamically rough boundary the individual roughness elements have heights greater than δ; ν therefore becomes irrelevant because the stress is transmitted by pressure forces in the wakes of the roughness elements (cf. §4.3.2). The velocity profile is now

$$u = \frac{u_*}{k} \ln \frac{z}{z_0}, \qquad (5.1.4)$$

where z_0, called the roughness length, is related to the geometry of the boundary.

Using (5.1.1) one can also evaluate the rate of production of mechanical energy per unit mass ϵ say (which in steady conditions, and when there is no flux of energy into or out of the region of interest, is also the rate of dissipation)

$$\epsilon = u_*^2 \frac{\mathrm{d}u}{\mathrm{d}z} = \frac{u_*^3}{kz}, \qquad (5.1.5)$$

and define an eddy viscosity or vertical transport coefficient for momentum

$$K_M \equiv \frac{\tau}{\rho} \bigg/ \frac{\mathrm{d}u}{\mathrm{d}z} \qquad (5.1.6)$$

which is equal to $ku_* z$ for the logarithmic profile (5.1.4). The relation between the flux and the gradient of a passive tracer (such as

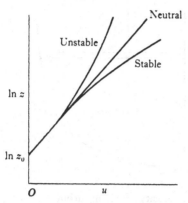

Fig. 5.2. The effect of a stable or unstable environment on
turbulent velocity profiles.

a small amount of heat or moisture) is predicted to good accuracy by assuming that the turbulent diffusivities (say K_H for heat) are the same as K_M, an idea which is called 'Reynolds analogy' (see §5.1.3). When the buoyancy associated with these properties becomes larger, however, the transfer mechanism itself is affected, the profiles change, and a separate discussion is needed. We will still assume initially that the boundary stress is dominant, i.e. that we are dealing with case (a) of fig. 4.19.

5.1.2. The effect of a buoyancy flux

The extra parameter of dynamical significance is the buoyancy flux, defined by

$$B = -g\overline{\rho'w'}/\overline{\rho}, \qquad (5.1.7)$$

where ρ' and w' are the fluctuations of density and vertical velocity and the bar denotes an average (over a horizontal area, or over time at a fixed point). When one is dealing with a heat flux H (positive upwards) in a gas, then $B = (g/\overline{\theta})(H/C_p\rho)$, where C_p is the specific heat and $\overline{\theta}$ the absolute temperature. For the purposes of discussion B, as well as u_*, is usually taken to be constant with height, but this can only be a good approximation in a layer of limited depth (see §7.3.3). Transports by any other mechanism (such as heat radiation) are neglected here.

Following Monin and Obukov (1954), we now show how the properties of the flow are determined when u_* and B are taken as the fundamental parameters. From these can be formed the 'Monin-Obukov length'

$$L = \frac{\bar{\rho}u_*^3}{kg\overline{\rho'w'}} = \frac{-u_*^3}{kgH/C_p\rho\bar{\theta}} = \frac{-u_*^3}{kB}, \qquad (5.1.8)$$

a scaling parameter for the flow which is negative in unstable conditions (e.g. heating from below) and positive in stable conditions. All dimensionless variables must now be functions of z/L. Equation (5.1.1) is replaced by

$$\frac{kz}{u_*}\frac{du}{dz} = \phi_M\left(\frac{z}{L}\right) \qquad (5.1.9)$$

(which implies that $K_M = ku_*\,z/\phi_M$), and the relations obtained by integrating (5.1.1) must be modified correspondingly.

The function ϕ_M has two basic forms, one in stable and the other in unstable conditions, and the description in terms of these universal functions receives support from measurements in the atmosphere (see fig. 5.3). Any further theoretical discussion, however, requires special assumptions about ϕ_M, going beyond the simple dimensional argument. If ϕ_M is expressed as a power series in z/L, and only the linear term is retained ($\phi_M = 1 + \alpha z/L$) then (5.1.9) leads to

$$u = \frac{u_*}{k}\left[\ln\frac{z}{z_0} + \alpha\frac{z}{L}\right], \qquad (5.1.10)$$

i.e. to a log-linear profile, where α has a value of about five (see Webb 1970). Such deviations from the logarithmic profile in opposite senses in stable and unstable conditions are observed (see fig. 5.2), the velocity increasing less rapidly than $\ln z$ in an unstable environment (z/L negative) because of the increased vertical mixing caused by convection.

In the same way a non-dimensional density (or potential temperature) gradient can be defined, such that

$$\frac{ku_*}{\overline{\rho'w'}}\frac{z}{dz}\frac{d\rho}{dz} = \frac{ku_*}{\overline{\theta'w'}}\frac{z}{dz}\frac{d\theta}{dz} = \frac{kz}{T_*}\frac{d\theta}{dz} = \phi_H\left(\frac{z}{L}\right), \qquad (5.1.11)$$

where $T_* = \overline{\theta'w'}/u_*$ is a convenient temperature scale. Another

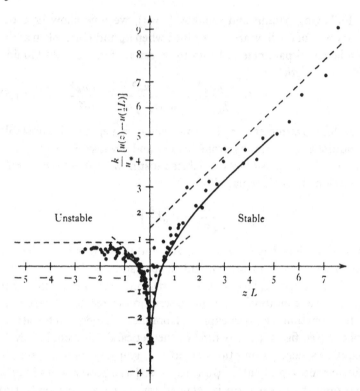

Fig. 5.3. Plot of experimental data in the finite difference form used by Monin and Obukov, showing the behaviour in stable and unstable conditions. (From Monin 1962a.)

function ϕ_I can be defined similarly for each transferred property, such as water vapour or a pollutant, and the use of 'Reynolds analogy' implies that the ϕs and hence the 'eddy transport coefficients' are equal. This seems to be a good assumption for heat, water vapour and dynamically neutral pollutants, but the experimental evidence described later suggests that momentum behaves differently; both ϕ_M^{-1} and K_H/K_M decrease as z/L increases (i.e. as the density gradient becomes more stable).

The gradient Richardson number Ri (1.4.3) can be used in this context as an alternative but slightly less convenient stability parameter. Of more direct physical significance is the flux Richardson number Rf, the ratio of the rate of removal of energy by buoyancy

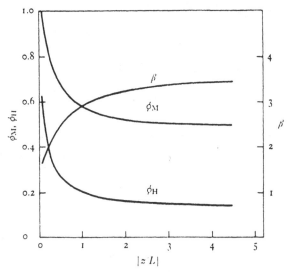

Fig. 5.4. The forms of the non-dimensional profile functions, and their ratio, derived from more recent analyses of observations in unstable conditions. (From Charnock 1967.)

forces to its production by the shear, which can be expressed in several equivalent ways:

$$Rf = \frac{K_H}{K_M} Ri = \frac{\overline{g\rho'w'}}{\bar{\rho}u_*^2 \, du/dz} = \frac{K_M}{ku_* L}. \qquad (5.1.12)$$

Thus near the limit of zero density gradient, where K_M and K_H are nearly equal, $Ri \approx Rf = z/L$.

In general Ri, Rf and K_H/K_M, as well as the non-dimensional profiles, are more complicated functions of z/L, whose forms are now known empirically from the observations in the lower atmosphere. Several other limiting cases are treated below.

5.1.3. Forced and free convection

It will be useful to digress from the main theme of this chapter to discuss briefly the effect of an unstable buoyancy flux on a turbulent shear flow (although convection problems where turbulence due to the mean flow is entirely absent are left until chapter 7). When $|z/L|$ is sufficiently small, the generation of turbulence by the shear dominates over generation by buoyancy; the deviation from neutral conditions is small, and the process of transfer of buoyancy (heat) is called 'forced convection'. The buoyancy flux can be calculated by putting $K_H = K_M = k^2 z^2 (du/dz)$; it follows that

$$\overline{\rho' w'} = K_H \frac{d\rho}{dz} = k^2 z^2 \left(\frac{du}{dz}\right)\left(\frac{d\rho}{dz}\right). \tag{5.1.13}$$

When this flux as well as u_* is constant with height, the density gradient is proportional to z^{-1} and the profile is again logarithmic.

'Free convection', on the other hand, occurs at large negative values of z/L, in a range where the contributions to the vertical transfer of momentum and (say) heat by mechanical turbulence can be neglected compared to those carried by convection. This does *not* mean that the mechanically generated energy itself is negligible – it may be comparable with that generated by buoyancy, and indeed over a smooth boundary the existence of the shear flow is implied if molecular conduction near the boundary is not to become important (see fig. 5.5, and also Townsend (1962), which is discussed in §7.1.3). It does mean that u_* and L are no longer relevant parameters, so that the density gradient or the dynamically more significant quantity $(g/\rho)(d\rho/dz)$ (cf. N^2; but the gradient is now unstable) should depend on height z and $B = -g\overline{\rho' w'}/\bar{\rho}$. (Over a rough boundary z_0 can in principle also be relevant.) A dimensional argument shows that the form of the relation between them must be

$$B \left(\frac{g}{\rho}\frac{d\rho}{dz}\right)^{-\frac{3}{2}} z^{-2} = H_*, \tag{5.1.14}$$

where H_* is the 'non-dimensional buoyancy flux', a constant in free convection with an experimental value close to unity. Rewriting (5.1.14) in the notation appropriate for a heat flux one obtains

$$-\frac{g}{\theta}\frac{d\theta}{dz} = H_*^{-\frac{2}{3}}\left(\frac{gH}{\theta C_p \rho}\right)^{\frac{2}{3}} z^{-\frac{4}{3}}, \\ -\frac{d\theta}{dz} = H_*^{-\frac{2}{3}}\left(\frac{H}{C_p \rho}\right)^{\frac{2}{3}} \left(\frac{g}{\theta}\right)^{-\frac{1}{3}} z^{-\frac{4}{3}}. \tag{5.1.15}$$

(The parameter g/θ has sometimes been regarded as a separate entity, but it is undesirable and sometimes misleading to do this.)

Integrating (5.1.15) with $\bar{\theta}$ and ρ regarded as fixed gives $\theta \propto z^{-\frac{4}{3}}$ (but note that $\ln \theta$ should replace θ when the temperature varies markedly through the range of interest). With the assumption of similar profiles for temperature and velocity, it follows for large negative z/L that

$$\frac{u}{u_*} \propto \left(-\frac{z}{L}\right)^{-\frac{1}{3}}. \qquad (5.1.16)$$

Both K_H and K_M become independent of u_* in the limit of free convection (and proportional to $z^{\frac{4}{3}}$), so $K_H/K_M = \beta$ approaches a constant and $\phi_M \propto Rf^{-\frac{1}{4}}$. These conclusions are supported by the observed forms of β, ϕ_H and ϕ_M in very unstable conditions, as summarized in fig. 5.4 (after Charnock (1967) who used a set of observations reported by Dyer (1965)).

The non-dimensional heat flux defined by (5.1.14) can also be evaluated for *forced* convection. In that case it follows from (5.1.13) that H_* is not constant, but a function of the stability parameters:

$$H_{*\text{forced}} = k^2 |Ri|^{-\frac{1}{2}} = k^2 \left(-\frac{z}{L}\right)^{-\frac{1}{2}}. \qquad (5.1.17)$$

These predictions about H_* are again supported by observations, which show that there is a transition from the $|Ri|^{-\frac{1}{2}}$ dependence of H_* to the constant value characteristic of free convection at a very small value of $|z/L|$, about 0.03. (See fig. 5.5.) At very large negative Ri, H_* tends to rise again over a smooth boundary as molecular effects make another lengthscale relevant near the boundary (Townsend 1962).

It is convenient for some purposes to be able to write down an explicit expression for ϕ_M which has the right limiting behaviour in extreme conditions and represents the observations reasonably well over the whole range. Ellison (1957) suggested the form

$$\frac{du}{dz} = \frac{u_*}{kz}(1 - \gamma Rf)^{-\frac{1}{4}} \qquad (5.1.18)$$

which can be rewritten using (5.1.9) and (5.1.12) as

$$\phi_M^4 - \gamma \frac{z}{L} \phi_M^3 = 1, \qquad (5.1.18a)$$

where γ is a constant, experimentally about 14. This has been taken

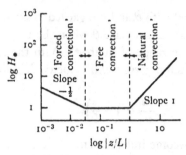

Fig. 5.5. Variation of the non-dimensional buoyancy flux H_* with z/L over the whole unstable range (schematic). (After Townsend 1962.)

up (and derived again independently) by others and widely used, but its status as an interpolation formula (rather than a strict theoretical deduction) tends to be forgotten. Neither this nor other formulae derived by more mechanistic arguments has been found satisfactory over the whole range of stability conditions, though it has been applied even in stable conditions by using a different value of γ in that range.

5.1.4. *Constant-flux layers in stable stratification*

Let us now return to the case of more immediate concern here, a stable density gradient produced (for example) by a downward (negative) buoyancy flux. In moderately stable conditions the observed velocity and density profiles are nearly similar to one another, and can again be described by the log-linear form (5.1.10) with nearly the same value of α (Webb 1970).

Using (5.1.12) we find

$$\phi_M = 1 + \alpha \frac{z}{L} = (1 - \alpha Rf)^{-1} \qquad (5.1.19)$$

which implies that Rf has a maximum (critical) value near $Rf = \alpha^{-1}$ ($\approx \frac{1}{5}$ experimentally), at which ϕ_M becomes very large and hence K_M small. From its definition, Rf cannot exceed unity, but here we have an indication (which is given further support in §5.2.3) that the value of Rf at which the turbulent transport is substantially altered can be much less than one. That is, most of the energy is always dissipated by viscosity, and only a small fraction by working against gravity.

In very stable conditions, the argument based on (5.1.19) will not be valid, but in the limit $z/L \to \infty$, Ellison (1957) has suggested that the *linear* dependence of ϕ_M on z/L is preserved. This result can be approached by a more direct dimensional argument, based on the following physical reasoning which is closely related to the views expressed by Stewart (1969). In a very stable gradient, the mixing motions must be so strongly damped that the vertical excursions of fluid parcels will be limited entirely by the work done against gravity. (This is case (*c*), referred to in §4.3.1.) All levels in the interior of the fluid must behave in the same way and will not be directly influenced by the boundaries, because mixing can occur only with adjacent fluid and cannot now extend right to the solid wall. (There will still be an indirect effect of the boundaries, where the assumed constant stress and buoyancy flux are applied; the extent to which it is *consistent* to retain all these assumptions will be taken up again in chapter 10.) It follows that any distance z (from a wall) becomes irrelevant in very stable conditions; L defined by (5.1.8) is the only lengthscale, and the velocity and density gradient must be determined by u_* and $B = -g\overline{\rho'w'}/\bar{\rho}$ alone. Thus on dimensional grounds

$$\frac{du}{dz} = k_1 \frac{-B}{u_*^2} = k_1 \frac{u_*}{L} = \frac{k_1}{k_2} N \qquad (5.1.20)$$

and

$$N^2 = -\frac{g}{\rho}\frac{d\rho}{dz} = k_2^2 \frac{B^2}{u_*^4} = k_2^2 \frac{u_*^2}{L^2}, \qquad (5.1.21)$$

where k_1 and k_2 are (as yet undetermined) constants. Both profiles are linear, as stated above, and the flow is in a kind of equilibrium, self-regulated state. This result receives support from observations in the atmosphere in very stable conditions, such as above Antarctic snowfields, though in such cases it is hard to justify the constant heat flux assumption. (For a discussion of the observations in this stability range, see Oke 1970 and Webb 1970.) The other quantities defined previously will in this stable limit also attain internally regulated values, which can be expressed in terms of u_*, B, k_1, k_2. The flux Richardson number $Rf = k_1^{-1}$, and is thus related to the fluxes and the velocity profile alone. We find also that

$$K_M = \frac{u_*^4}{k_1 B} = \frac{u_* L}{k_1}, \quad K_H = \frac{u_*^4}{k_2^2 B} = \frac{u_* L}{k_2^2} \qquad (5.1.22)$$

so that
$$\frac{K_H}{K_M} = \frac{k_1}{k_2{}^2}, \quad Ri = k_2{}^2/k_1{}^2 \qquad (5.1.23)$$

and the gradient Ri also has a constant ('equilibrium') value in a turbulent constant flux layer with a stable gradient.

It must be emphasized that this last result will only hold in practice under rather special circumstances, when the external conditions exactly match the internally regulated gradients. An apparently contradictory result has been suggested for more stable flows (with large overall Richardson number Ri_0, based on the whole depth), and it will be helpful to make the difference clear immediately. The latter case corresponds to (g) of fig. 4.19, in which the turbulence is so strongly damped that the motion is more appropriately described as a field of random, intermittently breaking gravity waves. While it may be possible to define a local gradient Ri through the individual turbulent patches which will have the same significance as $(5.1.23)$, any measured, time averaged value for the fluid as a whole will be more nearly akin to Ri_0. A critical value of a flux Richardson number (which is always imposed by energy considerations) does not, however, imply any limit on this 'average gradient Richardson number'.

Since the internal wave motions can readily transfer momentum through the action of pressure forces, but relatively little heat or matter even when the waves 'break', the ratio K_H/K_M can become small. Thus both transfer coefficients remain larger than the molecular values and some transfer persists to large values of 'average gradient Richardson number', while still satisfying $(5.1.12)$, with $Rf < 1$. (See also §§5.2.3 and 10.2.2.) Laboratory results consistent with the 'equilibrium' conditions are discussed in §6.2.4, and measurements made in a large scale flow of the second kind are described in §5.3.3.

5.2. Theories of turbulence in a stratified shear flow

Compared to the case of uniform density, little is known about the properties of turbulence in a stratified fluid. In this section we outline various theories which have been proposed to study the fluctuating motions themselves, their maintenance, and the diffusion they produce. Most of these invoke rather more explicit assumptions about the effect of stratification on the structure of the flow, but we begin with the results which follow from a simple extension of the similarity arguments used in §5.1. (See Lumley and Panofsky 1964.)

5.2.1. Similarity theories of turbulence and diffusion

Dimensional reasoning can be applied to the fluctuating quantities in just the same way as they were to the form of the mean profiles. Again u_*, B and z are the governing parameters, and variables like the variance of the vertical velocity fluctuations $\sigma_w = \langle w'^2 \rangle^{\frac{1}{2}}$ or of the buoyancy fluctuations $\sigma_\rho = g\langle \rho'^2 \rangle^{\frac{1}{2}}/\rho$ can be expressed as

$$\sigma_w = u_* f_1\left(\frac{z}{L}\right)$$

and

$$\sigma_\rho = \frac{B}{u_*} f_2\left(\frac{z}{L}\right). \qquad (5.2.1)$$

Here f_1 and f_2 are universal functions which can be related to ϕ_M and ϕ_H describing the profiles, and are constants in neutral conditions. In free convection (when u_* no longer enters explicitly)

$$\sigma_w \propto B^{\frac{1}{3}} z^{\frac{1}{3}} \qquad (5.2.2)$$

and the velocity fluctuations increase with height; similarly

$$\sigma_\rho \propto B^{\frac{2}{3}} z^{-\frac{1}{3}}. \qquad (5.2.3)$$

In stable conditions, the stratification is more important at larger z, and σ_w is a decreasing function of height. The forms of these functions (and related quantities such as 'gustiness' and the deviation of the wind direction from the mean) have been calculated using extensions of the log-linear theory and the interpolation formula (5.1.19), and agreement between theory and observation is reasonably good. The variation with stability of two of them is shown in fig. 5.6, after Monin (1962a). The variance of the horizontal velocity σ_u is relatively poorly predicted by similarity theory, since this is strongly influenced by larger scale motions, not just by the local properties.

The diffusion of a neutrally buoyant pollutant (such as smoke) in

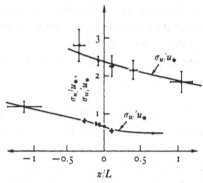

Fig. 5.6. The variances of vertical and horizontal velocity in the atmosphere near the ground, as functions of stability. (From Monin 1962a.)

a given stratified shear flow can also be approached by similar dimensional arguments. The Lagrangian vertical velocity

$$\mathrm{d}\bar{z}(t)/\mathrm{d}t = w_*,$$

say, following a typical particle released from the boundary into a constant stress layer, will be (Monin 1959, Ellison 1959)

$$w_* = bu_* f_3(z/L), \tag{5.2.4}$$

where f_3 is another universal function and b is a constant. The mean horizontal velocity is given by a relation like (5.1.10) so that the slope of the mean top of the smoke plume, $\mathrm{d}x/\mathrm{d}z = u/w_*$, will be a function of z/L (i.e. of stability) but not of mean velocity. In neutral conditions, the rate of spread becomes nearly linear some distance above the ground where u is changing slowly, and the concentration at the ground is inversely proportional to x. This result differs from the parabolic spread predicted by ordinary diffusion in a constant cross-stream flow, since it takes properly into account the increasing scale of the turbulence at greater heights. In unstable conditions the rate of spread increases, and in stable conditions it decreases (see fig. 5.7). A good summary of the predictions of similarity theory and the comparison with observations has been given by Klug (1968).

5.2.2. *The spectrum of nearly inertial turbulence*

Something should be said next about an idea which seemed promising when it was first introduced, but which is of limited

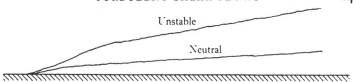

Fig. 5.7. The mean spread of the top of smoke plumes rising from the ground in neutral and unstable conditions. (After Monin 1959.)

practical significance for reasons which will become apparent. It is an extension of Kolmogorov's theory of 'local similarity' which is based on the idea of an energy cascade from small to large wavenumbers in a turbulent flow, and gives information about scales of motion much smaller than those of length scale l_1 and velocity scale u_1 at which energy is put into a constant density flow at high Reynolds number. A thorough discussion of this theory would take us too far afield here, and only the final result will be quoted; reference should be made to Batchelor (1953b, p. 114), Monin (1962b) and Phillips (1966a, p. 213) for the background to this problem.

Kolmogorov's second hypothesis leads to a description of the form of the spectrum, or the energy density function $E(k)$, in an 'inertial subrange' where viscous effects are unimportant. In this range $E(k)$ is determined entirely by two parameters, the magnitude of the wavenumber k and the rate of transfer of energy through the spectrum. The latter can be equated either to the rate of decay of the energy-containing eddies (which is of order u_1^3/l_1, cf. (5.2.15)) or to the eventual energy dissipation per unit mass ϵ_0 by the smallest eddies. The function $E(k)$ is related to the kinetic energy per unit mass $q^2 = \overline{u'^2} + \overline{v'^2} + \overline{w'^2}$ by

$$\tfrac{1}{2}q^2 = \int_0^\infty E(k)\,dk, \tag{5.2.5}$$

and the form predicted using a dimensional argument is

$$E(k) = A\epsilon_0^{\tfrac{2}{3}} k^{-\tfrac{5}{3}}. \tag{5.2.6}$$

This prediction has received firm support from experiments made in a tidal estuary by Grant, Stewart and Moilliet (1962), whose measurements also give a value for the numerical constant ($A \approx 1.5$). A '$-\tfrac{5}{3}$ power law' has been used to describe a wide variety of geophysical data, but (5.2.6) has quite unreasonably been elevated

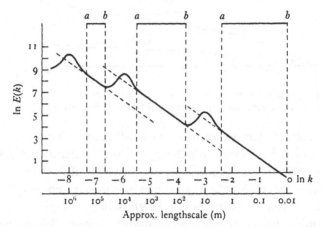

Fig. 5.8. Schematic diagram of the distribution of energy in the ocean as a function of wavenumber. Ranges a–b are those in which a '$\frac{5}{3}$ power law' may be applicable, though between them there is an energy input. (From Ozmidov 1965a.)

to the status of a general principle, without paying enough attention to the assumptions on which it is based (which include the condition that the turbulence should be isotropic in this range).

The arguments leading to (5.2.6) can break down if there is any substantial transfer of energy directly into or out of the 'inertial subrange'. Convective motions in the atmosphere can add energy at intermediate scales, for instance; but Ozmidov (1965a) has suggested that if this input is confined to a small range of wavenumbers, the $-\frac{5}{3}$ power law may still apply for values of k on either side of this. There will then be a jump in the E, k curve corresponding to each input (see fig. 5.8), but it can continue with the same slope at a higher level corresponding to the increased value of ϵ_0, the new rate of energy supply to higher wavenumbers.

Of more interest here is the possibility that, in a stratified fluid, there could be a systematic removal of turbulent kinetic energy over a range of wavenumbers, by working against buoyancy forces. This could only happen in a region of fully turbulent fluid (such as one of the turbulent patches referred to at the end of §5.1.4) in which the departure from neutral conditions is small, and it is still reasonable to think in terms of a cascade of energy through the spectrum. Regions where the Richardson number is large are therefore excluded; then the motions are more appropriately described as a set of weakly interacting waves (cf. §2.4.3), which can transfer energy selectively to remote wavenumbers. Moreover, it will become clear below that a significant buoyancy effect of the kind envisaged is only possible where energy is injected into a region from an external source, not where it is produced by the local shear.

The smallest scales of motion will remain unaffected in stable stratification, while buoyancy forces first become important for the larger length scales. Thus the discussion of the constant-flux boundary layer (5.1.10) indicates that buoyancy forces are significant at heights comparable with and greater than L (the Monin–Obukov length). The laboratory observations of Grigg and Stewart (1963), of turbulent vortices projected through a stratified fluid, show directly that the largest scales (i.e. the mean motion) are suppressed first, leaving the smaller scale turbulence practically unchanged. These results suggest that there could be a wavenumber k_b say, replacing L^{-1} in the present problem, at which one can further divide the wavenumbers previously comprising the inertial subrange. In the 'buoyancy subrange', such that

$$l_1^{-1} \ll k < k_b \qquad (5.2.7)$$

the loss of energy due to working against buoyancy can, it is argued, become comparable with that transferred. The physical parameters on which k_b can depend are ϵ_0 and N^2, and so on dimensional grounds

$$k_b = c_1 N^{\frac{3}{2}} \epsilon_0^{-\frac{1}{2}}, \qquad (5.2.8)$$

where c_1 is a positive constant. An equivalent argument was used by Ozmidov (1965b) to determine the critical length scale at which buoyancy forces become important in the oceanic boundary layer. The corresponding maximum vertical transfer coefficient can be expressed in the form

$$K_M \propto \epsilon_0 N^{-2}. \qquad (5.2.9)$$

Ozmidov also showed that in such a 'buoyancy subrange' the motion must become anisotropic because not only are the horizontal motions undamped by gravity, but some of the vertical energy is transferred to them.

Dimensional reasoning can take us a step further. In the 'buoyancy subrange', if it exists, the local rate of energy transfer

$\epsilon(k)$ through the spectrum will be much larger than the eventual dissipation rate ϵ_0, since the latter is by definition the residual rate of transfer after energy has been extracted by working against gravity at all larger scales. Thus ϵ_0 is not an important parameter in the buoyancy subrange,† and the form of the spectrum must be determined by N^2 and k, since $\epsilon(k)$ is itself a derived function. It follows that

$$E(k) = c_2 N^2 k^{-3}. \qquad (5.2.10)$$

Lumley (1964) used more explicit assumptions to derive a form for the complete buoyancy-inertial subrange, but in view of the rather special conditions under which it could be applicable, this will not be pursued further here.

The form (5.2.10) has received limited support from aircraft observations of atmospheric turbulence, though there must always be some doubt whether these were indeed made within fully turbulent air, as implied by the theory. Myrup (1968), for example, flew through a stable layer at the top of a convecting surface layer (cf. §9.2.3), and found a range of k, where $E(k) \propto k^{-3}$ approximately, changing to $k^{-\frac{5}{3}}$ at higher wavenumbers, in agreement with (5.2.6) for the inertial subrange. No buoyancy subrange has been detected in laboratory experiments.

Two major criticisms can be levelled against the interpretation of such observations in terms of a buoyancy subrange. Most important, the relation (5.2.7) is rarely satisfied in the atmosphere, so that there may be no range of scales in which the underlying assumptions are properly satisfied and the buoyancy effect dominant. Secondly, measurements of the kind quoted in the preceding paragraph could include an unknown contribution from internal waves. This is perhaps a less serious objection, since in a buoyancy subrange, the distinction between the non-isotropic turbulence described above and the forced, strongly interacting internal waves referred to at the end of chapter 2 becomes blurred. In fact the form (5.2.10) is identical to the condition (2.4.9), which ensures that strong interactions will dominate over the weak, selective wave interactions.

† This is quite different from the more usual situation where there is a local balance between production and dissipation of turbulent energy, with no external source. In the latter case, viscous dissipation will dominate over the buoyancy term even in stable stratification – see §5.2.3.

Whatever words are used to describe the process, each of these pictures implies a cascade of energy through the spectrum to higher wavenumbers, and relates the form of $E(k)$ to the same two physical parameters k and N^2. The 'turbulence' description is perhaps the more helpful, because it makes clear that working against gravity, and hence wave breakdown and mixing, is a necessary part of the mechanics of a buoyancy dominated subrange.

5.2.3. Arguments based on the governing equations

In this section we compare several attempts to deduce the structure of turbulence in a unidirectional stratified shear flow, which apply similarity hypotheses to the dynamical equations rather than directly to the external parameters. The flow is assumed to be fully turbulent and of high Reynolds number, and no molecular diffusion effects are included here. (See §9.1.3 for a discussion of this last point.) Arguments which also take wave propagation into account explicitly will be left until the final chapter.

For simplicity, horizontal homogeneity will be assumed here, though the results will be applied later (with some reservations) to developing flows such as wakes. In the absence of advection, the equations for the turbulent kinetic energy per unit mass

$$\tfrac{1}{2}q^2 = \tfrac{1}{2}(\overline{u'^2} + \overline{v'^2} + \overline{w'^2})$$

and the mean square density fluctuation may be expressed in the forms (derived from the Navier–Stokes equations):

$$\frac{1}{2}\frac{\partial q^2}{\partial t} + \frac{1}{2}\frac{\partial \overline{q^2 w'}}{\partial z} + \overline{u'w'}\frac{\partial u}{\partial z} + \frac{1}{\rho}\frac{\partial \overline{w'p'}}{\partial z} + \frac{g}{\rho}\overline{\rho'w'} + \epsilon_0 = 0, \quad (5.2.11)$$

$$\frac{1}{2}\frac{\partial \overline{\rho'^2}}{\partial t} + \frac{1}{2}\frac{\partial \overline{\rho'^2 w'}}{\partial z} + \overline{\rho'w'}\frac{\partial \bar{\rho}}{\partial z} - \kappa\overline{\rho'\nabla^2\rho'} = 0. \quad (5.2.12)$$

In the steady state, and neglecting small 'diffusion' terms like $\partial(\overline{q^2 w'})/\partial z$ (or alternatively, regarding u' etc. as vertically averaged values), (5.2.11) reduces to

$$\tau\frac{\partial u}{\partial z} = -\overline{u'w'}\frac{\partial u}{\partial z} = \frac{g}{\rho}\overline{w'\rho'} + \epsilon_0, \quad (5.2.13)$$

where the terms are respectively the production by the Reynolds stress working against the mean velocity gradient, the rate of working against buoyancy, and viscous dissipation.

By considering the equations for the separate components (see, for example, Stewart (1959)) it can be shown that the energy is put initially into the $\overline{u'^2}$ component, that in the direction of mean motion, but is redistributed by pressure fluctuations among all three turbulent velocity components. The loss to buoyancy affects only $\overline{w'^2}$ directly, whereas viscosity affects all three. Moreover, the mechanism which tends to make the turbulence isotropic is known (from experiments in flows of constant density) to be inefficient compared to the decay mechanism. Thus we can already see qualitatively how the buoyancy term, though making only a small direct contribution to the energy dissipation, can have a large effect on the turbulence. Even in a stratified fluid ϵ_0 remains the dominant term on the right hand side of (5.2.13), so that the maximum value of Rf (defined by (5.1.12)) will be much less than unity.

The corresponding equation to (5.2.13) for the density fluctuations may be written

$$\overline{w'\rho'}\frac{\partial\overline{\rho}}{\partial z} + \chi = 0, \tag{5.2.14}$$

where the terms are respectively the rates of production and dissipation of mean square density fluctuations. Further progress depends on the assumptions made at this point about the forms of ϵ_0 and χ; those due to Townsend (1958) will be considered first. He supposed that results which have been used to relate ϵ_0 and χ to the large scale properties of a constant density flow (Batchelor 1953 b, p. 102) will continue to be valid in stable stratification. Explicitly, he put

$$\epsilon_0 = q^3/L_u, \quad \chi = \overline{\rho'^2}q/L_\rho \tag{5.2.15}$$

where L_u and L_ρ are characteristic integral scales which are assumed to remain in a fixed ratio even in a stratified flow. Substituting in (5.2.13) and (5.2.14) and rearranging gives

$$\overline{w'\rho'} = -k_\rho^2 L_\rho q \frac{\partial\overline{\rho}}{\partial z} \tag{5.2.16}$$

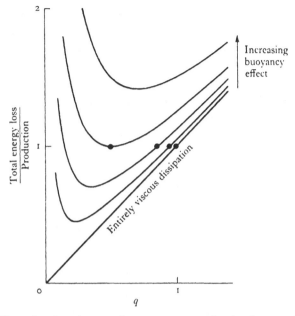

Fig. 5.9. The ratio of total energy loss to energy production in a stratified flow, as a function of the turbulent intensity for various ratios of the buoyant and viscous contributions to the energy dissipation. (From Townsend 1958.)

and
$$-\overline{u'w'}\frac{\partial u}{\partial z} = -k_u q^2 \frac{\partial u}{\partial z} = -\frac{g}{\bar{\rho}}k_\rho^2 L_\rho q\frac{\partial \bar{\rho}}{\partial z}+\frac{q^3}{L_u}, \qquad (5.2.17)$$

where
$$k_\rho^2 = \frac{(\overline{w'\rho'})^2}{q^2\overline{\rho'^2}}, \quad k_u = \frac{\overline{u'w'}}{q^2}. \qquad (5.2.18)$$

To complete the solution, Townsend also had to assume that k_ρ^2 and k_u are constants (averaged across the flow section and independent of stability) and that the velocity and density gradients are also nearly constant.

The ratio of the total energy loss to the production (the ratio of the right hand to the left hand side of (5.2.17)) is plotted in fig. 5.9 as a function of the turbulent intensity q, for several ratios of the two dissipation terms. Possible equilibrium points are marked; clearly no equilibrium intensity exists for curves whose minimum value exceeds unity, i.e. for which the assumed buoyancy contribution is too large. At each minimum, the buoyant and viscous terms are equal on this model, and so $Rf_{max} = \frac{1}{2}$. (This numerical value can change with different detailed assumptions, but the essential features of the model are preserved provided one of the terms on the right of (5.2.17) increases and the other decreases with increasing q.) The turbulent intensity is finite in the critical flow, and a sudden collapse of the turbulence occurs as the limit is passed.

By solving (5.2.17) for q and substituting the result in (5.2.16), Rf can be expressed in terms of gradients as

$$Rf = \frac{1}{2}\left[1 - \left(1 - 4\frac{L_\rho k_\rho^2}{L_u k_u^2}Ri\right)^{\frac{1}{2}}\right]. \qquad (5.2.19)$$

Thus Townsend's assumptions lead to the conclusion that the gradient Richardson number Ri will also have a critical value

$$Ri_{crit} = \frac{1}{4}\frac{L_u k_u^2}{L_\rho k_\rho^2} \qquad (5.2.20)$$

above which no real solutions to the balance equations are possible. This is in agreement with the simple deduction made in §5.1.4, and it is supported also by the later theories of Monin (1965) and Ellison (1966), to which further reference is made below. No satisfactory theoretical estimate of Ri_{crit} has yet been made, but the experimental results to be discussed in §5.3 suggest that $Ri_{crit} \approx 0.1$.

The most questionable of the assumptions made by Townsend are that k_u, and especially k_ρ, are constants independent of stability. This can certainly not remain true in the final stages of decay when the vertical motions are strongly damped, for much the same reasons as those outlined at the end of §5.1 to show that K_H can become much less than K_M. It is also not clear how the averages are to be taken, especially in a boundary layer where such properties will be strong functions of the height.

With the constant-flux boundary layer in mind, Ellison (1957) proposed a method which avoids some of these difficulties. This was the first attempt to use simplifications of the dynamical equations to study the effect of stratification, and it still merits careful consideration, though that author has since developed a more ambitious model. He again used (5.2.13) and (5.2.14), but replaced (5.2.15) by

$$\epsilon_0 = q^2/T_2, \quad \chi = \overline{\rho'^2}/T_1, \qquad (5.2.21)$$

i.e. he defined decay times T_1 and T_2 such that (in the absence of production) turbulence would begin to destroy $\overline{\rho'^2}$ and q^2 at rates T_1^{-1} and T_2^{-1}. (Compare also with the full equations (5.2.11) and (5.2.12).) Instead of deriving an expression for $\overline{w'\rho'}$ from these (which led previously to a constant k_ρ), he used an additional equation for the change in the density flux. With the same approximations as before, and introducing a separate timescale T_3 for the decay of $\overline{w'\rho'}$, this is

$$\overline{w'^2}\frac{d\overline{\rho}}{dz} + \frac{\overline{\rho'^2}g}{\overline{\rho}} + \frac{\overline{w'\rho'}}{T_3} = 0. \tag{5.2.22}$$

The three equations involving T_1, T_2 and T_3 can be rearranged to express the ratio of eddy transport coefficients (defined by (5.1.6), and equivalently for buoyancy) in the form

$$\frac{Rf}{Ri} = \frac{K_H}{K_M} = \frac{q^2\overline{w'^2}\left[1 - Rf\left(1 + \frac{q^2}{\overline{w'^2}}\frac{T_1}{T_2}\right)\right]}{(\overline{u'w'})^2(T_2/T_3)(1-Rf)^2}$$

$$= \frac{b(1 - Rf/Rf_{crit})}{(1-Rf)^2}, \tag{5.2.23}$$

where b is a constant. If it is supposed that $T_1 \approx T_2$ and $q^2/\overline{w'^2}$ changes little from its measured value in neutral conditions (which is about 5.5), this expression predicts that $K_H/K_M \to 0$ when $Rf = Rf_{crit} = 0.15$. Again this is much smaller than unity, in agreement with the view that the mixing rate is greatly altered before buoyancy forces have much effect on the overall energy balance (since only the vertical velocity component is directly affected). Equation (5.2.23) can also be regarded as a relation between K_H/K_M and Ri; and it is clear that according to this theory the low limit to Rf puts no similar restriction on Ri, which may become large provided $K_H/K_M \to 0$.

A second important consequence of Ellison's assumptions is that

$$\frac{(\overline{w'\rho'})^2}{\overline{\rho'^2}\,\overline{w'^2}} = \frac{1 - Rf\left[1 + \frac{q^2}{\overline{w'^2}}\frac{T_1}{T_2}\right]}{(T_1/T_3)(1-Rf)} \tag{5.2.24}$$

which decreases towards zero in very stable conditions, instead of remaining constant as Townsend supposed. This shows that inefficiency of the density transport is due not to the suppression of the velocity and density fluctuations themselves, but to a reduction of the mean product $\overline{w'\rho'}$. (In the limit of pure internal wave motions, ρ' and w' are $90°$ out of phase, so this correlation tends to zero. See §2.2.1.)

The later theory of Monin (1965) applied a related but more detailed scheme of simplifications to the dynamical equations for all the second moments of the four variables u', v', w' and ρ', instead of just the three considered above. Ellison (1966) (quoted by Yaglom 1969), showed how the resulting system of algebraic equations can be solved to give each of the quantities of interest as a function of Rf. The numerical values of five dimensionless characteristics of turbulence in a fluid of constant density now need to be specified, instead of just the two that were required in (5.2.23). With these stronger constraints, not only does Rf have a critical value, but so does K_H/K_M, the limiting values predicted by Ellison in very stable conditions being $Rf = 0.15$ and $K_H/K_M = 0.5$.

In the following section, experimental and observational results are presented which give support to each of the above conclusions but in rather different circumstances. A critical value of K_H/K_M (and hence of the gradient Richardson number Ri) applies in wake-like flows, which are maintained by Reynolds stresses generated at the edge of the flow itself; once these stresses have become small there is no mechanism for the further generation of turbulent energy (cf. fig. 4.19b). Ellison's (1957) theory, which suggests that Ri is unlimited, seems to describe flows over a solid boundary which approximate to the assumed constant-flux layer, and into which there is a continuing supply of turbulent energy (cases (a) and (g) of fig. 4.19). For this to be possible the relations (5.2.23) and (5.2.24) must be taken to apply to flows which are only intermittently turbulent (§5.1.4), so that the gradients entering into the definitions of Ri are time averaged values, not instantaneous local gradients. This is not inconsistent with the weaker assumptions used by Ellison (1957), whereas the later theory refers more explicitly to a region of fully turbulent fluid.

5.3. Observations and experiments on stratified shear flows

5.3.1. *The generation and collapse of turbulent wakes*

In §§3.1.4 and 4.3.2 we referred to the drag associated with the separation of the flow behind a blunt obstacle and the formation of a turbulent wake. This flow will now be examined in more detail. We begin with a general description of the structure of the wake as it develops downstream under various conditions, both in a continuously stratified fluid and at the interface between two layers, before making the more detailed comparisons with theory.

Immediately behind a cylinder moving horizontally with high Reynolds number (say greater than 10^3, based on the diameter), and moderate Richardson number (say $Ri^{\frac{1}{2}} = \kappa = 0.4$ in the notation of §3.1.4), the wake in a density gradient behaves much as it does in a uniform fluid. Its thickness increases, a small scale turbulent structure is observed, and the shape of the velocity defect profile is little changed. At about ten diameters downstream, however, the rate of growth is clearly being affected, and shortly afterwards the wake collapses as the largest turbulent eddies are rather abruptly suppressed by the stratification. Later the small scale turbulence is also damped out, and the residual motion 100 diameters downstream consists almost entirely of larger scale periodic motions, i.e. of internal waves. This behaviour is illustrated in fig. 5.10 pl. XIII by a sequence of shadowgraph pictures due to Pao (1968b). This author has also studied the whole range of wake effects from slow viscous flows (§3.3.3) to the fully turbulent wakes of interest here. At intermediate Reynolds numbers regular vortex shedding can occur as in a uniform fluid, but with stratification the vortices are flattened, and they also act independently to generate internal waves. Such vortices are quite distinct from the rotors which are produced by local overturning of the stratification (by large amplitude lee waves), in regions remote from the obstacle (see §4.3).

Schooley and Stewart (1963) carried out experiments (on a three-dimensional wake behind a self-propelled body) in which they paid special attention to the mechanism of wave generation. They pointed out that the collapsing motion is of just the right kind to generate gravity waves efficiently; it too is driven by gravity, and must take place at a rate comparable with the vertical velocity in internal waves. An experiment of Wu (1969), in which he studied the collapse of a semi-cylindrical region of uniform fluid in a density gradient, is also relevant here (though there was no small scale turbulence present initially and the later stages of flow, an advancing nose like a gravity current (§3.2.5) and finally a viscously dominated motion (cf. §3.3.4), do not concern us here). He showed that waves are generated which propagate ahead of the advancing front with an initial peak in the energy density at an angle of about 55° to the horizontal (corresponding to a frequency of 0.8 of the buoyancy frequency, §2.2.2). At later times the line of the crests moves toward the horizontal.

The turbulent wake produced by a flat plate towed normal to its plane along an interface between uniform layers of fresh and salt water has been studied more quantitatively by Prych, Harty and Kennedy (1964). Density profiles were measured by withdrawing samples from a rake of probes mounted at various distances behind the plate (and moving with it). Velocity profiles were obtained similarly using a propeller type velocity meter, mostly in a uniform fluid but also, for comparison, in the stratified cases. The measured velocities were in good agreement with a theory which allowed for a backflow arising because of the finite length (10 m) of the tank; the more accurate results for the uniform fluid will be used in the later quantitative calculations. The drag force was also measured directly in all runs. The drag coefficient C_D (defined as in §3.1) changed little with density difference from its value of nearly two in a uniform fluid, so that form drag and not wave drag dominated the behaviour at low Ri and high Re in these two-layer experiments. (Contrast this with the results for a continuously stratified fluid described at the end of §3.1.4, where both contributions to the drag were seen to be significant.)

The strong effect of stratification on the mixing is clearly shown in the density profiles. Prych *et al.* expressed their results in terms of the parameter

$$J^2 = g \frac{\Delta\rho}{\rho} b_0 \bigg/ C_D U^2 \qquad (5.3.1)$$

a kind of overall Richardson number based on the half-width b_0 of the plate and its velocity, but with the drag coefficient added (this

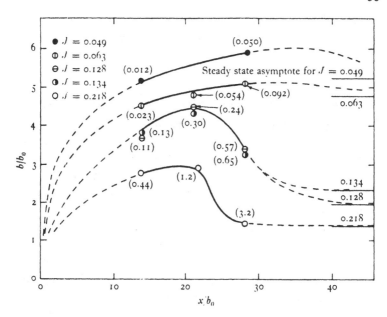

Fig. 5.11. Variation of the width of a wake with distance x behind a plate (of width $2b_0$) towed along an interface between two layers of different density. Curves are shown for various values of the parameter J, and the numbers in brackets are minimum gradient Richardson numbers computed using the data given by Prych, Harty and Kennedy (1964).

enters into their estimate of the work done by the body on the fluid). Numerically, J is very nearly equal to the parameter κ defined in §3.1.4. The rate of spread, and the extent of the later collapse, depend systematically on J as shown in dimensionless form in fig. 5.11, where the separation between two arbitrarily chosen concentrations is plotted against distance downstream.

An overshoot and strong collapse are prominent at the higher values of J. A visual examination of the wake, with one layer dyed, showed that this is the result of large globules of fluid being transferred across the interface by vortices shed by the plate. If the turbulence is damped before much mixing has taken place (as it is when J is large), these settle back to an equilibrium position much closer to the centre of the wake; the final equilibrium thicknesses are shown on the right of fig. 5.11. It is also worth remarking that the final density profile is not far from linear across the centre of the tank. Stirring in this apparently crude way is a surprisingly effective way of producing a uniform gradient over nearly the whole depth.

5.3.2. *The suppression of turbulence at an interior shear layer*

As our first test of Townsend's predictions (§5.2.3) we will use the data just described. There is some uncertainty in the interpretation for developing flows, since a failure of the regulating mechanism of the turbulence does not cause mixing to stop instantly; nevertheless one can deduce a range of conditions over which buoyancy effects are becoming significant.

Minimum gradient Richardson numbers, computed using the (nearly linear) density profiles and the maximum gradients on the corresponding velocity profiles, have been added (in brackets) to the experimental points shown in fig. 5.11. Consider first the experiments with $J = 0.128, 0.134$ (which incidentally give good support to the choice of J as a modelling parameter, since they were carried out with both the density difference and velocity varied). At the position of maximum width the turbulence must already have been strongly affected since the wake collapsed so strongly after this point; thus the values 0.24 and 0.30 are certainly larger than the Ri_{crit} defined in Townsend's sense. The results at lower values of J, which approach their asymptotic values more smoothly, suggest that Ri_{crit} is less than 0.1 and may be as low as 0.05.

Townsend (1957) made similar deductions from observations of a jet of fluid of intermediate density injected along an interface. A complication here was the more rapid spread horizontally than vertically, but he was able to show that mixing of outside fluid into the jet had almost ceased at $Ri = 0.3$. The first signs of an effect of the density gradient on the jet development occurred at about $Ri = 0.05$; the 'critical' value must lie between these extremes, and is probably less than 0.1.

Further examples of the same turbulence suppression mechanism are to be found in flows where the density distribution is stabilizing over a limited range of heights. This often occurs near a boundary, and though superficially the flow then seems more like the cases considered in the following section, the distinguishing feature of 'interior' layers is that the turbulence must be generated elsewhere in the flow, not directly by the shear at the adjacent wall. The laboratory experiment of Nichol (1970) is related to Reichardt's investigation of the stability of a boundary layer under the heated roof of a wind tunnel (discussed in §4.2.2). Instead of heating the whole roof, however, Nichol began heating at some point downstream, and examined the effect of the growing temperature boundary layer on an established turbulent wall layer. He observed a sudden and nearly complete collapse of the turbulence near the wall at some distance behind the leading edge of the heated section, measuring very small values of turbulent intensity and wall stress, and a highly inflected velocity profile with low velocities near the wall. The inside of the wall layer is disconnected from its energy source, though dissipation must still occur near the wall. Townsend (1957) used Nichol's (then unpublished) profiles to give further support to the view that Ri just before collapse is less than 0.1.

A natural example of this phenomenon is the 'dying away' of the wind at night, or more spectacularly, the sudden calm which has been observed during an eclipse of the sun (Townsend 1967). Clearly only the wind near the ground can be changed, since the duration of an eclipse is too short for the geostrophic flow at great heights to be affected, and the wind can remain strong even at tree-top height. When the solar heating is removed, the ground cools by radiation and heat is transferred downwards out of the air towards the ground. This produces a layer with a stable density gradient, whereas previously the whole profile was convectively unstable. As the depth of the stable layer grows, its top reaches a height where the velocity gradient is small, while the density gradient there changes only slowly with the depth. Eventually the value of Ri rises to a value of about 0.1 and the turbulence will die away. (Radiation effects will change the critical value somewhat, but the principle of the argument remains unchanged.) Without any turbulent Reynolds stress, the motion higher up (which has been driving the whole flow) cannot any longer affect the air below, and so the friction at the ground will cause the wind to die away at low levels. (An associated change in the intensity of temperature fluctuations has been documented by Okamoto and Webb (1970).) In time, however, this layer can be accelerated by the overall pressure gradient, and may again become turbulent for a short while during the night – until the same process begins again and the wind dies down a second time. The last result recalls again the observations referred to in passing in §4.1.4. Mittendorf (1961) showed that, after the Kelvin–Helmholtz mechanism had led to the breakdown of an interface between two accelerating layers in an inclined tube, the turbulence in the mixing region was suppressed before the transition layer had spread out very far. In one experiment he observed three successive breakdowns of the interfacial region, separated by quiescent periods. While turbulence is present the drag on the layers increases and the velocity falls, but when it is suppressed the flow is accelerated again by gravity. This behaviour can now be interpreted in terms of an increase of an appropriate Richardson number to a stable value, because of the decrease of the velocity gradient due to mixing. In §10.2.3 it will be shown that the supply of energy to the turbulent interface will be cut off when an overall Richardson number based on its thickness reaches a value of order one. During the whole of the thickening process, however, the gradient Ri can be much smaller, allowing turbulence to persist according to Townsend's criterion. (Compare with §6.2.4.)

Finally, it is convenient to mention here a contrasting (but related) natural phenomenon, the wind-driven layer at the surface of the sea. In conditions of strong surface heating, a well-mixed warmer (and therefore lighter) layer is formed, which is of limited depth because the stabilizing density distribution inhibits vertical mixing with the deeper water. (See §9.2.1 for a fuller discussion of the mixing problem.) At the bottom of this surface layer is a strong density gradient where (according to the ideas developed in this section) the turbulence is suppressed and the Reynolds stresses are small. A given wind stress at the surface can thus accelerate the water to produce stronger surface currents in this case compared to the unstratified ocean for two reasons: the depth of the layer involved is smaller, and the retarding stress below it is reduced. The recognition of conditions likely to produce such a 'slippery sea' played a large part in the success of the winning Olympic sailing team at Acapulco in 1968 (Houghton and Woods 1969). (We should however, not forget the 'dead water' effect described in §2.1.2, which could slow a ship down because of the extra wave drag generated at the interface.)

5.3.3. *Stratified flows in pipes, channels and estuaries*

The flow just discussed also has much in common with those lying at one extreme of the range of enclosed flows to be considered next. Another example is a turbulent layer of salt water flowing under a deep layer of fresh water in conditions where the bottom stress is much greater than that at the interface (which is so when a shallow layer flows along a horizontal bed – see §6.2.3). The intrusion of a saline wedge into an estuary against a fresh water flow (§3.2.5) is also of this kind. The distinguishing feature is that buoyancy effects become dominant somewhere in the interior of the channel, and mixing generated by the shear at the boundary is almost entirely suppressed across a sharp interface bounding one or more well-mixed layers (cf. case (*d*) fig. 4.19).

At the other extreme, the turbulence generated at the boundaries can be so strong that the fluid is thoroughly mixed over the whole depth of the flow (fig. 4.19*a*). This too occurs in natural estuaries, when the tidal current is much larger than the river flow; the salinity is then nearly constant at any section, but decreases gradually with distance upstream. For intermediate conditions, and especially near the turn of the tide, there is a measurable vertical gradient as well as a horizontal one, and the density gradients are clearly having an effect on both the rate of vertical mixing and the longitudinal motion and diffusion.

It was in an 'intermediate' type of flow that the often quoted measurements of velocity and density profiles were made in a tidal channel (the Kattegat) by Jacobson. Taylor's (1931*b*) analysis of these results showed that vertical mixing (at much greater than molecular rates), can persist at time-averaged 'gradient' Richardson numbers (based on measurements made $2\frac{1}{2}$ m apart) of ten or more. The ratio of eddy diffusivities for salt and momentum K_S/K_M deduced from several sections was about 0.03, so that Rf was still less than unity and there was no violation of the energy condition. The discussion given in §5.1.4 suggests however, that such an interpretation is consistent only if the turbulence is very intermittent, not continuous throughout the fluid (case (g), fig. 4.19). The stress at the boundary remains an important parameter governing the whole flow, but a large fraction of the momentum flux will be carried by internal waves. (See also § 10.2.)

The whole range of steady two-dimensional pipe or channel flows can also be characterized in an overall way by two parameters, the slope of the pipe and a 'pipe Richardson number'. The latter is most simply defined by

$$Ri_p = \frac{|\rho_m - \rho_0|}{\rho_0} \frac{gd\cos\theta}{U^2}$$

$$= A\cos\theta / U^3 \qquad (5.3.2)$$

where d is the total depth, U is the mean (discharge) velocity, ρ_m is the density when the flow is fully mixed across a section, and ρ_0 is the density of the ambient flow (before any buoyant fluid is added). Thus $A = g(\Delta\rho/\rho)dU$ is the buoyancy flux per unit width, which is constant in many problems of interest.

A related, but at first sight very different parameter used by civil engineers to correlate model experiments and field data, is the *estuary number* E_s, defined for a tidal channel to be

$$E_s = P_t F_0^2 / Q_f T. \qquad (5.3.3)$$

Here P_t is the 'tidal prism' (the volume of sea water entering the estuary on the flood tide), $F_0 = U/(gd)^{\frac{1}{2}}$ is an ordinary Froude number based on the tidal mean velocity and the depth, Q_f is the fresh water discharge and T the tidal period. (See for example Harleman and Ippen (1967).) From its definition, $P_t = UbdT$, where b is the width of the estuary, so that

$$E_s = \frac{P_t U^2}{gdQ_t T} = \frac{U^3 b}{gQ_t}. \qquad (5.3.4)$$

Writing $A = g(\Delta\rho/\rho)Q_t/b$ and comparing (5.3.2) and (5.3.4) gives finally

$$Ri_p = E_s^{-1}\frac{\Delta\rho}{\rho}. \qquad (5.3.5)$$

The density ratio factor is a constant for a particular estuary (and about 0.02 for all estuaries), so Ri_p and E_s are just inversely related. According to Harleman and Ippen, the division between 'stratified'

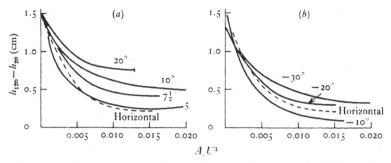

Fig. 5.12. Mean experimental curves showing the increase in thickness of a layer of salt solution as it flows between two sections 1 m apart under an uphill flow of fresh water. (a) Fully reversed layer. (b) Layer flowing downhill, with gravity but against the main flow. (From Ellison and Turner 1960.)

and 'well-mixed' cases occurs in the range $E_s = 0.03-0.3$, increasing values corresponding to increasingly mixed conditions.

The laboratory experiments of Ellison and Turner (1960) support this choice of overall governing parameters. They measured the rate of spread of a layer of salt solution, injected into the bottom of an inclined pipe of rectangular section through which there was a steady turbulent flow of fresh water. When the pipe was horizontal or the flow was downhill, the mixing rate was small, and when the flow was uphill two cases were distinguished. At low 'ventilating' velocities a thin layer of salt solution crept back along the pipe against the main flow, but at higher flow velocities all the salt was immediately carried uphill, in a layer which thickened with distance. (The criterion for reversal is discussed in §6.2.) As shown in fig. 5.12, the increase of depth of the salt layer in the 'fully reversed' cases depends systematically on θ and Ri_p, with quite a strong dependence on slope. At low values of Ri_p all curves come together, as they must when density differences are negligible. Buoyancy effects are already important at $Ri_p = 0.005$, by which time the rate of spread at small slopes has decreased by a factor of three.

Bakke and Leach (1965) have applied the results of these and similar experiments to describe the behaviour of layers of methane gas flowing along the roof of a coal mine. They defined a 'layering number', equivalent to $Ri_p^{-\frac{1}{3}}$ in the notation of (5.3.2), and were able to recommend minimum values of this parameter for various slopes, such that explosive concentrations of gas will be confined to acceptably short sections of a mine. This application is discussed further in §6.2.3.

We now turn to an interpretation of the same laboratory experiments in terms of the detailed local properties, which led to a test of Ellison's (1957) theory. Ellison and Turner (1960) showed that the distortion of the velocity profile in a tilted pipe depends mostly on Ri_p. Choosing cases where the profile changed least with distance, and combining the velocity measurements with salinity profiles at two sections, they were able to estimate K_S and K_M for salt and momentum, as a function of position in the flow and over a range of stability condition. A consistent correlation with local values of Ri were obtained at two heights intermediate between the wall and the centre of the channel, where the salt flux and Ri were changing least rapidly. K_M is affected rather little by the density gradient, whereas K_S/K_M is a strong function of Ri, as shown in fig. 5.13. It is by no means obvious that such a dependence on a strictly local parameter should emerge under the conditions of this experiment, when Ri is varying and the turbulent eddies cover a considerable range of heights. (Refer to the discussion on p. 150.)

On fig. 5.13, which is plotted using a logarithmic scale to cover a large range of Ri, is also shown the family of theoretical curves predicted using (5.2.23). The form of each of these is fixed once one specifies the value of Rf_{crit} and $b = 1.4$, the measured value of K_H/K_M (or of K_S/K_M, since no molecular effects are taken into account here) in neutral conditions. The experimental points are best fitted by the curve corresponding to $Rf_{crit} = 0.15$, in agreement with the estimate made earlier on purely theoretical grounds. At higher Ri, the theoretical curves are also broadly in agreement with Taylor's values (mentioned above) but the comparison is less sensitive in this range as well as the theory being more questionable.

A related experiment using air as the working fluid was carried out by Webster (1964). He investigated the detailed structure of a turbulent stratified shear flow, using a specially designed wind tunnel with differential heating to produce the density gradient, and comparing his results with the theories of Ellison and Townsend. He showed that various fluctuating quantities such as the mean

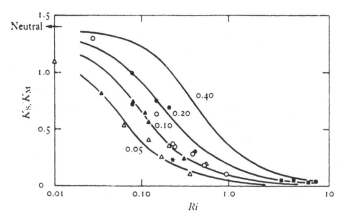

Fig. 5.13. Ratio of the transport coefficients for salt and momentum plotted against Richardson numbers deduced from time-averaged profiles, on a logarithmic scale. Various laboratory measurements reported by Ellison and Turner (1960) are compared with the theory of Ellison (1957); values of Rf_{crit} are marked on the theoretical curves. The Kattegat results, as analysed by Taylor (1931b), are also plotted as the squares on the far right of the diagram.

square temperature fluctuation (normalized with the local temperature gradient) and the ratio of the turbulent transports of heat in the downstream and vertical directions, as well as the ratios K_H/K_M deduced from correlation measurements, depend systematically on Ri. The trend of decreasing K_H/K_M with increasing stability was confirmed, and the existence of a critical value of Rf supported, but the absolute values were high compared with other experiments (even in near-neutral conditions). The discrepancy can be attributed to the development of the flow, which probably never reached equilibrium and still depended strongly on the properties of the decaying wakes behind the grid bars which produced the shear. A much larger wind tunnel is needed to remove these ambiguities; recent developments in this field have been reported by Arya (1972).

5.3.4. Longitudinal mixing and advection

Another important property of the estuary flows described above is the longitudinal spread of the region of mixing between the fresh and salt water. When an estuary is not far from the well-mixed state, and the fresh water flow is constant, the transition region can be described reasonably well using a one-dimensional diffusion equation, with a longitudinal dispersion coefficient which is determined by the mean flow over a long time. The shape of the salinity distribution changes little in one tidal cycle, but just shifts back and forth between low and high tide marks; the fresh water flow may however change relatively rapidly, and this can give rise to difficulties in evaluating the dispersion coefficient from observations. (See Ward and Fischer 1971.) The model can be adapted to include flows having considerable vertical salinity gradients, provided the diffusivity (and certain other parameters specifying the boundary conditions) are allowed to be empirical functions of E_s defined by (5.3.3).

Physically it is clear that a strictly one-dimensional model is an oversimplification. Firstly, the motions arising because the fresh water tends to flow on top of the salt must play a role in any channel which is not perfectly mixed. When this density driven flow is accompanied or followed by vertical mixing, the result is an apparent increase in the 'longitudinal diffusivity'. The longitudinal spread will be largest when the rate of vertical mixing is small, since then the shear will have its maximum effect. (Compare this with the result of Taylor (1954), who showed how the non-uniform velocity profile enhances the longitudinal spread in a turbulent pipe flow.) Secondly, Fischer (1972) has shown that the non-uniformity of the velocity in the lateral directions, due to variations in depth, can have an even larger effect on longitudinal dispersion than do the vertical shears. The combination of small scale, vertically controlled mixing with the mean horizontal shear is very effective because the widths of estuaries are usually so much larger than their depths.

An experiment which isolates the effect of the density-driven flow from that of the tidal motion has been performed by Harleman, Jordan and Lin (1959). They stirred mechanically over the whole length and depth of a large laboratory tank (to simulate mixing due to many tidal excursions), and measured the horizontal concentration profiles as a function of time after a central barrier separating fluid of different densities had been removed. With no density difference the effective diffusivity K was approximately proportional to the stirring velocity. When the density difference was 1 %, the diffusivity (K' say) was the same as K at high stirring rates, but

$K' - K$ was inversely proportional to turbulent intensity over the range of conditions they studied.

Most attempts at solving even a fully two-dimensional problem have fallen short of being complete theories, but they can nevertheless provide a useful framework for the comparison of observations and for the evaluation of the dependence of vertical and longitudinal diffusivities on other parameters. A notable contribution is that of Hansen and Rattray (1965), who surveyed the previous work and themselves gave solutions which cover a wide range of observed types of flow in an estuary. Using equations derived by averaging over a tidal cycle, they distinguished modes of motion associated with the mean fresh water flow, the differential motion due to density gradients (and a third caused by surface wind stress, which will not be discussed further here). Note that nearly uniform salinity in the vertical does *not* imply the absence of gravitational effects; especially in deep channels there can be a reversal of mean longitudinal velocity with depth even with only slight stratification. (See also Abbott 1960.)

Finally, it is instructive to present in this context the similarity solution used by Phillips (1966b) to describe the buoyancy-driven circulation in an almost enclosed sea, such as the Red Sea. There are some important differences and simplifications here: the flow can be regarded as steady, and both the longitudinal density differences driving it and the turbulent stirring are produced in another way, by surface cooling and evaporation. If there is a sill (of depth d) far from the origin, the flow has the form shown in fig. 5.14, inwards at the surface and out again below, with little motion below the sill depth. The surface density must increase with distance from the entrance, while the density profiles must remain stable throughout.

The only external parameters defining the flow are the length-scales x and d and the downward buoyancy flux B defined by (5.1.7), which is assumed independent of position. The expression of B in terms of the sensible heat flux H must now, however, be modified to take into account density changes produced by the increase of salinity S due to an evaporation rate E, i.e.

$$B = \frac{-g}{\bar{\rho}}\left[(H - LE)\frac{\alpha}{C_p} - ES\right],$$ (5.3.6)

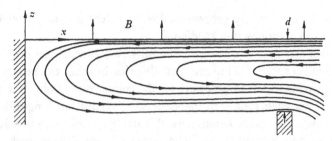

Fig. 5.14. Sketch of the streamlines in a convective motion driven by a uniform surface buoyancy flux, with a sill of depth d far from a vertical boundary at $x = 0$. (From Phillips 1966b.)

where L is the latent heat and α the coefficient of expansion. Using dimensional reasoning with the constraints imposed by the governing equations, Phillips showed that the longitudinal velocity and the density distribution must have the forms

$$U = (Bx)^{\frac{1}{3}} f\left(\frac{z}{d}\right), \qquad (5.3.7)$$

$$g\frac{\rho'}{\bar{\rho}} = B^{\frac{2}{3}}x^{\frac{2}{3}}d^{-1}j\left(\frac{z}{d}\right). \qquad (5.3.8)$$

(Compare these with (5.2.2) and (5.2.3); dimensionally the arguments are closely related.) There are no current data which allow (5.3.7) to be tested, but measurements of the longitudinal density variations in the Red Sea agree with the $\frac{2}{3}$ power law predicted by (5.3.8). Vertical buoyancy profiles scaled in this way all fall close to a single curve, giving further strong support to the solution and permitting the empirical evaluation of the dimensionless function j.

We should also note that in this kind of turbulent convective motion the turbulence adjusts itself so that there is everywhere a balance between the mean shear, the buoyancy and the Reynolds stress terms: all are of the same order of magnitude. It follows from (5.3.7) and (5.3.8) that the mean gradient Ri must be independent of B and x (but it may still be a function of z/d), and the same is true of Rf. In view of the prediction made in (5.1.23), one can speculate that the case considered here is another kind of internally determined flow, for which there will be unique constant values of Ri and Rf (or K_ρ/K_M).

CHAPTER 6

BUOYANT CONVECTION FROM
ISOLATED SOURCES

The buoyancy effects discussed so far have for the most part been stabilizing, or have been assumed to produce a small modification of an existing turbulent flow. Now we turn to convective flows, in which buoyancy forces play the major role because they are the source of energy for the mean motion itself. The usual order of presentation will be reversed: it is convenient to set aside for the present the discussion of the mean properties of a convecting region of large horizontal extent, and flows near solid bodies, and in this chapter to treat various models of the individual convective elements which carry the buoyancy flux. (See Turner (1969 a) for a review of this work and a more extensive bibliography.)

Such models can be broadly divided into two groups, those which assume the motion to be in the form of 'plumes' or of 'thermals'. (See fig. 6.1.) In both of them motions are produced under gravity by a density contrast between the source fluid and its environment; the velocity and density variations are interdependent, and occupy a limited region above or below the source. *Plumes*, sometimes called buoyant jets, arise when buoyancy is supplied steadily and the buoyant region is continuous between the source and the level of interest. The term *thermal* is used in the sense which has become common in the meteorological literature to denote suddenly released buoyant elements. The buoyancy remains confined to a limited volume of fluid, which, as it rises, loses its connection with the source which produced it. Other models have some features in common with both of these. (See §6.3.3 and fig. 6.1.)

Both plumes and thermals can be axisymmetric or two-dimensional, and each case will be discussed here with varying degrees of detail. The Boussinesq approximation will be used throughout, neglecting (potential) density differences except where they occur in

[165]

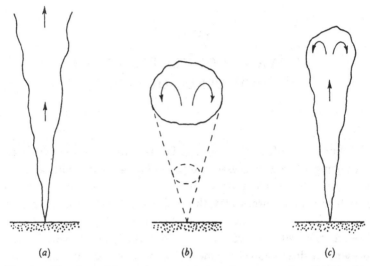

(a)   (b)   (c)

Fig. 6.1. Sketches of the various convection phenomena described in this chapter: (a) plume, (b) thermal, (c) starting plume. The arrows indicate the direction of mean motion. (From Turner 1969a.)

the buoyancy terms. We will discuss the motion with buoyancy taken as positive upwards, i.e. as if heated air were rising through the atmosphere, though of course the direction of motion is immaterial to the dynamics, and laboratory experiments often for convenience use plumes of salt water descending through a tank of fresh water.

Most of the results described here are similarity solutions of the kind described by Batchelor (1954a). Dimensional arguments are very powerful tools for investigating the overall properties of such flows, and when combined with numerical values of parameters obtained from laboratory experiments, they allow one to make predictions about the flow pattern and density distribution over a very wide range of scales. Detailed theories of these motions are much less well developed, however, and there are important natural flows (e.g. large convective clouds – see §6.4.1) which are not adequately described by similarity theory but for which there is at present no alternative model.

6.1. Plumes in a uniform environment

The major parameter governing the flow above a small axisymmetric source of buoyancy can be defined as

$$F_0 = 2\pi \int_0^\infty wg'r\,\mathrm{d}r, \qquad (6.1.1)$$

where w is the vertical velocity, $g' = g\rho'/\rho_0$ and r is the radial distance from a vertical line above the source. Thus $\rho_0 F_0$ is the total weight deficiency produced per unit time at the source, and F_0 must remain constant with height z in an environment of constant density ρ_0 (though w will decrease and the radial lengthscale b increase with increasing distance z above the source). Such flows can stay laminar only for very low values of F_0 – even the plume from a cigarette becomes unstable after rising a short distance – and attention will be concentrated here on the practically more important turbulent plumes.

6.1.1. *Axisymmetric turbulent plumes*

Basic to the understanding of all free turbulent flows is the process of 'entrainment' or mixing of outside fluid into the plume. It is observed that (like jets) turbulent plumes have a sharp boundary separating nearly uniform turbulent buoyant fluid from the surroundings, as pictured in the photograph of fig. 6.2 a pl. xiv. This boundary is indented by large eddies and the mixing process takes place in two stages, the engulfing of external fluid by the large eddies, followed by rapid smaller scale mixing across the central core. The vertical velocity and the turbulence measured at a fixed point off the axis have an intermittent character, since the plume is waving about and a measuring instrument spends part of its time in the fully turbulent fluid, and part outside it. Though an instantaneous profile across a plume is sharp-edged, the time averaged profiles of velocity and temperature are smoother, and can be well fitted by Gaussian curves.

A detailed theory of the mechanism of entrainment (and a comparison between different free flows) has been given by Townsend (1970), but the above simplified picture is sufficient for the present purpose. The *similarity* of the profiles at all heights above the source is the main assumption made here; no information is obtained from similarity theory about their form, and the particular choice made is a matter of convenience only. In the following we will often use the simplest 'top hat' shape (i.e. the properties have one constant value inside the plume, and another outside it). This is equivalent to replacing the mass and momentum fluxes by mean values defined as integrals across the plume, and does not imply any physical assumption. The rate of spread is governed by the large scale structure of the turbulence generated by the motion of the plume itself, and the assumption of similarity means that this must have the same relation to the mean flow whatever the scale of motion. Provided the Reynolds number $Re = wb/\nu$ is large enough, neither the molecular properties ν and κ of the fluid nor Re itself can enter directly into the determination of the overall properties of a turbulent plume.

The parameters defining the behaviour of an axisymmetric turbulent plume can therefore be taken as F_0, z and r. A dimensional analysis shows immediately that

$$
\left.
\begin{aligned}
w &= F_0^{\frac{1}{3}} z^{-\frac{1}{3}} f_1\!\left(\frac{r}{b}\right), \\
g' &= F_0^{\frac{2}{3}} z^{-\frac{5}{3}} f_2\!\left(\frac{r}{b}\right), \\
b &= \beta z,
\end{aligned}
\right\}
\tag{6.1.2}
$$

where β is a constant to be defined for particular profiles. The turbulent plume thus has a conical form, with apex at the source. Such a point source is of course physically unrealistic, since (6.1.2) implies that both w and g' become infinite as $z \to 0$. At some distance above a real source, however, the flow behaves as if it came from a 'virtual point source' below, and the point source remains a useful theoretical concept, provided (6.1.2) are applied to plumes which rise to a height of many diameters above the source.

The actual form of f_1 and f_2 must be obtained using either more detailed theories (all of which entail some questionable assumptions about the turbulence distributions across the plume), or directly by experiment. For example, the experiments of Rouse, Yih and Humphreys (1952), made using a heat source in a large room, both

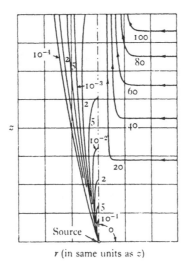

Fig. 6.3. Mean isotherms (left) and streamlines (right) for turbulent convection from a maintained point source. The numbers on the isotherms are relative values of $\Delta\rho/\rho_0$, and those on the streamlines are relative values of the Stokes stream function (from Rouse, Yih and Humphreys 1952).

confirmed the above forms of dependence on z and indicated that the mean profiles were closely Gaussian. The results can be represented to the experimental accuracy by

$$\left.\begin{aligned} w &= 4.7 F_0^{\frac{1}{3}} z^{-\frac{1}{3}} \exp\left(-96 r^2/z^2\right), \\ g' &= 11 F_0^{\frac{2}{3}} z^{-\frac{5}{3}} \exp\left(-71 r^2/z^2\right). \end{aligned}\right\} \tag{6.1.3}$$

The corresponding mean isotherms and streamlines in the plume are reproduced in fig. 6.3; these show that the width of the region containing turbulent buoyant fluid is always expanding, but the *visible* region (containing concentrations above a certain minimum in a plume marked by smoke, for example) will at first expand and then contract and disappear. The value of β found experimentally (and used in the plot) will be discussed in the next section. Note that (6.1.3) implies a slightly greater (mean) spread of buoyancy relative to momentum.

The above solutions show that there is a flow from the environment into the turbulent plume, since the mass flux is clearly increasing with height. This radial flow at large r is not taken into

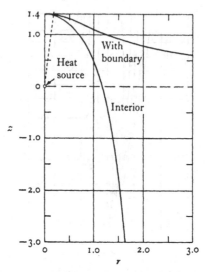

Fig. 6.4. Streamlines for inflow into a plume rising from a point heat source located on the horizontal boundary or in the interior of a large region of uniform fluid. (From Taylor 1958.)

account in (6.1.3) which refer only to the narrow region of strong vertical motion, but it may be found using a method introduced by Taylor (1958). As far as the external fluid is concerned, a plume (or jet) can be replaced by a line of sinks along the axis, and the inflow calculated approximately by potential flow theory. The sink distribution must be chosen to match the calculated inflow at each level, and varies with the kind of flow; for axisymmetric buoyant plumes, the increase in mass flux with z implied by (6.1.2) gives a sink strength proportional to $z^{\frac{2}{3}}$. The forms of typical streamlines computed for the two cases where the point source is on a plane, or in the middle of an unbounded region of uniform fluid, are reproduced in fig. 6.4; they show that the fluid entering the plume may have come from very different parts of the environment, according to the shape of the region in which convection is taking place.

6.1.2. *The entrainment assumption*

As Batchelor (1954a) pointed out, this increasing vertical flow in a plume also implies that there is a mean inflow velocity across the boundary which varies as $z^{-\frac{1}{3}}$. That is, the linear spread of radius with height implies that the mean inflow velocity across the edge of the plume is proportional to the local mean upward velocity (6.1.2). This statement, regarded now as the basic assumption, was first introduced by Sir Geoffrey Taylor, and was used by Morton, Taylor and Turner (1956) to discuss cases where similarity solutions of the above form do not exist. It is in fact another type of similarity assumption, which implies that there is the same kind of turbulent structure and balance of forces at each height. It is simple and powerful, and is to some extent justified by its success in a variety of contexts, but its range of validity must always be examined carefully. It has been argued by Morton (1968), for example, that for flows where the equilibrium structure has not been attained, the entrainment should be related not to the mean velocity but to the Reynolds stress, and Telford (1966, 1970) has proposed that the local level of turbulence should be used as the velocity scale. The detailed analysis of the entrainment process by Townsend (1970) shows that none of these is precisely satisfied, but nevertheless they remain useful approximations.

When the simplest entrainment assumption is made, so that the inflow velocity is taken to be some fraction α of the upward velocity, the equations of conservation of mass, momentum and buoyancy can be reduced to the form

$$\left.\begin{array}{l} \dfrac{\mathrm{d}(b^2\overline{w})}{\mathrm{d}z} = 2\alpha b\overline{w}, \\[2mm] \dfrac{\mathrm{d}(b^2\overline{w}^2)}{\mathrm{d}z} = b^2 g', \\[2mm] \dfrac{\mathrm{d}(b^2\overline{w}g')}{\mathrm{d}z} = -b^2\overline{w}N^2(z). \end{array}\right\} \qquad (6.1.4)$$

The neglect of a pressure term in the second of these equations can be rigorously justified provided the plume remains narrow. With a later use in mind, an arbitrary density variation in the environment has been allowed for; $g' = g(\rho_0 - \rho)/\rho_1$ is calculated using the local density difference between the plume ρ and its environment ρ_0 at the height z and some standard density ρ_1 in the environment, and $N^2 = (-g/\rho_1)(\mathrm{d}\rho_0/\mathrm{d}z)$ is the square of the local buoyancy frequency. The velocity $\bar{w}$ and width b are defined by integrating the mass and momentum fluxes across the plume:

$$\bar{w}b^2 = 2\int_0^\infty w(r)r\,dr, \quad \bar{w}^2b^2 = 2\int_0^\infty w^2(r)r\,dr, \qquad (6.1.5)$$

i.e. an equivalent 'top hat' profile is implied in (6.1.4).

When the environment is of uniform density, $N = 0$, so the third equation in (6.1.4) is a formal statement of the fact that the buoyancy flux is constant, $b^2\bar{w}g' = F_0/\pi = F$ say. The solution of the other two equations, with the boundary conditions that the mass and momentum fluxes are zero at the source, is entirely equivalent to the similarity solution (6.1.2). The multiplying constants such as β are now, however, all given explicitly in terms of the 'entrainment constant' α:

$$\left.\begin{aligned} b &= \tfrac{6}{5}\alpha z, \\[2mm] \bar{w} &= \frac{5}{6\alpha}(\tfrac{9}{10}\alpha F)^{\frac{1}{3}}z^{-\frac{1}{3}}, \\[2mm] g' &= \frac{5F}{6\alpha}(\tfrac{9}{10}\alpha F)^{-\frac{1}{3}}z^{-\frac{5}{3}}. \end{aligned}\right\} \qquad (6.1.6)$$

Allowance can be made for different widths of the velocity and buoyancy profiles with little increase in complication.

It will be seen later (§6.4) how other more general kinds of plumes, having different rates of variation of buoyancy with height, can be treated by the same method. For the present it will suffice to comment on two of the consequences of the equations (6.1.4), which hold regardless of the form of $N^2(z)$ and g'. The mass continuity equation implies that the *proportional* rate of entrainment or change of mass flux (or, in the Boussinesq approximation, the volume flux V) is inversely proportional to the radius, i.e.

$$V^{-1}(dV/dz) = 2\alpha/b. \qquad (6.1.7)$$

The first two equations may be combined to express the rate of spread as

$$db/dz = 2\alpha - (bg'/2\bar{w}^2). \qquad (6.1.8)$$

Thus the half angle of spread ($\tfrac{6}{5}\alpha$) of a plume in a uniform environment is proportional to the entrainment constant, but is less than the angle 2α for a non-buoyant jet (with $g' = 0$), for the same value of α. (See fig. 6.5, which is discussed further in the following section.)

The numerical value to be chosen for α cannot be obtained theoretically (though comparisons between different kinds of flow have been made), and it must be taken from laboratory experiments. Morton (1959b) adopted a value of 0.116 for top hat profiles, based on measurements in non-buoyant jets (for Gaussian profiles, if the scale of w is defined to be the velocity on the axis, and b the radial distance to the point where the velocity is reduced by a factor e^{-1}, then it follows from the definitions (6.1.5) that the corresponding entrainment constant $\alpha_G = 2^{-\frac{1}{2}}\alpha$).

Ricou and Spalding (1961) have found a value of α for jets closer to 0.08; they used a direct method of measuring the inflow which avoids the errors introduced by integrating across measured profiles. Their experiments also suggested that the entrainment constant for buoyant plumes is somewhat greater than for jets. A value of $\alpha = 0.12$ for plumes is deduced from the profiles (6.1.3), but direct results of comparable accuracy to those in jets are not yet available for plumes. Since α does not enter sensitively into many calculations, a fixed value of say $\alpha = 0.10$ (or $\alpha_G = 0.07$) might be adopted in all cases until more definitive measurements become available.

6.1.3. *Forced plumes*

Although it is often useful to interpret observations on plumes from finite sources in terms of one arising from an equivalent virtual point source of buoyancy alone, it is not necessary always to do so. The equations (6.1.4) can be solved with any initial values of radius and velocity (always bearing in mind the reservations about the validity of the entrainment assumption in the 'developing' part of the flow). Another important case is the turbulent flow generated by a finite source which emits fluxes of mass and momentum at a steady rate, as well as buoyancy. Morton (1959a) has shown that this general flow, which he has called the forced plume (and includes non-buoyant jets and pure plumes as special cases) can be related to the flow from a virtual point source of buoyancy and momentum only.

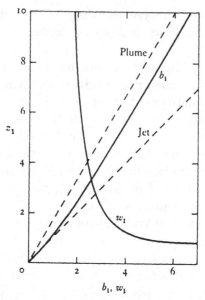

Fig. 6.5. The behaviour of a forced plume in a uniform environment, showing the non-dimensional radius b_1 and vertical velocity w_1 as functions of height. The initial spread is like a jet, but later approaches that characteristic of plumes. (From Morton 1959a.)

Two physical parameters are now specified at the source, the rates of discharge of buoyancy $\rho_1 F_0$ and momentum $\rho_1 M_0$ say. All the variables of the problem can be scaled in terms of these, and one can write

$$\left. \begin{aligned} z &= \alpha^{-\frac{1}{2}} |M_0|^{\frac{3}{4}} |F_0|^{-\frac{1}{2}} z_1, \\ b &= \alpha^{\frac{1}{2}} |M_0|^{\frac{3}{4}} |F_0|^{-\frac{1}{2}} b_1, \\ \bar{w} &= \alpha^{-\frac{1}{2}} |M_0|^{-\frac{1}{4}} |F_0|^{\frac{1}{2}} w_1, \end{aligned} \right\} \qquad (6.1.9)$$

where z_1, b_1 and w_1 are non-dimensional variables. The equations (6.1.4) reduce to a non-dimensional pair in the variables $(b_1 w_1)$ and $(b_1{}^2 w_1)$, with boundary conditions $b_1 w_1 = \pm 1$ and $b_1{}^2 w_1 = 0$ at $z = 0$.

This formulation includes the possibility that F_0 and M_0 can have either sign; two cases will be discussed as examples. When both the buoyancy and momentum fluxes are directed upwards, the non-dimensional solutions for radius and velocity are shown in fig. 6.5. The rate of spread is at first like that in a pure jet, but as the momentum generated by buoyancy dominates over the initial momentum,

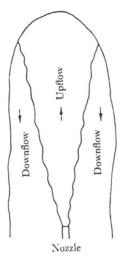

Nozzle

Fig. 6.6. Sketch of the up and downflow regions in a heavy salt jet projected upwards. The outer boundary is approximately to scale, but the inner boundary is schematic only. (From Turner 1966.)

the spread (for given α, of course) becomes smaller and approaches that of a plume. This behaviour must be kept in mind in any experiments designed to *measure* α in plumes, since any extra momentum at the source could have an effect that might be interpreted as a larger α. It will also be of importance in a stable environment ($\S 6.4.3$).

When the source fluid is *heavier* than the environment, but its initial momentum is upwards, then the momentum is continually being decreased by the buoyancy forces, until it becomes zero and then begins to increase downwards. The solution of the equations derived from (6.1.4) can be used to estimate the maximum height to which fluid first rises, but after that the model becomes unrealistic. Experiments (Turner 1966) show that in this case the heavy fluid falls back in an annular region surrounding the upflow (see fig. 6.6), and the simple theory does not allow for the mixing between the two streams moving in opposite directions. It is observed, however, that in the steady state the plume top fluctuates slightly about a height z_m which is a constant fraction (0.70) of that to which it first rises, and experiments using salt water injected into fresh have confirmed the relation $z_m = 1.85 M_0^{\frac{3}{4}} F_0^{-\frac{1}{2}}$, which is the form to be expected from (6.1.9).

There is a useful alternative interpretation of these results, which is related to the analysis given in §6.2.2. In an actual experiment fluid is ejected from a nozzle of radius r_0 with velocity U_0, so $M_0 = \pi r_0^2 U_0^2$ and $F_0 = \pi g_0' r_0^2 U_0$. For the experiment with heavy fluid, the maximum height z_m may be written as

$$\frac{z_m}{r_0} \propto \frac{U_0}{(g_0' r_0)^{\frac{1}{2}}} = F = Ri_0^{-\frac{1}{2}}. \tag{6.1.10}$$

That is, the ratio of the height of rise to the radius of the nozzle is directly proportional to an internal Froude number F based on the initial properties of the negatively buoyant jet.† The Froude number is also relevant, of course, in the case where both buoyancy and initial momentum are directed upwards. A pure plume in neutral surroundings has a particular *constant* value of the Froude number (with the definitions and scaling used in (6.1.6), this is $(\frac{8}{5}\alpha)^{-\frac{1}{2}}$, but the actual value is less important than the fact that it exists). If fluid is ejected from a finite source with this value of F, it will behave exactly like a plume originating from a point source below; if it comes out with different properties, these will change so as to approach the value of F appropriate to a plume. A jet-like flow initially has $F \to \infty$ and gives the most rapid mixing of an effluent with its environment, whereas a plume accelerating fluid from rest ($F = 0$ at the source) results in a more rapid removal of plume fluid from the neighbourhood of the source.

6.1.4. *Vertical two-dimensional plumes*

Analogous results can be obtained for line sources of buoyancy in uniform surroundings, using similarity arguments (based now on the buoyancy flux per unit length, $\rho_0 A$ say) or the equivalent entrainment assumption. The rate of spread of the transverse dimension x is again linear with height, and the profiles nearly Gaussian; dimensional arguments show that the velocity scale is independent of height z and the density difference varies inversely as z. The full

† The use of the same symbol F for Froude number and buoyancy flux is unfortunate here; in no other section do they occur together.

solutions, with the profile shapes suggested by the experiments of Rouse, Yih and Humphreys (1952) are

$$\left.\begin{aligned} w &= 1.8A^{\frac{1}{3}}\exp(-32x^2/z^2), \\ g' &= 2.6A^{\frac{2}{3}}z^{-1}\exp(-41x^2/z^2). \end{aligned}\right\} \qquad (6.1.11)$$

The accuracy of such experiments is not high enough to be sure that the suggested greater spread of velocity relative to buoyancy is real, and indeed the waving about of these plumes makes the estimate of an entrainment constant from the profiles very inaccurate. A direct estimate of entrainment is available here too (from the experiments of Ellison and Turner (1959) which will be described more fully in the following section); these give the value of $\alpha = 0.08$ for top hat profiles.

Line plumes arise in several applications of practical importance. Priestley (1959) has suggested that the sources of warm air near the ground get drawn out into lines down wind, so that typical convection elements are like two-dimensional symmetric plumes. Artificial plumes formed by lines of burners have been used in attempts to disperse fog from aircraft runways. Examples in the ocean are the discharge of sewage from a manifold laid on the bottom, and the 'bubble breakwater' invented by Sir Geoffrey Taylor. This last example arose from the suggestion that if a curtain of air bubbles is forced from a horizontal pipe some distance below the surface of the sea, then waves will be damped by the surface outflow of water driven upwards by the bubbles. The point of interest here is that the flux of density deficiency due to the small bubbles can be treated in exactly the same way as that due to fresh water rising through salt water, or to heat in a turbulent plume in air, since in each case the molecular diffusivity is irrelevant. The relevant flux can be calculated from the total volume of air released; if the bubbles are rising at terminal velocity, the whole of their buoyancy is transmitted to the water around them, and provided they are small they are transported about by the turbulence in the plume. In the same way sediment, if its concentration is not too great, can drive 'turbidity currents' by changing the buoyancy of the water containing it. This last example will be treated at greater length in the following section.

6.2. Inclined plumes and turbulent gravity currents

We now come to a phenomenon which can be regarded as a generalization of the one just treated – a two-dimensional turbulent plume, but now on a slope at an arbitrary angle θ to the horizontal (see fig. 6.7). Related topics have already received some attention, the frictionless flow of a heavy fluid under a lighter one, and its relation to hydraulic theory in chapter 3, and the laminar viscous flow and its stability in chapter 4. The new problems introduced when the flow is turbulent and can mix with the environment, assumed to be non-turbulent, are of sufficient importance to require extra discussion here. There is also a close connection with the turbulent stratified flows mentioned in §5.1, whose discussion we have deferred till this section so that both the driving and the stabilizing aspects of the density gradients could logically be considered together.

There are many situations in nature where such flows can arise. Katabatic winds in the atmosphere occur when air cooled by contact with cold ground flows downhill. Similar currents arise in the ocean, for example when cold water from Arctic regions flows along the bottom under warmer water. Heavy layers can also be formed because of the extra weight of suspended solids; turbidity currents in the ocean (Johnson 1963) and 'density currents' in reservoirs arise from this cause. In the atmosphere too, some dust storms and powder snow avalanches are driven by the weight of the suspended material (see fig. 3.14 pl. VI) though in these cases the advance of the 'front', or leading edge of the disturbance, is of more importance than the steady flow behind. Gravity currents can equally well be formed by a lighter layer under a sloping roof, as in the example of a layer of methane flowing along the roof of a mine, which was mentioned in §5.3.3.

6.2.1. A modified entrainment assumption

When a two-dimensional plume, or gravity current, is inclined to the vertical there will be a component of gravity acting normal to the entraining edge, so that gravity will inhibit mixing into the plume as well as driving it along the slope. A measure of the stabilizing effect of the density gradient relative to the shear is again an overall

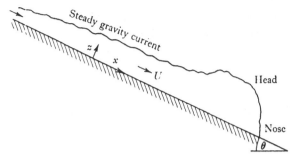

Fig. 6.7. Sketch of a gravity current on a slope, with an
initial head followed by a steady layer flow.

Richardson number, which can be defined for the present purposes
as
$$Ri_0 = \frac{g'h\cos\theta}{U^2} = \frac{A\cos\theta}{U^3}, \qquad (6.2.1)$$

where U is the mean velocity down the slope and h the depth,
defined by integrals analogous to (6.1.5)
$$Uh = \int_0^\infty u\,dz, \quad U^2h = \int_0^\infty u^2\,dz, \qquad (6.2.2)$$

and $A = g'hU$ is the buoyancy flux per unit width (cf. §5.3.2).

The new assumption required when the plume is inclined
(Ellison and Turner 1959) is that the entrainment parameter is not
now the constant α appropriate to vertical plumes, but that it can be
a function $E(Ri_0)$ of the Richardson number (6.2.1). If the assumption of similarity of profiles is retained, relations of the form (6.1.11)
will apply, but with the magnitudes modified because of the slope.
Gravity currents will still spread linearly with distance x, their
mean velocity will be constant and proportional to $A^{\frac{1}{3}}$, and the
density difference will be inversely proportional to x.

The mass continuity equation (with the ambient fluid at rest) is
$$\frac{d(Uh)}{dx} = EU, \qquad (6.2.3)$$

where E is now a function of Ri_0. In the simplest case, when one can
follow the method used for vertical plumes and regard bottom
friction as unimportant compared with the stress at the edge due to
entrainment, the momentum equation is just
$$\frac{d(U^2h)}{dx} = g'h\sin\theta. \qquad (6.2.4)$$

(For the moment a 'profile constant', which could arise because of the form of the density distribution, is ignored.) It follows from the last two equations that for a flow dominated by the turbulent entrainment

$$E = Ri_0 \tan \theta. \qquad (6.2.5)$$

Thus E is a strong function of slope, and is zero (except for the effects of bottom friction) at a finite value of Ri_0 when the bed becomes horizontal.

6.2.2. *Slowly varying flows*

In the general case, the momentum equation can be written

$$\frac{d(U^2h)}{dx} = -C_D U^2 - \frac{1}{2}\frac{d(S_1 g'h^2 \cos\theta)}{dx} + S_2 g'h \sin\theta, \quad (6.2.6)$$

where C_D is the drag coefficient (cf. (3.2.7)), and S_1 and S_2 are profile constants defined by integrating across the plume. The terms on the right represent respectively the turbulent frictional drag on the bottom, the pressure force on the layer due to its changing depth (a term which was ignored in (6.2.4) and which is relatively less important), and the force of gravity accelerating the layer. If any variation of S_1, S_2 and θ with x is neglected (6.2.6) can be combined with (6.2.3) to give separate equations for dh/dx and $d Ri_0/dx$. These are:

$$\frac{dh}{dx} = \frac{(2 - \frac{1}{2}S_1 Ri_0)E - S_2 Ri_0 \tan\theta + C_D}{1 - S_1 Ri_0}, \qquad (6.2.7)$$

$$\frac{h}{3Ri_0}\frac{dRi_0}{dx} = \frac{(1 + \frac{1}{2}S_1 Ri_0)E - S_2 Ri_0 \tan\theta + C_D}{1 - S_1 Ri_0}. \qquad (6.2.8)$$

Equation (6.2.7) adds the effects of mixing to those earlier considered in deriving (3.2.8) using the 'hydraulic' assumptions. In the previous case, however, h and Ri_0 (or F) were directly connected through the equation of continuity; there was a certain depth at which $Ri_0 = 1$ and another depth for which $dh/dx = 0$, $dRi_0/dx = 0$. In the present case where mixing can take place, the behaviour of Ri_0 is little different from that in ordinary hydraulics, but h is no longer connected uniquely with it. There is a particular value of

Ri_0 called the *normal* value Ri_n, determined by the slope and also by friction, for which the right-hand side of (6.2.8) vanishes and Ri_0 is constant along the slope; to determine Ri_n one must therefore know S_1 and S_2 and use measured values for E as a function of Ri_0.

The analogy with hydraulic theory may be pursued further. The forms of solution of (6.2.7) and (6.2.8) depend on the relative magnitudes of $S_1 Ri_0$ at the origin, $S_1 Ri_n$ and 1. The distinction can still be made between tranquil (subcritical) flow in which $S_1 Ri_0 > 1$ and the velocity of long internal waves on the layer is greater than the flow velocity, and shooting (supercritical) flow in which $S_1 Ri_0 < 1$ and disturbances are unable to propagate upstream. Experimentally, it appears that entrainment becomes small at $Ri_0 \approx 1$, so that all flows on small slopes may be treated approximately as if they did not mix (as was done in chapter 3). At the other extreme, on steep slopes both $S_1 Ri_0$ and $S_1 Ri_n$ are less than unity, and the flow is supercritical throughout. The adjustment to the 'normal' state is quite rapid; if the flow starts too slowly, gravity accelerates it, and if it starts too fast, increased mixing occurs until Ri approaches Ri_n. At this time the spread is linear with

$$\frac{dh}{dx} = E(Ri_n). \qquad (6.2.9)$$

6.2.3. *Laboratory experiments and their applications*

The form of the entrainment function $E(Ri_0)$ must, as in the case of the entrainment constant for vertical plumes, be found by experiment. Two kinds of laboratory experiment have been used. The first consists of a horizontal turbulent surface jet above a heavier fluid, or a bottom current on a horizontal floor under a lighter layer, which is started with high velocity and low Ri_0 (and therefore in a 'supercritical' state). As it mixes and spreads out, its Richardson number increases towards a value where the flow is subcritical, and by assuming local equilibrium, E can be evaluated from the rate of spread at a number of intermediate values of Ri_0; the most important feature is a rapid fall of E with increasing Ri_0. (See fig. 6.8.) The relation between the velocity and depth during this process is

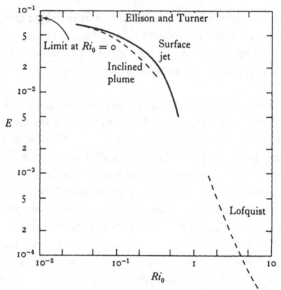

Fig. 6.8. Rate of entrainment into a turbulent stratified flow as a function of overall Richardson number, for two types of experiment described by Ellison and Turner (1959), and the experiments of Lofquist (1960).

obtained by integrating the momentum equation (6.2.6) with $\theta = 0$, $C_D = 0$ to give (when $S_1 = 1$)

$$\frac{h(1 + \frac{1}{2}Ri_0)}{Ri_0^{\frac{2}{3}}} = \text{const.} \qquad (6.2.10)$$

Wilkinson and Wood (1971) have shown that it is useful to interpret this region of flow adjustment as a kind of hydraulic jump, in which some of the kinetic energy lost from the mean flow is responsible for mixing. There is the important difference that the flow conditions on each side of the mixing region are not now uniquely related; a range of possible states can be attained downstream (for given values of h and Ri_0 upstream), depending on the nature of the downstream control. The maximum total entrainment occurs when $Ri_0 = 1$ downstream (or rather less when S_1 and S_2 are included), by which stage E has become negligible. If the flow downstream is controlled by a weir whose height is gradually increased, Ri_0 increases, a region of reversed flow moves upstream and the total mixing is reduced, until the jump is completely submerged and there is no entrainment at all. (See fig. 6.9 pl. xiv.) In the unsteady

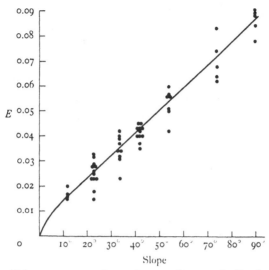

Fig. 6.10. Measurements of entrainment into an inclined plume as a function of slope. (From Ellison and Turner 1959.)

case, when the layer flow is preceded by a nose, the front of the flow itself acts as a control which keeps Ri_0 close to unity. (See §3.2.4.)

The second kind of experiment is the one illustrated in fig. 6.7, the inclined plume of salt water on a slope, to which the theory applies more directly. Here a steady plume can be set up, and the inflow of fresh water necessary to make up for the entrainment is measured as a function of slope. The results obtained in this way by Ellison and Turner are shown in fig. 6.10; the general dependence of E on slope is clear, but there is considerable experimental scatter attributed to the fact that the Reynolds number was not always large, and a layer of heavy fluid near the wall could sometimes remain non-turbulent.

The mean experimental results for E as a function of Ri_0 are plotted on logarithmic scales in fig. 6.8. At the smaller values of Ri_0 the surface jet results and those for inclined plumes obtained by Ellison and Turner are seen to be in good agreement. At the higher values of Ri_0, the curve is that due to Lofquist (1960), who studied an undercurrent on a much larger scale. Since this was maintained in a turbulent state by bottom friction, some measure of the roughness must also be a relevant parameter here. The two sets of results do seem to be consistent with one another (though there is a gap not covered by any experiments), and with the assumption that a stability parameter like Ri_0 is the major factor determining the entrainment rate in both cases. We must, however, sound here the warning that at high values of Ri_0 molecular diffusion may play a significant role in the mixing process (see §9.1.3).

The laboratory results may be used confidently to make predictions of the velocity of flow of a larger scale turbulent gravity current on a steep slope. All that is required is an estimate of the rate of output of light or heavy fluid A; small changes in Ri_n due to Reynolds number effects or changes in friction coefficient are unimportant since U is proportional to $(A/Ri_n)^{\frac{1}{3}}$. (The constant of proportionality can be obtained as a function of slope from the laboratory experiments.) At low slopes, E becomes very small so that even at large Reynolds numbers there should be negligible friction at the interface. The weight of the layer must then be balanced solely by friction at the solid wall and

$$S_2 \frac{A}{U^3} \sin \theta = C_D. \qquad (6.2.11)$$

On intermediate slopes, both entrainment and bottom friction may be important in determining the friction and flowrate.

The results of the previous section can easily be extended to calculate the opposing (non-turbulent) flow U_a which would be required to reverse a gravity current, for example, the magnitude of the upper wind which could destroy a katabatic wind at the surface, or the ventilating flow needed in a mine to prevent a dangerous layer of methane flowing uphill along the roof of a sloping roadway (see §5.3.3). If U now denotes the difference between the layer velocity and the external flow (with Ri_0 again defined using U), the buoyancy flux becomes $A = g'h(U+U_a)$, and the 'normal' state is described by the following generalization of (6.2.8):

$$\left(1 + \tfrac{1}{2} S_1 Ri_0 \frac{U}{U+U_a}\right) E = S_2 Ri_0 \tan \theta - C_D \frac{(U+U_a)^2}{U^2} \operatorname{sgn}(U+U_a).$$

$$(6.2.12)$$

Note that the last term (the drag) depends on the sign of $(U+U_a)$, the velocity of the layer, which may be in either direction, with or against U_a.

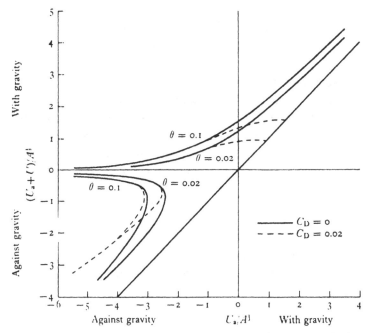

Fig. 6.11. Calculations of the velocity $(U+U_a)$ of an inclined plume entraining a moving ambient stream of velocity U_a. The empirical relation from fig. 6.8 has been used, and curves are shown for two slopes and two friction coefficients. (From Ellison and Turner 1959.)

For given values of θ and C_D and experimental values of $E(Ri_0)$, (6.2.12) may be solved numerically to give $(U+U_a)/A^{\frac{1}{3}}$ as a function of $U_a/A^{\frac{1}{3}}$, i.e. the non-dimensional layer velocity as a function of that in the ambient fluid, as shown in fig. 6.11. When U_a opposes the natural flow and is greater than about $3A^{\frac{1}{3}}$ (twice the downhill layer velocity in the absence of an opposing flow), the direction of layer flow can be reversed. The numerical constant varies only slowly with slope and friction coefficient, and this value has been verified by direct laboratory experiment. The Richardson number and hence the entrainment function and rate of spread do vary more markedly as the slope is changed. Small values of Ri_0 are relevant when the ambient flow is acting against gravity, whatever the magnitude of the slope, and the greatly increased mixing and stress in this state accounts for the sudden reversal of direction which is often observed. (Compare with §3.2.5.)

6.2.4. *Detailed profile measurements*

Though the experimental results of Ellison and Turner (1959) reported above were designed originally to obtain an 'overall' understanding of inclined plumes or bottom currents, they are sufficiently detailed to allow in addition an interpretation in terms of the local gradients. At moderately steep slopes the stress depends mostly on the entrainment at the outer edge of the layer and little on bottom friction (6.2.6). The turbulence effecting the entrainment is therefore produced in the same region as it is used to do work against buoyancy forces, and in this respect the flow is like the wakes discussed in § 5.3. Because of the component of gravity which tends to accelerate the flow if the stress becomes too small, one might expect in addition that this outer edge could be *maintained* in a marginally stable state. The theoretical argument of § 5.1.4 can be recast in terms of the local velocity and buoyancy differences across the edge of the plume to give

$$\frac{du}{dz} = k_1 \frac{g'}{\Delta U}, \quad \frac{g}{\rho}\frac{d\rho}{dz} = k_2 \left(\frac{g'}{\Delta U}\right)^2. \qquad (6.2.13)$$

Again it predicts linear profiles, essentially because no external lengthscale is relevant here.

Examination of the detailed measurements (such as those in fig. 6.12) shows that the outer edges of both the mean velocity and salinity profiles are indeed very closely linear. The whole of this outer region can be characterized by a single gradient Richardson number (corrected for the slope). For fifteen experiments in which reliable profiles were measured (with $\theta = 14°$ and $23°$) the range of values of $Ri \cos \theta$ was 0.044–0.077, with a mean value of 0.058. To the accuracy of the experiments they therefore support the existence of a critical gradient Richardson number, less than 0.1, at which turbulence is strongly damped.

6.3. Thermals in a uniform environment

6.3.1. *Dimensional arguments and laboratory experiments*

Similarity solutions can also be written down to describe the motion of axisymmetric thermals, or suddenly released regions of buoyant

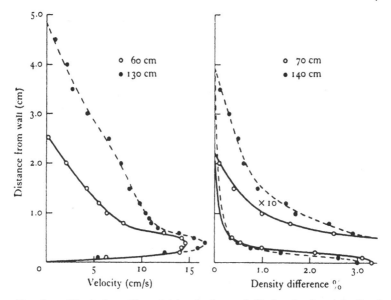

Fig. 6.12. Typical profiles of (a) velocity and (b) density in an inclined plume on a slope of 14°. (From Ellison and Turner 1959.)

fluid, as they rise and mix with their surroundings of constant density ρ_0. (It should be noted immediately that such solutions are less likely to be strictly applicable to real flows than is the case for plumes, since thermals typically grow to only a few times their original diameter and therefore do not have time to adjust to an equilibrium state.) Underlying these solutions is the assumption of similar distributions of velocity and buoyancy relative to the moving centre, and the sole parameter governing the motion is the total weight deficiency or buoyancy say $F_* = g(\rho_0 - \rho)V_0/\rho_0$, where ρ and V_0 are the initial density and volume of source fluid. The solutions for the typical horizontal dimension r_*, the mean vertical velocity w and g' in terms of powers of x and F_* (which is a constant in uniform surroundings) are

$$
\left.
\begin{aligned}
r_* &= \alpha z, \\
w &= F_*^{\frac{1}{2}} z^{-1} f_1\!\left(\frac{\mathbf{r}}{r_*}\right), \\
g' &= F_* z^{-3} f_2\!\left(\frac{\mathbf{r}}{r_*}\right).
\end{aligned}
\right\}
\tag{6.3.1}
$$

The profiles are functions of the vector position $\mathbf{r}$, which have so far only been obtained by experiment.

Scorer (1957) checked various consequences of this solution in laboratory experiments, using thermals of heavy salt solution falling through fresh water. He expressed his results in terms of the direct relation between the velocity $w_c = \mathrm{d}z_c/\mathrm{d}t$ of the cap of the thermal, the extreme horizontal radius b, and g' as

$$w_c = C(g'b)^{\frac{1}{2}}, \qquad (6.3.2)$$

and also verified that

$$b \propto t^{\frac{1}{2}} \quad \text{and} \quad w_c \propto t^{-\frac{1}{2}} \qquad (6.3.3)$$

which follow by writing (6.3.1) in terms of time instead of height. He found that the typical shape of thermals is a slightly oblate spheroid, with $m = V/b^3 = 3$. In any given experiment C (which has the form of a Froude number – compare (6.1.12)) and α ($= b/z_c$) remain constant with time, but there can be large variations in angle of spread from one experiment to another; the mean values given were $C = 1.2$ and $\alpha = 0.25$. It can be shown (Turner 1964b) that there is in fact a relation between C and α, which is determined using only the shape of the thermal and the assumption of a potential flow around it, without requiring any information about the flow inside.

More detailed experiments have shown that the mean motion inside a thermal does remain approximately similar at all heights and is rather like that in a flattened Hill's spherical vortex, with an upflow at the centre (in a rising thermal) and descent near the edges (see fig. 6.13 pl. xv). The kinematic consequences of applying the exact spherical vortex model are illustrated in fig. 6.14. The surrounding flow is instantaneously the same as that in a perfect fluid round a solid sphere, but superimposed on this is a constant rate of inflow across the sharp boundary separating the turbulent interior from the surroundings. No 'skin friction' is possible here because any boundary layer is immediately incorporated into the thermal; the only drag arises because of the external fluid which must be accelerated from rest, either by being added to the thermal or displaced around it. The figure shows both the particle paths relative to the boundary and the shapes into which originally horizontal surfaces are distorted as the vortex overtakes them. Mixing may be enhanced at the front where the entrained fluid is

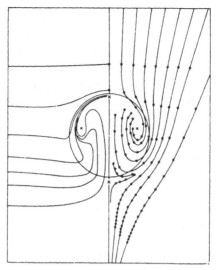

Fig. 6.14. The motion of particles into an expanding Hill's spherical vortex (right) and the distortion of initially horizontal planes of fluid (left) plotted in coordinates in which the spherical boundary is fixed in size. The tangent of the half angle of spread is $\alpha = \frac{1}{4}$. (From Turner 1964a.)

swept sideways in a thin layer (because the density contrast is statically unstable there), but there is also a broader region of inflow at the rear. Another result, which follows when the explicit spherical vortex velocity profiles are added to the similarity solutions, is worth a mention here. Only a fraction $(\frac{5}{14})$ of the work done by buoyancy appears as kinetic energy of mean motion, distributed between the 'solid' forward motion $(\frac{1}{6})$, the internal circulation $(\frac{3}{28})$ and the displaced external fluid $(\frac{1}{12})$; the balance $(\frac{9}{14})$ must go into turbulence and is eventually dissipated. (See Turner 1972.)

6.3.2. Buoyant vortex rings

A thermal released from rest can be regarded as a special case of a buoyant vortex ring (Turner 1957). For a thermal, the circulation is generated entirely by buoyancy, whereas in a vortex ring the initial buoyancy F_* and circulation K_0 can be specified separately at the source. When a volume of buoyant fluid is forcibly ejected upwards into uniform surroundings at rest, a vortex ring is formed in which both the vorticity and light fluid are contained within a sharp core.

This can either be laminar, and have a constant volume, getting thinner as the ring expands sideways, or turbulent, in which case the radius of its cross-section increases and remains nearly proportional to the radius R of the ring. (See fig. 6.15 pl. xv.) The cores are, however, carrying along with them a spheroidal region of non-buoyant fluid, rather like a thermal in shape.

One can take a circuit passing along the axis of the ring which contains only particles of fluid of constant density, so the circulation $K_0 \propto wR$ will remain constant in time. (This is true also for thermals, from the solutions (6.3.1). In that case it can be explained in terms of the balance between the generation of vorticity by buoyancy along the axis, and its destruction by mixing of vorticity of opposite signs across the centre-line.) The total impulse P of a vortex and its surroundings is

$$P = \pi \rho K_0 R^2. \qquad (6.3.4)$$

If we suppose that this is being increased by the action of the constant buoyancy force F_*, then the momentum equation may be written as

$$\frac{dP}{dt} = \pi \rho K_0 \frac{dR^2}{dt} = \rho F_* \qquad (6.3.5)$$

which gives

$$R^2 - R_0^2 = \frac{F_* t}{\pi K_0}. \qquad (6.3.6)$$

With the additional assumption that the velocity distribution remains similar at all heights, so that $w = cK_0/R$, it follows that

$$R = \alpha z = \frac{F_*}{2\pi c K_0^2} z. \qquad (6.3.7)$$

Thus the angle of spread of a vortex (and the special case of a thermal) is proportional to F_*/K_0^2. The release of a thermal from rest generates, during the process of acceleration, a relatively small value of K_0 and therefore a large angle of spread. Small differences in the manner of release can produce a different K_0, and a range of different angles, but once the similarity state is attained F_*/K_0^2 stays constant. Increasing K_0 for a given F_* (by giving extra momentum at the source) leads to a smaller angle of spread and hence less mixing with the environment. This is a very different behaviour from that of plumes, where extra momentum was seen to make the flow more jet-like and the spread greater (see fig. 6.5).

It is sometimes useful to formulate the dynamics of thermals and buoyant vortex rings in terms of the entrainment assumption (though as we have seen above, the total rate of addition of external fluid follows from a consideration of the overall dynamics, without mentioning the detailed mixing). If the inflow velocity is assumed proportional to the mean upward velocity of the centre w_0 and to the surface area, then one obtains a set of equations analogous to (6.1.6), which for a spherical thermal are

$$\left.\begin{array}{l} db^3/dt = 3\alpha b^2 w_0, \\ d(b^3 w_0)/dt = \tfrac{2}{3}b^3 g', \\ d(b^3 g')/dt = -b^3 w_0 N^2. \end{array}\right\} \tag{6.3.8}$$

In writing these forms the thermal has been treated as if it were a well-mixed sphere and the 'virtual mass coefficient' appropriate to a sphere has been used, but only the multiplying constants will be changed if the properties are defined by integrating over observed distributions.

The first of (6.3.8) shows immediately that $b = \alpha z$, i.e. the entrainment assumption plus the equation of continuity alone implies a linear spread (whatever the properties of the environment); moreover for a spherical thermal the entrainment constant and the angle of spread are identical. (Contrast this result with (6.1.8).) If V is the volume of the thermal, it also follows that

$$V^{-1}\frac{dV}{dz} = \frac{3\alpha}{b}. \tag{6.3.9}$$

As for plumes (6.1.7) the proportional rate of entrainment depends inversely on the radius, but the rate of dilution of a thermal is much larger than it is for plumes, since α is larger.

6.3.3. 'Starting plumes'

As mentioned in the introduction to this chapter, and sketched in fig. 6.1, a phenomenon related to both plumes and thermals is observed when a steady source of buoyancy is suddenly turned on, producing what has been called a 'starting plume'. This flow pattern stays similar at all times, but with increasing size; well behind the front the motion is just that in a steady plume, and at the front it is like a thermal. (See fig. 6.16 pl. xv.) At first sight it seems unlikely that these flows could stay together, since the velocity dependence on height in (6.1.6) and (6.3.1) is quite different; but the appropriate thermal solution is not that for constant total buoyancy, since allowance must be made for the buoyancy and momentum fed into the cap from the plume behind. Here it is appropriate to use Gaussian profiles in the plume since this form gives a good match to that at the bottom of a spherical vortex.

With these modified assumptions one can obtain a detailed understanding of the similarity solution near the cap in terms of the separate plume and thermal dynamics, using the vortex ring equation (6.3.5) (Turner 1962). Again, certain parameters have only been obtained by experiment. The cap spreads with a half angle of $\alpha_0 = 0.18$ (defined as the maximum radius divided by the corresponding height), much less than the mean value of $\alpha_0 = 0.30$ defined in the same way for ordinary thermals. To preserve similarity the cap velocity must follow the power law appropriate to a plume, and it has a mean value of 0.61 times the velocity at the centre of the established plume at the same height. Although it may look like an isolated thermal from above, a plume cap dilutes only about one third as rapidly for a given size, because about half the fluid entering it comes from the plume below. The density difference will follow the plume dilution law, rather than that in a thermal.

6.3.4. *Line thermals and bent-over plumes*

Two-dimensional thermals have also received some attention, and they have a practical significance, since a section of a plume bent over by a (non-turbulent) wind will behave rather like a line thermal (provided we are able to ignore the differential velocity in the downwind direction which caused the plume to bend over in the first place). Richards (1963) performed laboratory experiments with line thermals released from rest which supported a similarity assumption, including a linear spread with height of the turbulent buoyant region (at a much greater angle than axially symmetric thermals). The physically relevant parameter for a bent over plume is now the buoyancy per unit length F_1, which is related to the buoyancy flux $F_0 = \pi F$ from the source by $F_1 = F_0/U$ where U is the wind velocity (supposed constant with height). Dimensional arguments show that the height of rise z in time t is therefore

$$z \propto F_1^{\frac{1}{3}} t^{\frac{2}{3}}. \tag{6.3.10}$$

This can be put in terms of F and the distance $x = Ut$ downwind, giving

$$z = 1.6 F^{\frac{1}{3}} U^{-1} x^{\frac{2}{3}}. \tag{6.3.11}$$

The numerical constant has been added using the summary of observations in the atmosphere given by Briggs (1969), which also support the functional form (6.3.11) in an intermediate range of the total rise of a bent-over plume.

Alternatively one can approach the problem using a theory of buoyant vortex pairs analogous to the vortex ring case, assuming that the circulation K_0 as well as the buoyancy per unit length F_1 remain constant in time (Turner 1960a). Because of the different dimensions of F_1 this is not now equivalent to the similarity assumption; another length must enter the problem. It follows using the momentum equation

$$\frac{dP_1}{dt} = 2\rho K_0 \frac{dR}{dt} = \rho F_1 \tag{6.3.12}$$

that R spreads linearly in time, and therefore with distance downwind. Adding the expression for the velocity of a pair vortex $w = K_0/4\pi R$ shows that the radius increases exponentially with height

$$\frac{R}{R_0} = \exp\left(\frac{2\pi F_1}{K_0^2} h\right), \tag{6.3.13}$$

where h is the height between states with radii R_0 and R.

The apparent conflict between these two ideas was resolved by Lilly (1964) who showed (using a numerical model) that the behaviour depends on the relative magnitudes of turbulent diffusion and the rate of increase of R. If buoyant fluid and vorticity mix across the centre of the thermal, K varies but a linear spread of R with z is obtained; if they stay separate in two distinct cores, K remains constant and the vortex pair theory is appropriate. Both laboratory plumes and occasionally smoke from chimneys are observed to split sideways as they bend over away from the outlet, and in these cases the second model should be used.

Experiments have also been carried out with line thermals travelling along a solid boundary, either in isolation or as 'noses' at the front of turbulent gravity currents. When the wall is vertical the latter is equivalent to a two-dimensional starting plume (Tsang 1970) for which a similarity solution can again be found. The cap moves at only 0.38 times the maximum velocity of the steady plume behind it, and (unlike the axisymmetric case) spreads at a much faster rate than the plume behind, with $\alpha = 0.33$. As the slope of the boundary is decreased towards zero, a turbulent front persists, still moving rather more slowly than the flow behind, but not now so closely connected with it (see §3.2.4). The nose is flattened and spreads out backwards along the top of the steady flow and can be mixed into it again. Little attention has been given to this problem of mixing near turbulent fronts, compared to the inviscid (hydraulic) theories of chapter 3 or the mixing of the steady gravity currents described in §6.2.

6.4. The non-uniform environment

6.4.1. *Motions in an unstable environment*

Similarity arguments (or the equivalent entrainment assumption) can also be used to derive some of the features of convection from small sources in non-neutral conditions. Batchelor (1954a) showed that power law solutions can again be obtained for an environment whose density gradient is of the form

$$\frac{g}{\rho_1}\left(\frac{d\rho_0}{dz}\right) = Cz^p. \tag{6.4.1}$$

The constant C must be positive for this kind of solution to hold, implying that the environment must be unstable; the solution for stable surroundings will be considered separately. If a plume is supposed to arise from a small source at $z = 0$, then the variations of plume properties with height, consistent with (6.4.1) and the entrainment equations (6.1.6), are

$$\left.\begin{array}{l} \bar{w} = C^{\frac{1}{2}}(4+p)^{-\frac{1}{2}}(4+\frac{3}{2}p)^{-\frac{1}{2}}z^{(1+\frac{1}{2}p)}, \\ g' = C(4+\frac{3}{2}p)^{-1}z^{(1+p)}, \\ b = 2\alpha(3+\frac{1}{2}p)^{-1}z. \end{array}\right\} \tag{6.4.2}$$

Various special cases can be recovered by choosing the value of p. Notice that for all $p > -\frac{8}{3}$, the buoyancy flux at $z = 0$ is zero, so plumes of this kind can arise spontaneously from small disturbance in the environment, and no extra parameter specifying the initial buoyancy flux is needed. When $p = -\frac{8}{3}$, the equations can only be satisfied with $C = 0$, in which case the whole analysis reduces to that for a uniform environment (and the buoyancy flux is a relevant parameter). With $p = -\frac{4}{3}$ one obtains the solutions for plumes in conditions of 'free convection' (see §5.1.3) which can be used (as in §7.3) to give a detailed interpretation of the flux from a heated plane.

The discussion of thermals in unstable power law gradients can be carried through in the same way, though it is less interesting here: because disturbances can grow spontaneously, a continuing plume is more likely to develop than an isolated thermal. Line plumes are, however, of some practical importance in the atmosphere under unstable conditions since observations suggest that plumes from local hot spots tend to get drawn out into lines downwind.

The power law solutions (6.4.2) for plumes and the corresponding ones for thermals can also be used in some situations where the environment is conditionally unstable, for example because of the release of latent heat in the atmosphere. The important feature is that the density *difference* is changing in a certain way, and this may be achieved by a process of internal generation of buoyancy, even when the environment is uniformly or stably stratified. (A necessary qualification in stable stratification is that the momentum carried away by internal gravity waves should be small; see §6.4.3.) Using the chemical release of gas bubbles in a liquid to simulate latent heat, Turner (1963 a) produced laboratory thermals which can be described in this way. These were unstable in the sense that they accelerated, but they did not lead to the formation of a continuing plume.

The processes of generation of buoyancy due to condensation and dilution due to mixing with the environment are basic to the understanding of the dynamics of clouds. Much of the work on convection in conditionally unstable surroundings has been carried out with this application in mind, and many of the theoretical models used by meteorologists have been based on, or can be reformulated in terms of, the entrainment assumption. The original application of the simple 'thermal' experiments was to small clouds, but the practical situations are so complex that simple exact solutions do not take us far, and the details of more complicated models have been worked out numerically. (See, for example, Ogura (1963) for a conditionally unstable thermal calculation.) It is outside the scope of this book to pursue this subject far, but some results of more general interest should be mentioned.

In the case of a simple plume, Morton (1956) added an equation of continuity for moisture to the other three equations (6.1.4) and showed how the solution for a point source of various strengths can be modified by the presence of condensing water. His ideas were applied by Squires and Turner (1962) to a steady plume model of a cumulonimbus cloud. They specified the mass flux over a finite area at cloudbase, with given environmental properties, and calculated the vertical velocity, the liquid water concentration and height of penetration. This model may be applicable to individual updraughts within the cloud, but it is unlikely to be valid (as originally suggested) if the whole cloud is treated as single plume. A recent evaluation by Warner (1970) has shown that this and related steady-state 'entrainment' models are inadequate when they are tested against measurements in clouds. It is impossible to predict both liquid water concentrations and heights of penetration simultaneously, and there is some doubt whether the dilution depends simply on size in the manner suggested by (6.1.7). In retrospect, it seems clear that clouds (which are typically nearly as broad as they are tall) will not achieve a fully developed state, and are poor candidates for the application of similarity theories which describe the steady flow many diameters above a small source. 'Starting plume' models (§6.3.3) allow to some extent for mixing at the top as well as the sides, but this is not enough; a fully time-dependent approach is needed which treats a cloud as a whole and takes the non-equilibrium lateral entrainment properly into account.

6.4.2. *Plumes in a stable environment*

It is clear that the ordinary kind of similarity solution, expressing plume properties in powers of the height above the source, cannot be found in a stable environment, since the buoyant fluid must come to rest at some finite height. There is no such limitation on the equations (6.1.6) based on the entrainment assumption. A useful case to discuss in detail (following Morton, Taylor and Turner 1956) is again the point source, with initial buoyancy flux $\rho_1 F_0$, in a constant stable density gradient $G = N^2 = -(g/\rho_1)(d\rho_0/dz)$.

The buoyancy flux will now of course be decreasing with height, but the equations can be put into a non-dimensional form using the two governing parameters, F_0 at the source and G (or N). Denoting non-dimensional functions corresponding to height z, radius b, vertical velocity w and buoyancy parameter g' by a subscript 1, the following transformations

$$\left.\begin{aligned}
z &= 0.410\alpha_G^{-\frac{1}{2}}F_0^{\frac{1}{4}}N^{-\frac{3}{4}}z_1, \\
b &= 0.819\alpha_G^{\frac{1}{2}}F_0^{\frac{1}{4}}N^{-\frac{3}{4}}b_1, \\
w &= 1.158\alpha_G^{-\frac{1}{2}}F_0^{\frac{1}{4}}N^{\frac{1}{4}}w_1, \\
g' &= 0.819\alpha_G^{-\frac{1}{2}}F_0^{\frac{1}{4}}N^{\frac{5}{4}}\Delta_1,
\end{aligned}\right\} \qquad (6.4.3)$$

reduce (6.1.6) to the convenient non-dimensional form

$$\frac{dm_1}{dz_1} = v_1, \quad \frac{dv_1^4}{dz_1} = f_1 m_1, \quad \frac{df_1}{dz_1} = -m_1. \qquad (6.4.4)$$

The numerical factors in (6.4.3) are those appropriate to a value of α_G defined for Gaussian profiles, and the new functions in (6.4.4) are related to the physically relevant mass, momentum and buoyancy fluxes by

$$m_1 = b_1^2 w_1, \quad v_1 = b_1 w_1, \quad f_1 = m_1 \Delta_1. \qquad (6.4.5)$$

The numerical solutions of the equations (6.4.4), with the boundary conditions $m_1 = 0$, $v_1 = 0$, $f_1 = 1$ and $z_1 = 0$ (corresponding to a virtual point source of buoyancy alone) are shown in fig. 6.17(a). Up to $z_1 = 2.0$ the plume spreads nearly linearly, and the solutions are little different from those in a neutral environment, but above this height it spreads more rapidly sideways. In this region the entrainment formulation becomes more questionable, and the plume must also overshoot and fall back; a fuller discussion of this behaviour is given below. Ignoring these difficulties for the moment, values for two heights are obtained from the solutions: $z_1 = 2.13$

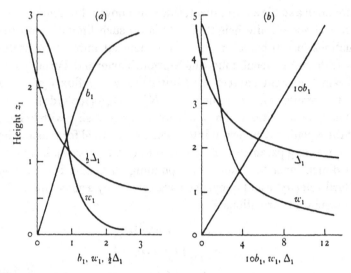

Fig. 6.17. The non-dimensional solutions for the width b_1, the vertical velocity w_1 and the buoyancy parameter Δ_1, as a function of height z_1 for (a) a plume and (b) a thermal in a linearly stratified environment. (After Morton, Taylor and Turner 1956.)

where the density difference first vanishes, and $z_1 = 2.8$ where the vertical velocity vanishes. (These are also marked on fig. 6.2.)

The top of the layer of fluid flowing out sideways will lie somewhere between these levels; substituting in (6.4.3), with the experimentally determined value of α_G (see §6.1), gives a numerical estimate for the final height. Uncertainties about the multiplying constant can be removed by direct experiment. As shown in fig. 6.18, Briggs (1969) has found that the formula

$$z_{max} = 5.0 F^{\frac{1}{4}} G^{-\frac{3}{8}} \qquad (6.4.6)$$

gives good agreement with observations over a wide range of scales, from the laboratory to a plume above a large oil fire. Note that $F = F_0/\pi$ has been used here (instead of F_0 as in (6.4.3)) and of course potential density gradients are implied for all the measurements in the atmosphere.

Detailed observations near the top of a plume in a stable environment have been made by Abraham and Eysink (1969), and they can be interpreted using a model related to our fig. 6.6. As in a uniform

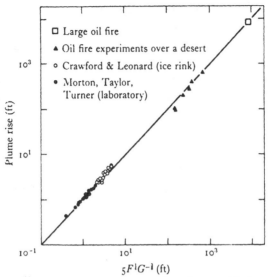

Fig. 6.18. Measurements of plume rise in calm stratified surroundings compared with the relation (6.4.6). (From Briggs (1969), where the various sources of data are discussed more fully.)

fluid, excess momentum causes the plume to continue rising in the centre, past the level of zero buoyancy. It forms a dome on top, then falls back in an annular region surrounding the upflow and finally spreads out horizontally (see fig. 6.2 pl. xiv). At all levels above the height where the plume density first equals that of the environment, the density on the centreline remains substantially constant, showing that the spreading layer protects the core of the plume from further mixing with the environment.

In this context of a stable environment, we should also refer to a different formulation of the problem by Priestley and Ball (1955), which leads to essentially the same predictions for the final height though it is based on rather different assumptions. Instead of entrainment, they assumed similarity of the profiles of shear stress and a quadratic dependence of stress on mean velocity. An acceleration-dependent term is thereby added to the mass-continuity equation (the first of (6.1.6)), but otherwise the equations are the same. (See Priestley 1959, p. 79.) Both theories break down near the plume top, though the limiting behaviour predicted by them is different. In the case of the Priestley and Ball model, the shape of the plume depends on the form of the profile adopted. For example, the spread is linear if the profiles are Gaussian, whatever the density gradient in the environment, whereas 'top hat' profiles give an increased sideways spreading in stable surroundings. When the entrainment assumption is made, however, this latter behaviour is predicted (perhaps more realistically) as an integral property of the flow, regardless of the particular profile chosen.

6.4.3. *Forced plumes and vortex rings in a stable environment*

The theory of forced plumes described in §6.1.4 can be extended to include the case of a linear density stratification. Since there are now three governing parameters, the momentum and buoyancy fluxes M_0 and F_0 at the source, plus the stratification parameter G, not all of these can be removed by making the equations non-dimensional. Morton (1959a) took $\sigma = GM_0{}^2/(F_0{}^2 + GM_0{}^2)$ as the additional non-dimensional parameter, and solved a series of problems for virtual point sources of various kinds over the range $0 < \sigma < 1$.

His solutions are shown in fig. 6.19. The most striking result is that for upward directed buoyancy and momentum fluxes, the addition of some extra momentum to a buoyant plume can actually slightly decrease the total height to which it will rise in stable conditions, because of the greater mixing produced near the source when the flow is more like a jet (see (6.1.8)). Only at much larger values of M_0 (as $\sigma \to 1$) is this effect reversed so that the plume top will rise again; the result suggests that the continuous discharge of chimney gases at high velocities is not a practical way to increase plume heights (or to reduce pollution).

A similar analysis to that leading to fig. 6.17(a) has been carried out, based on the entrainment assumption, for thermals released into a linearly stratified environment. (For details see Morton, Taylor and Turner (1956).) The non-dimensional solutions found in this way are shown in fig. 6.17(b). The final height is of the form

$$z_{\max} \propto \alpha^{-\frac{3}{4}} F_*{}^{\frac{1}{4}} G^{-\frac{1}{4}} = 2.66 F_*{}^{\frac{1}{4}} G^{-\frac{1}{4}}. \qquad (6.4.7)$$

The multiplying constant, taken from laboratory experiments, is in reasonable agreement with the theoretical predictions using the

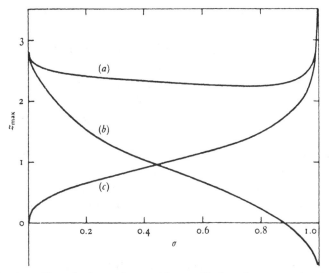

Fig. 6.19. Forced plumes in a stably stratified environment showing the final heights achieved as a function of the parameter σ for sources with (a) positive buoyancy and upward momentum, (b) positive buoyancy and downward momentum and (c) negative buoyancy and upward momentum. (From Morton 1959a).

value of α previously obtained for thermals in a neutral environment.

For thermals, the equivalent of the forced plume is a buoyant vortex ring with arbitrary initial buoyancy F_* and circulation K_0. With the similarity assumption, the angle of spread remains constant and proportional to F_*/K_0^2 (6.3.7) in a stratified environment. Substituting this in (6.4.7) shows that

$$z_{max} \propto F_*^{-\frac{1}{2}} K_0^{\frac{3}{4}} G^{-\frac{1}{4}}. \qquad (6.4.8)$$

The curious feature of a decrease in final height with increasing buoyancy, for fixed K_0, has been confirmed by laboratory experiment.

The most important property of (6.4.8) is the sensitive dependence on the initial circulation. Increasing K_0 (and hence reducing the angle of spread, and the rate of mixing) can lead to very large increases in the final height to which buoyant material could rise in stable surroundings. Such increases have indeed been observed for large explosion clouds, and the result is the basis for suggestions

(Turner 1960b, Fohl 1967) that waste gases should be ejected from chimneys intermittently rather than continuously in very stable conditions. The contrast between plumes, for which extra momentum has an adverse effect (fig. 6.19) and vortex rings, where there is in principle a great advantage to be gained, is very marked.

The same remarks do not apply to plumes bent over by a (non-turbulent) cross wind, where the evidence suggests that an extra rise can be produced by increasing the efflux velocity. This too can be interpreted in terms of the 'line thermal' model of a bent-over plume. Following the argument used in §6.3.4 for neutral surroundings, the total rise of a bent-over plume driven by buoyancy alone must depend on $F_1 = F_0/U$ and G. On dimensional grounds it must be of the form

$$z_{max} \propto (F_0/UG)^{\frac{1}{3}}. \tag{6.4.9}$$

Extra vertical momentum injected during the first part of the bent-over phase can decrease the angle of spread and increase the total rise as before.

The simple entrainment assumption gives a poor representation of the behaviour of thermals near the final height, but it can be replaced by separate physical assumptions about the variations of F_* and K_0 with height (Turner 1960a). These predict an increased spreading in stably stratified surroundings if the circulation falls to zero before the momentum, or a collapse if the momentum vanishes first. The latter behaviour seems to correspond to the phenomenon of 'erosion' observed when laboratory thermals (or cloud towers) come to rest in a stable gradient. Models which relate entrainment to the mean velocity certainly cannot take account of the effect observed by Grigg and Stewart (1963), and discussed in §5.2.2. They showed that a thermal can be brought to rest before the small scale motions are very much affected, and the latter will continue to produce some mixing even in the absence of a mean upward velocity. Another process which is left out of account in all these models is the wave drag, which becomes especially important when the buoyancy of the thermal reverses and it is oscillating about its equilibrium height (Warren 1960, Larsen 1969b).

6.4.4. Environmental turbulence

So far in this chapter the environment has been supposed to be at rest and completely unaffected by the passage of the buoyant elements through it. Surrounding fluid has been entrained into the plume or thermal by the self-generated turbulence, but no transfer has been contemplated in the other direction. This assumption will now be relaxed to allow for the removal of buoyant fluid by turbulence in the surroundings, though we still concentrate on the behaviour of the buoyant elements themselves, leaving the discussion of the complementary changes in the environment to §7.3.

Priestley (1953) introduced the concept of the *open parcel*, of fixed size, over whose (spherical) surface interchange of fluid is taking place at the same rate in each direction. Turbulence in the environment is thus assumed to dominate the whole of the mixing process, and if the region of interest is so large that its properties remain unchanged, the momentum and buoyancy equations can be written, following Priestley (1959, p. 74), in the form

$$\left.\begin{aligned} \frac{dw}{dt} &= g' - k_1 w, \\ \frac{dg'}{dt} &= -Gw - k_2 g'. \end{aligned}\right\} \qquad (6.4.10)$$

The last term in each of these is a simple (but arbitrary) representation of the effect of the turbulent interchange, which is assumed to reduce w and g' at a rate proportional to their magnitude; k_1 and k_2 are defined as 'mixing rates' for momentum and buoyancy. These parameters have dimensions t^{-1} in this formulation and, Priestley argued, they can be regarded as inverse measures of the size of the parcel, since with a given level of turbulence it will take longer for the properties of the larger elements to change. The solutions of (6.4.10) can take three forms according to the relative magnitudes of the coefficients, which can be assumed constant for a given parcel. Small enough elements (large k_1 and k_2), starting with finite velocity or excess buoyancy, travel a finite distance whatever the sign of the density gradient G, and approach their final height asymptotically. When the parcel exceeds a critical size, two different modes of motion become possible. In stable conditions there is an oscillltion about the final equilibrium height, in damped harmonic motion with period

$$T = 2\pi |G - \tfrac{1}{4}(k_1 - k_2)^2|^{-\frac{1}{2}}. \qquad (6.4.11)$$

When $k_1 = k_2$ this is just $2\pi/N$, but when the mixing rates for momentum and buoyancy are different, the period is increased. In an unstable gradient large parcels accelerate, and solutions with exponentially increasing velocities are relevant.

The model just described takes no account of the entrainment due to self-generated turbulence which was the basis of the earlier sections. Both environmental turbulence and that due to the motion can be combined in an 'entrainment' type of model if it is assumed that the outflow velocity is constant, proportional to a characteristic velocity u_0 in the environment, while the inflow is again proportional to the mean vertical velocity. The radius b is retained as a variable, and an extra term describing the outflow is added to each of (6.3.8). The continuity equation gives immediately

$$b - b_0 = \alpha z - u_0 t, \qquad (6.4.12)$$

so with these assumptions the increase of radius due to entrainment is reduced by the existence of an outflow.

The modified entrainment equations can be made non-dimensional using the initial total buoyancy F_* and u_0 as scaling parameters. The effect of stratification (described by constant $G = N^2$ say) is contained in the non-dimensional parameter $\gamma = -\tfrac{4}{9}\alpha G F_* u_0^{-4}$, and solutions have been obtained for point sources and a range of values of γ (Turner 1963 b). When $G = \gamma = 0$, the radius increases to a maximum, then decreases to zero at

$$z_{max} = 0.83 F_*^{\frac{1}{2}} u_0^{-1}. \qquad (6.4.13)$$

This maximum height is in reasonable agreement with the results of laboratory experiments, provided u_0 is taken to be comparable with the r.m.s. turbulent velocity. Solutions for other values of γ are shown in fig. 6.20. In stable and also slightly unstable conditions they are like the neutral case; no oscillatory motion occurs now (in contrast to the result (6.4.11) obtained for the 'open parcel' model). At a value of γ of about 145, the behaviour changes markedly,

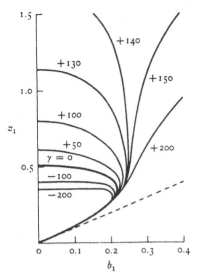

Fig. 6.20. The calculated behaviour of buoyant thermals in turbulent surroundings, for various values of the parameter $\gamma = -\frac{4}{9}\alpha G F_*/u_0^4$; positive values of γ correspond to unstable environments. (From Turner 1963 b.)

and absolute instability becomes possible, just as with the 'open parcel'. This second formulation does, however, emphasize the sensitivity of the criterion for exponential growth to the level of turbulence in the environment, through the parameter $\gamma \propto u_0^{-4}$.

Another kind of buoyant element which has received some attention in a turbulent environment is the bent-over plume. Priestley (1956) considered this in two stages, the first when entrainment dominates (leading to (6.3.11)), with an abrupt transition to a state where the turbulence is everywhere the same as that in the environment, so that a two-dimensional form of 'open parcel' argument can be used. Slawson and Csanady (1967) suggested that it is useful to subdivide this latter range further. In the intermediate phase, turbulence in the inertial subrange is supposed to dominate the mixing, and the spreading rate depends on the plume radius b and on ϵ, the rate of energy dissipation (see §5.2.2 and Batchelor 1952). These assumptions imply an effective inflow velocity proportional to the velocity scale of eddies of radius b, and lead to the prediction that the plume will level off at an asymptotic height. In the final phase the eddy diffusivity is essentially constant, and the influx velocity decreases as b increases. The asymptotic path of the plume is a straight line of fixed slope, a prediction which seems to agree with observations in the atmosphere in neutral conditions. These authors have also extended the model to stable and unstable stratification, and have shown that the intermediate phase can sometimes be important; this suggests that it may always be necessary to specify the scale of the motions considered, and not just a characteristic turbulent velocity of the environment.

CHAPTER 7

CONVECTION FROM HEATED SURFACES

The problems which arise when the buoyancy sources are distributed over a surface, rather than being localized, are rather
different from those treated in the previous chapter. Often the temperature difference between the fluid and the surface or between
two surfaces is now specified, and the heat flux is not given but is a
derived quantity which it is desired to predict. Attention is shifted
from the individual elements to the properties of the mean flow and
buoyancy fields, since it is not clear at the outset what form the
buoyant motions will take. As we shall see later in this chapter,
however, the details of the observed flows can only be properly
understood if we again consider the buoyant elements explicitly
and try to understand how their form and scale is determined
by the boundary conditions (which in many cases are nominally
uniform over the solid surface).

There is also the new feature of stability to be discussed here. A
point source of buoyancy in an otherwise unconstrained perfect
fluid always gives rise to convective motions above or below it, with
an associated production of vorticity, and the same is true of a
horizontal interface with the lighter fluid below (see § 1.2). In a real
fluid, however, the criterion based on (1.2.1) is no longer sufficient
to determine whether a horizontal layer of fluid is gravitationally
unstable or not; even when the fluid is lighter below, molecular
viscosity and diffusion can act to damp out small disturbances, so
that no overturning occurs.

We begin with a discussion of the linear stability problem for
convection between horizontal planes, proceeding straight to the
case where both viscosity and diffusion are important, and go on to
discuss theories describing this process when the amplitude is finite.
The related experimental evidence will then be presented, together
with the results of numerical experiments which complement the

laboratory results. The nature of the individual convective elements, and the way in which they interact with, and in some cases determine, the mean properties of their environment will be discussed next. Finally, a few problems relating to convection in different geometries will be presented briefly.

7.1. The theory of convection between horizontal plates

7.1.1. *The governing parameters*

The classic problem of thermal convection is to determine the motion of a layer of fluid contained between horizontal planes, uniformly heated below and cooled above. This continues to attract a great deal of attention, both because of its relation to the heat transfer at the earth's surface and the resemblance between convection patterns observed in thin sheets of cloud and in the laboratory, and because of its inherent mathematical and experimental interest to successive generations of workers in this field. This interest dates back to the experiments of Bénard (1901), who showed that a thin layer of fluid becomes unstable when the temperature difference exceeds a minimum value (which depends strongly on the depth and less strongly on the properties of the fluid). As discussed in more detail in §7.2, the motion at first takes place in a regular steady pattern, but with larger depths and temperature differences it can become turbulent. The meaning of the major parameter determining the behaviour can be appreciated by considering the following simple mechanistic model of the convection process (which is of course equivalent to a scale analysis of the governing equations).

Suppose that a small laminar buoyant element, with fixed characteristic dimension δ, leaves the neighbourhood of one boundary with density difference ρ_0' from the surrounding fluid, and that it subsequently has a density difference $\rho'(t)$. (In terms of temperature, $\rho_0'/\rho \approx \frac{1}{2}\alpha\Delta T$, where α is the coefficient of expansion and ΔT the temperature difference between the plates.) It is driven upwards (in the z-direction) by buoyancy forces but retarded by viscosity, and neglecting inertia effects there is the local balance

$$g\rho'\delta^3 \sim \mu\delta w \sim \mu\delta\frac{\mathrm{d}z}{\mathrm{d}t}. \qquad (7.1.1)$$

Because of diffusion, the density difference ρ' will be decreasing from its initial value at a rate which depends on the surface area and is approximately

$$\frac{d\rho'}{dt} \sim -\frac{\kappa \rho'}{\delta^2}, \qquad (7.1.2)$$

where κ is the thermometric conductivity, so that

$$\rho' \sim \rho_0' \exp(-\kappa t/\delta^2).$$

Combining (7.1.1) and (7.1.2) gives the total distance travelled between $t = 0$ and $t = \infty$ as

$$z = \frac{g'\delta^4}{\kappa \nu}. \qquad (7.1.3)$$

If convection is to transport heat to a second plane a distance d from the first, the right-hand side of (7.1.3) must be greater than d. For fixed ρ_0', κ and ν conditions are most favourable when δ is as large as possible, i.e. is of order d, so marginal stability corresponds to a particular value of the parameter

$$Ra = \frac{g'd^3}{\kappa \nu} = \frac{g\alpha\Delta T d^3}{\kappa \nu}, \qquad (7.1.4)$$

called the Rayleigh number.

The Rayleigh number remains the most important parameter over the whole range of more unstable conditions. As is clear from the above model, it expresses the balance between the driving buoyancy forces and the two diffusive processes which retard the motion and tend to stabilize it. Another interpretation following from (7.1.1) is that

$$Ra = w\delta/\kappa = Pe, \qquad (7.1.5)$$

i.e. a Péclet number based on the parameters of the laminar motion set up by the buoyancy forces. The second parameter needed to describe the state of the fluid completely (for all problems except the initial instability) is the Prandtl number $Pr = \nu/\kappa$. The combination $Ra/Pr = Gr$ is the Grashof number, which arises (as in §4.2.3) in problems where the density difference is given and diffusion need not be taken explicitly into account. (See also §7.4.) The Nusselt number Nu, a non-dimensional heat flux, is defined as the ratio of the actual heat (or more generally, buoyancy) transport to the purely diffusive flux which would occur through a linear temperature gradient between the two boundaries, and it must be a function of

Ra and *Pr*. The Richardson number cannot enter here, because there is no externally imposed velocity, independent of the buoyancy forces.

7.1.2. *Linear stability theory*

A more detailed discussion of the methods used will be given in chapter 8, in the context of two-component convection; but the main results of historical importance should at least be mentioned here. The stability problem was first formulated and solved by Lord Rayleigh (1916), for the idealized case of free conducting boundaries with a linear temperature gradient. He used a perturbation expansion of the linearized Boussinesq equations, assuming a convection pattern which varies sinusoidally in the horizontal (*x* and *y*) directions (i.e. 'cells' of rectangular planform), and showed that there is a critical value of *Ra*, the non-dimensional ratio now called after him, below which a fluid heated from below is stable to small disturbances. At a value of $Ra_c = \frac{27}{4}\pi^4 \approx 657$, cells which fill the gap between the boundaries become unstable in a steady mode. There will be disturbances growing monotonically (in fact to a steady state, see §7.1.3) when $Ra > Ra_c$; at successively higher values of *Ra* other modes can appear. The Prandtl number ν/κ does not enter into this time-independent problem (cf. (7.2.2)). Jeffreys (1926, 1928) derived the differential equation for the small density perturbation ρ', whose eigenvalues determine the state of marginal stability (cf. (4.1.1)). In non-dimensional form this is

$$\nabla^6 \rho' = -Ra\left(\frac{\partial^2 \rho'}{\partial x^2} + \frac{\partial^2 \rho'}{\partial y^2}\right), \tag{7.1.6}$$

and an equation of the same kind follows for the velocity perturbation. Jeffreys again restricted himself to rectangular cells, but added the more realistic feature of rigid boundaries. He obtained solutions by numerical methods and showed that the most unstable two-dimensional cells (bounded by solid boundaries and by neighbouring upward and downward currents) are nearly square in cross section. The critical value of *Ra* (corrected by later work) is 1708, and with one rigid and one free boundary it is 1108; the constraining effect of rigid boundaries clearly makes the flow more stable. Jeffreys (1928) also pointed out that the presence of a small shear will promote the stability of all modes except the two-dimensional one for which the cell boundaries are aligned in the direction of the shear. This in agreement with laboratory observations and with the recent detailed calculation of Gage and Reid (1968) for plane Poiseuille flow, to which reference has already been made in §4.2.2.

Pellew and Southwell (1940) relaxed the assumption of rectangular cells, and treated also circular, triangular and hexagonal boundaries. They showed that the horizontal and vertical variables in (7.1.6) can be separated by writing $\rho' = F(z)f(x,y)$, say, to give

$$[(D^2 - a^2)^3 + a^2 Ra] F = 0 \qquad (7.1.7)$$

and

$$\frac{\partial^2 f}{\partial x^2} + \frac{\partial^2 f}{\partial y^2} + a^2 f = 0, \qquad (7.1.8)$$

where $D = \partial/\partial\zeta$, $\zeta = z/d$ and x and y are non-dimensional horizontal coordinates (also scaled with the depth of the fluid d). The number a^2 is a characteristic number of the horizontal structure, related to the cell shape; for rectangular cells with wavenumbers l and m it is $a^2 = (l^2 + m^2)d^2$.

The value of Ra corresponding to marginal stability is again found from the eigenvalues of (7.1.7), subject to the appropriate boundary conditions. In the case of rigid boundaries, the minimum value of Ra for the first mode in the vertical is represented as a function of a by the lower line in fig. 7.1. The minimum value for any a is the critical Ra_c, and this occurs at $a = 3.13$ compared with $a = \pi/\sqrt{2} = 2.22$ for two free boundaries. Note that this result is obtained without specifying the horizontal form of the cell, and indeed linear theory gives no information about the most unstable form. When a shape is specified, however, the most unstable size follows and the corresponding eigenfunctions of (7.1.8) give the initial form of the density distribution (and of the motion) satisfying the given boundary conditions. Exact solutions have been found for hexagonal and for annular cells, both for the case of two free boundaries.

7.1.3. Finite amplitude convection

The state of knowledge of the linear theory is now virtually complete (see Chandrasekhar 1961) and recent interest has concentrated on

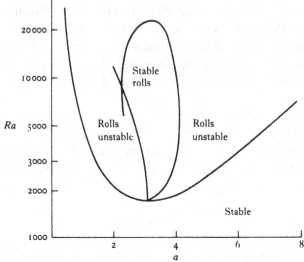

Fig. 7.1. Stability diagram for convection between rigid horizontal planes: the Rayleigh number plotted against horizontal wavenumber, showing the critical value for the first instability, and the region where two-dimensional rolls are stable according to Busse (1967).

the non-linear problems, associated with convection at higher Rayleigh numbers.

To obtain information about the steady amplitude and preferred planform of the finite motions established above Ra_c, one must take into account the advection of buoyancy and momentum by the flow field generated by the buoyancy, and therefore by definition must use the non-linear equations of motion. Malkus and Veronis (1958) showed, however, that the non-linear equations can be expanded as a sequence of linear equations whose solutions follow from those of the linear problem, and can thus again be found for any planform. Their perturbation method is based on an expansion parameter which is effectively $(Ra/Ra_c - 1)^{\frac{1}{2}}$, so it is valid only for relatively small departures from marginal stability. They suggested that the relevant solution will be the one for which the mean-square temperature gradient is a maximum, and on this basis predicted that square cells are preferred just above Ra_c in ordinary fluids. (There is no firm physical basis for this particular assumption, but some extra condition must always be used to close the problem when the boundary conditions at the sides are left unspecified.) The initial heat flux due to convection depends nearly linearly on Ra on their model, up to $Ra = 10\,Ra_c$, and the preferred horizontal wavenumber increases with increasing Ra.

Schlüter, Lortz and Busse (1965) extended this expansion technique to investigate the stability of each of the possible steady finite amplitude solutions. They showed for both rigid and free boundaries that all three-dimensional cellular patterns in a fluid with fixed properties can be unstable with respect to infinitesimal disturbances. There is a range of parameters for which only two-dimensional rolls, having a more restricted range of wavelengths than suggested by the linear theory, are stable, and the boundary of this region (determined by Busse (1967) with the assumption of infinite Pr) is shown in fig. 7.1. Analyses in which the fluid properties are allowed to be functions of temperature, thus permitting some vertical assymmetry in the layer, suggest on the other hand that the hexagonal pattern can be stable just above Ra_c.

The application of these and related theories is limited to a small range of Ra near Ra_c; numerical methods developed to extend such calculations to higher Ra will be described in §7.2. In the other limit of very large Ra, when the flow is certainly no longer steady, different arguments must be sought. Some predictions can be made on the basis of dimensional reasoning, whose validity of course depends on the underlying physical assumptions being satisfied. Let us now assume that the buoyancy flux depends on the conditions very near the boundaries and is independent of the plate separation d and of Pr. It follows from the definitions of Nu and Ra that

$$Nu = c\,Ra^{\frac{1}{3}}, \qquad (7.1.9)$$

since this is the only form which removes the dependence on d. The constant c can still depend on the boundary conditions, however.

With the assumption that d is no longer relevant, the distinction between two parallel plates and a single horizontal surface disappears. In both cases, the relation (7.1.9) may be rewritten, using the definition of Ra, in the form

$$\alpha H/\rho C_p = 2^{\frac{4}{3}}c(g\kappa^2/\nu)^{\frac{1}{3}}(\alpha\Delta T_{\frac{1}{2}})^{\frac{4}{3}} = A(\alpha\Delta T)^{\frac{4}{3}}, \qquad (7.1.10)$$

where $(\Delta T_{\frac{1}{2}})$ refers now either to the temperature difference between one plate and the far environment or to *half* the temperature difference between two plates. One might expect then that the similarity arguments of §5.1 could be used to find the distribution of density and the fluctuating quantities at a distance z above the lower plate; in the 'free convection' regime where the direct influence of molecular properties can be assumed negligible,

$$d\rho/dz \propto z^{-\frac{4}{3}},$$

which is the form derived on p. 134.

We have deferred until now, however, the point raised by Townsend (1962), which is especially relevant close to the heated plates in the laboratory experiments and in the atmosphere in the complete absence of a wind. In these circumstances the initial organization of the flow produced in the boundary layer (thickness z_b say) dominates the behaviour in the turbulent region outside; z_b must depend only on the buoyancy flux B and the conductivity κ and so must have the form

$$z_b \sim \kappa^{\frac{3}{4}} B^{-\frac{1}{4}}. \tag{7.1.11}$$

Because of this second relevant lengthscale all we can say on purely dimensional grounds is that the non-dimensional heat flux H_* must be some function of z/z_b. To go further, one must turn to more detailed theoretical arguments. Using the results of Malkus (1954b), it appears that H_* is a *linear* function of z/z_b, corresponding to $d\rho/dz \propto z^{-2}$ (in agreement with the results of Townsend's (1959) experiments, to be described in §7.2).

The theory of Malkus mentioned above has had a strong influence on later work, but an adequate treatment of its rather complicated structure would take us too far afield here. We will just refer to a review of the subject by Howard (1964), and summarize some of the results which have been obtained for large Ra. In addition to the form of the mean gradient, Malkus calculated the r.m.s. temperature and velocity fluctuations and obtained a numerical estimate for the constant c in (7.1.9). This is in reasonable agreement with experiments at large Pr, but no dependence on Pr is predicted by the theory.

Kraichnan (1962) extended the similarity arguments to study this dependence on Pr explicitly. For large Pr he distinguished three regions of the flow; close to the boundary both molecular conduction and viscosity are dominant, somewhat further out molecular viscosity remains important but heat is transported by convection, and at greater distances eddy processes control both transports. Using a 'mixing length' type of analysis (but paying special attention to numerical factors in order to make quantitative predictions), he showed that the temperature gradient in the fully turbulent region is proportional to $z^{-\frac{4}{3}}$, and in the intermediate region to z^{-2}. For the Nusselt number, Kraichnan predicted in the two limiting cases

$$Nu = 0.089 Ra^{\frac{1}{3}} \quad \text{at high} \quad Pr, \qquad (7.1.12)$$

$$Nu = 0.17 (Pr \cdot Ra)^{\frac{1}{3}} \quad \text{at low} \quad Pr. \qquad (7.1.13)$$

The second result is valid only when $(Pr \cdot Ra)^{\frac{1}{3}}$ is appreciably greater than 6; otherwise $Nu = 1$ and the transport is purely by conduction. Note that this form follows from the assumption that only the boundary layer flow is important, and that the power law is changed if interaction with larger eddies (and therefore a definite lengthscale) is taken into account.

Mention should also be made of the more recent theoretical work which seeks to put upper bounds on the heat flux as a function of Ra using variational methods. The bounds on the power law and the multiplying constant originally obtained by Howard (1963) were rather weak, but the addition of more physically realistic constraints on the possible motions greatly improves the estimates (Busse 1969.) The final limiting form at high Rayleigh number, with many horizontal wavenumbers taken into account, is again $Nu \propto Ra^{\frac{1}{3}}$, though with only one or a few wavenumbers the exponent is reduced and there is also a weak dependence on $\ln Ra$.

7.2. Laboratory and numerical experiments on parallel plate convection

More detailed predictions about convection between parallel plates at intermediate Rayleigh numbers have been made using high speed computing techniques to solve the coupled non-linear equations of motion and diffusion. Both because this work has been strongly influenced by the laboratory results, and because the numerical approach itself represents a kind of experimenting with the system to determine the relative importance of the various terms, it is convenient to group them together here.

7.2.1. *Observations of laminar convection*

Many laboratory studies have followed that of Bénard. Much of the early qualitative work was aimed at reproducing the regular patterns often observed in convective clouds; Phillips and Walker (1932), for example, showed that a wide variety of forms of cellular structure can be obtained by suitable variation of the rate of shear. Brunt (1952, p. 221) sought to relate such experiments to the atmospheric observations, replacing molecular conduction and viscosity by the joint effects of radiation and turbulence, but the analogy is far from exact and a detailed explanation of the patterns is still not available. The tendency for long lines of cloud to form with their axes aligned along the direction of the shear can, however, be explained in just the same way as in the laminar case: the downstream modes are the only ones not inhibited by the shear. An example of the last effect is shown in fig. 7.2 pl. XI; compare this with the clouds resulting from the K–H mechanism (fig. 4.14 pl. X) which are aligned across the shear.

Quantitative experiments were first reported by Schmidt and Milverton (1935), who confirmed the predicted value of the critical Rayleigh number, and heat transfer measurements were begun by Schmidt and Saunders (1938). We can refer again to Chandrasekhar (1961, ch. 2) for the history of this work, and concentrate here on describing some of the more recent developments.

Koschmieder (1967) carefully repeated Bénard's experiments in silicone oil, using various shapes of convection chamber with free and solid top boundaries, and showed that the initial cellular pattern to form when the bottom is gradually and uniformly heated is the two-dimensional roll. The orientation of these rolls depends, however, on the shape of the sidewall boundary, and in a circular chamber, for example, they appear as a series of concentric annular rings (fig. 7.3 pl. XVI). With a solid top boundary this initial configuration persists at larger temperature differences, but with air above, the ring pattern breaks down into hexagonal cells – very regular if the air above is still, and more nearly random if air motions are not suppressed. It seems most likely now that variations of surface tension with temperature are an essential element in these and Bénard's original experiments with a free surface

(Pearson 1958). The measured wavelengths are in good agreement with the predictions of theories such as that of Nield (1964) which take this effect as well as the density instability into account.

It is possible to set up cells of arbitrary shape and scale with non-uniform heating of the fluid layer, as demonstrated by Chen and Whitehead (1968) using temperature perturbations imposed with radiant heating. In the range $Ra_c < Ra < 2.5 \, Ra_c$ which they covered, they showed that when the disturbance was removed, the convection pattern approached that predicted by linear theory. Foster (1969), on the other hand, showed in a numerical experiment that a different finite amplitude disturbance can persist in a range where another solution is most unstable.

Krishnamurti (1968) has devised an elegant optical method of examining the plan-form of convection in a chamber having solid metal top and bottom boundaries. She traversed a beam of light across one side of the chamber and photographed in the perpendicular direction in the horizontal plane to pick out the cross sections of cells marked with aluminium powder. She has shown that roll cells (again aligned by the boundaries) occur at Rayleigh numbers just above critical when the mean temperature of the layer is fixed (fig. 7.4 pl. XVI). However, hexagons can be realized (under conditions where the dependence of fluid properties on temperture must be small) when the mean temperature is increasing steadily (at a rate η), and the temperature profile is thus non-linear. This observation is in agreement with her theory based on an expansion in both η and the amplitude, which also shows that there is a finite amplitude instability at a lower value of the Rayleigh number than the usually accepted one. This gives rise to stable hexagons in which the direction of motion (left undetermined in the linear Boussinesq theory) depends on the sign of η, being downwards if η is positive. Both these further predictions were verified experimentally (with values of η corresponding to a change in temperature of only a couple of degrees per hour), and they go a long way towards explaining discrepancies in previous experiments which were not so carefully controlled, or in which temperature differences were produced by heating or cooling at only one boundary.

Tritton and Zarraga (1967) investigated the cellular convection patterns produced by the instability of a layer of fluid heated uni-

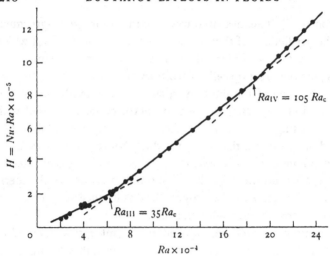

Fig. 7.6. Heat flux $H = Nu \cdot Ra$ as a function of Rayleigh number, showing the third and fourth transitions in a fluid with $Pr = 100$. (After Krishnamurti 1970.)

formly through its interior by an electrolytic current, and cooled from above. Hexagonal cells were formed in which the motion was downwards in the centre (in agreement with Krischnamurti's criterion), but more surprisingly the horizontal scale was much larger than with boundary heating. This observation becomes relevant when we wish to compare geophysical convection phenomena, especially in the earth's mantle, with laboratory results.

Heat flux measurements have been made by many experimenters, and the results are well summarized in a critical review by Rossby (1969). Recent experiments by Krishnamurti (1970), in which she has taken great care to study a series of steady states with fixed heat flux, have introduced a new order of precision into this field. She has shown that just above Ra_c the non-dimensional heat flux expressed as $H = Nu \cdot Ra$ is a linear function of Ra, with slope about 2.7 for water and slightly larger for fluids of higher Pr. At a second critical value Ra_{II} about $12\,Ra_c$ there is a discrete change to a higher slope, which was identified with a change in planform from two-dimensional to three-dimensional flow caused by a finite amplitude instability. Busse and Whitehead (1971) have also carried out experiments in this range, and have observed two distinct types of transition, which are shown in fig. 7.5 pl. XVII.

Further changes of slope in the heat flux curves at higher Ra are associated with transitions to time-dependent flows of various kinds (see fig. 7.6 and the next section). Malkus (1954a) had earlier obtained similar results (in unsteady experiments) but interpreted the transitions in terms of successive instabilities of higher order modes in the vertical.

7.2.2. Measurements at larger Rayleigh numbers

A feature of convection at high Rayleigh number, which has only recently become clear as experiments have been made in silicone oils of high viscosity and in liquid metals, is the strong dependence on Prandtl number. The transition to turbulence does not depend only on Ra reaching a value of order 10^5, as had previously been suggested; according to Rossby (1969) it occurs at $Ra \approx 14,000 \, Pr^\alpha$, where $\alpha \approx 0.6$ for $Pr \gg 1$. When Pr is large, *steady* convection can persist to quite high Ra. The flux is then carried by large convection cells across which the velocity is gradually varying, but with temperature anomalies concentrated in narrow vertical regions at the cell boundaries which are fed directly by the boundary layer at the walls. As Ra is increased (for a given finite Pr) the motion becomes time dependent; remarkably regular oscillations are first observed, and these increase in number and frequency until finally the flow becomes more disordered and turbulent. The various regimes observed are shown in fig. 7.7 as a function of Ra and Pr. At small Pr, on the other hand (in mercury, for example, with $Pr = 0.025$), no steady flows are ever observed above Ra_c and flows at high Ra are always turbulent.

Various heat flux measurements show that some Prandtl number dependence persists even at large Pr and Ra (i.e. that the limit (7.1.12) is not approached in practice). Silveston (1958) and Globe and Dropkin (1969) respectively suggested the empirical relations

$$Nu = 0.10 Ra^{0.31} Pr^{0.05}, \atop Nu = 0.069 Ra^{\frac{1}{3}} Pr^{0.074}, \qquad (7.2.1)$$

to describe their experimental results for $Ra > 10^5$ and Pr from 0.02 to 8750. Rossby (1969) on the whole agrees with their individual measurements at the larger Pr, but questions the usefulness of

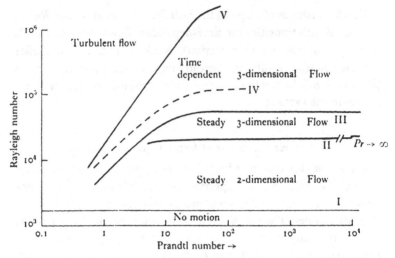

Fig. 7.7. Diagram showing the various forms of convection observed in a horizontal layer of fluid, as a function of Rayleigh and Prandtl numbers. (From Krishnamurti 1970.)

condensing all the results into a single formula, since as $Pr \to \infty$ the dependence on Pr should disappear. He also shows that the results for low Pr are quite different (cf. (7.1.13)). The power law dependence on Ra which he found using mercury, water and a silicone oil was consistently less than $\frac{1}{3}$, and the range of validity of this limiting value therefore remains in some doubt.

Thomas and Townsend (1957) and Townsend (1959) have made detailed measurements in air, recording mean and fluctuating quantities at various distances above a heated horizontal plate. Their measured heat fluxes imply a value of $C = 2^{\frac{4}{3}}c = 0.193$ in (7.1.10). (The corresponding value of c in (7.1.9), with Ra based on the whole temperature difference between two plates, is $c = 0.08$, slightly lower than the best value for water, $c = 0.09$.) Outside a purely conductive region, the mean temperature profiles were close to $T \propto z^{-1}$, the form predicted by the theory of Malkus (referred to in §7.1.3) which includes the direct effect of molecular diffusion near the wall. Their records of temperature fluctuations revealed periods of strong activity (which were more frequent close to the wall), alternating with relatively quiescent periods. These were interpreted as evidence for the intermittent detachment of buoyant elements

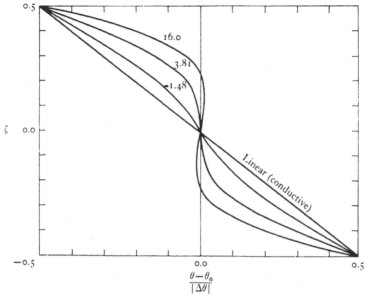

Fig. 7.8. Observed mean temperature profiles in a convecting fluid layer. (From Gille 1967.) Note the reversals of gradient at the larger values of $\lambda = Ra/Ra_c$, which are marked on the curves.

from the conductive boundary layer, and their subsequent erosion in the (still turbulent) surrounding air. Such observations point to the need to consider explicitly the nature of the buoyant elements and their interaction with the environment, which will be done in §7.3.

Other studies have concentrated on obtaining reliable mean profiles of the measured quantities, and comparing these with the numerical models. Deardorff and Willis (1965) compared the behaviour of a system constrained to convect in two-dimensional rolls with one in which three-dimensional, unsteady motions were permitted. In the latter case they showed that the heat flux was reduced by the unsteadiness and that averaging for a very long time is necessary before temporal and spatial estimates of the variance are the same; the former is crucially dependent on the slow shifting about of the convection pattern. In a later experiment (Deardorff and Willis 1967) they adopted the technique of spatial averaging and concluded that their observations were consistent with a thermal structure dominated by plumes extending most of the distance between the plates.

A feature of some of the detailed measurements, especially those at high Pr, is a reversal of the mean density profiles, which are slightly stable across the centre of the convecting region at high Rayleigh numbers. A particularly convincing experimental demonstration of this effect (in air with Ra about $16\,Ra_c$) was given by Gille (1967) using an optical method of measuring refractive index, which therefore did not require the insertion of any probes into the flow (fig. 7.8). This effect too can be explained (§7.3) in terms of the individual buoyant elements.

7.2.3. *Numerical experiments*

The predictions made using high speed computing techniques now far outweigh in detail those obtained from direct laboratory experiments, but only a few examples can be given here. They are all based on the numerical solution, with various approximations, of the coupled non-linear Boussinesq equations of motion and diffusion in two dimensions, which can be written in the non-dimensional form (Deardorff 1964)

$$\left(\frac{\partial}{\partial t}+u\frac{\partial}{\partial x}+w\frac{\partial}{\partial z}-Pr\nabla^2\right)\zeta = Pr\cdot Ra\frac{\partial T}{\partial x} \qquad (7.2.2)$$

and
$$\left(\frac{\partial}{\partial t}+u\frac{\partial}{\partial x}+w\frac{\partial}{\partial z}-\nabla^2\right)T = 0. \qquad (7.2.3)$$

Here $\zeta = \nabla^2\psi$ is the non-dimensional vorticity component in the y-direction, ψ is a streamfunction and T a non-dimensional temperature (proportional to the density).

Deardorff (1964) and Fromm (1965) retained the full form of the equations, and obtained solutions for just a few values of Pr and Ra. They imposed an initial disturbance on fluid at rest having an initially linear temperature gradient, and followed the development towards a steady state. The model is an essentially laminar one, leading at large times to the steady cellular convection pattern shown in fig. 7.9, which has the plume-like temperature structure characteristic of high Pr, high Ra convection (though the parameters used were $Pr = 0.71$ and $Ra = 6.75 \times 10^5$, corresponding to Townsend's experiments in air). A later model (Deardorff 1965) introduced extra assumptions to allow to some extent for three-

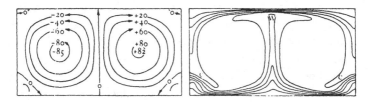

Fig. 7.9. Steady-state streamlines and isotherms calculated by Deardorff (1964) for a layer of fluid with $Pr = 0.71$, $Ra = 6.75 \times 10^5$. Relative values of the stream functions are given in the left-hand diagram; W denotes warm and C denotes cold regions in the right-hand diagram.

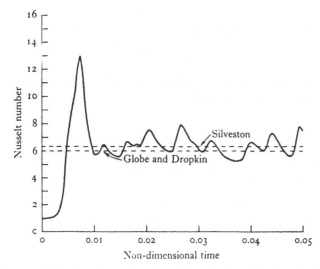

Fig. 7.10. The time-dependent heat flux calculated by Deardorff (1965), compared with laboratory measurements.

dimensional effects, and did permit fluctuations to persist. Fig. 7.10 shows the Nusselt number at $Pr = 0.71$, calculated by taking the horizontal average of the non-dimensional temperature gradient at the upper and lower boundaries. The computed mean heat flux was close to the measured values (also shown on the figure), and various other space-averaged quantities, such as mean density and the standard deviations of density and velocity were in rough agreement with observations. A wider range of steady conditions has been studied recently by Cabelli and de Vahl Davis (1971), who also included the effect of surface tension.

Herring (1963, 1964) introduced a model which greatly simplifies the analysis, and has also led to a clearer understanding of the important physical processes. His 'mean field equations' are obtained by averaging (7.2.2) and (7.2.3) over horizontal planes, retaining those non-linear terms which describe the interaction of the mean temperature field with a single velocity or temperature fluctuation, but omitting all interactions between fluctuating quantities. He was interested in the statistically steady state, and the limit $Pr = \infty$; a single, dominant wavenumber k was also assumed in the horizontal, which reduced the problem to one in a single space dimension (z). (Herring's calculations were based on a truncated Fourier representation of the variables in the vertical, but this should be regarded as a particular numerical technique, rather than an essential feature of the model.) For the case of rigid boundaries, the calculated heat fluxes at large Ra (corresponding to values of k chosen to maximize the heat flux in each case) were interpreted by fitting the form

$$Nu = 0.115 Ra^{\frac{1}{3}} \qquad (7.2.4)$$

to the individual points. This is rather larger than the experimental values, and other functional relationships are not ruled out by his data; nevertheless the agreement between this (and other predictions of the model, such as the mean profiles) and the laboratory data are surprisingly good in view of the strong approximations made.

Elder (1969 a) has shown that the mean field equations can be used to provide an adequate description not only of the steady state but of the temporal development of high Rayleigh number convection. He used an iterative finite difference method, with $Pr = 1$, and his paper contains an excellent discussion of the underlying assumptions, as well as a careful comparison with related work. At an intermediate stage of the calculation, a pronounced negative temperature gradient develops in the central region, and persists to the final state; this feature appears also (at the higher values of Ra) in all the numerical solutions described above. Observational evidence for such a reversal has been given in fig. 7.8, and the reasons for it will be discussed in the following section. Another feature of both Herring's and Elder's results is a small bump in the temperature profile at the edge of the unstable boundary layer. Though both of them attributed this to errors in their solutions, it may again represent a real physical effect (§7.3.4).

A finite amplitude solution for roll cells which retains the complete set of non-linear interactions has been obtained by Veronis (1966). For numerical reasons the number of terms must be truncated; but he has obtained in this way solutions valid for $Ra \gtrsim 30Ra_c$ and over a range of Pr from 0.01 and 100. The stable solutions have the form of a single steady cell, which, however, accounts surprisingly well for the measured heat flux. A related model, which involves the expansion of the horizontal structure of all flow variables as a sum of solutions of the linear problem, has been proposed for high Ra and a general Pr by Gough, Spiegel and Toomre (1975). They average the resulting equations in the horizontal to obtain equations in z and t for the mean and fluctuating quantities. Retaining only the first terms in the series is equivalent to Herring's mean field approximation and leads as before to steady solutions. Including more terms, however, can produce *unsteady* solutions, some of which are periodic or aperiodic oscillations resembling those observed in Krishnamurti's laboratory experiments.

Information about turbulent convecting flows in a viscous fluid has also been obtained from numerical models designed primarily to study convection in a porous medium (Elder 1967). There is now the simplification that the dominant processes are the generation of vorticity by the horizontal gradient of the buoyancy forces (cf. (1.3.5) and (2.2.4)) and the transfer of heat by diffusion and advection, only the last being non-linear. By comparing such results with similar calculations which include the extra effects of diffusion and advection of vorticity (and again, from (7.2.2), only the former enters when Pr is large), Elder has determined how important these two processes can be; and in many cases the qualitative behaviour is little changed by them. We refer the reader to Elder's papers for a description of his numerical method, but will present some of his results in the context of the next section.

7.3. The interaction between convective elements and their environment

In the discussion of turbulent parallel-plate convection it has several times been remarked that certain properties of the mean distributions (of density for instance) can only be properly understood if one considers explicitly the behaviour of the buoyant elements themselves. In this section we therefore turn attention from the averaged distributions to the detailed observations, and the interpretation of these in terms of simple theoretical models. The ideas and terminology used have something in common with those of chapter 6, but the major difference here is that we can certainly no longer neglect the effect of the buoyant elements on the environment. On the contrary the distinction between the two is not clearcut; the properties of the 'environment' are modified or even produced entirely by the convection elements, and there is a complex interaction between the two.

7.3.1. The formation of plumes or thermals near a horizontal boundary

Townsend's (1959) detailed observations, as well as common experience in the lower atmosphere, make it clear that the flux from a heated boundary is intermittent rather than steady. Buoyant fluid slowly accumulates and then breaks away, either as a thermal, or as an unsteady plume which wanders about the surface (unless there is some preferentially heated spot to which this can remain attached). All the theories presented so far fail to bring out this feature, but Howard (1964) has proposed the following model which considers it explicitly.

Suppose that a fluid layer of depth d is initially at rest and has a constant temperature $T = 0$. At time $t = 0$ the upper and lower boundaries have temperatures $-\frac{1}{2}\Delta T$ and $\frac{1}{2}\Delta T$ applied to them, and the fluid heats up by thermal conduction, producing error function profiles of the form

$$T = \tfrac{1}{2}\Delta T \operatorname{erfc}(z/2\sqrt{(\kappa t)}). \tag{7.3.1}$$

The thickness $\sqrt{(\pi\kappa t)}$ of the boundary layer increases, until at time t_*

the Rayleigh number Ra_δ based on ΔT and $\delta = \sqrt{(\pi\kappa t_*)}$ reaches a critical value, of order $Ra_\delta \sim 10^3$. At this time the buoyant fluid in the boundary layer suddenly breaks away as a discrete entity. With Pr of order unity and larger Howard has shown that this can happen in a time small compared with t_* (the model does not apply to low Pr, since conduction across the whole depth d will then be dominant). Thus for most of the time the process of transfer near the boundary is one of conduction, followed by a comparatively short interval during which the conditions are locally restored to the original uniform state by the removal of the buoyant fluid as a plume or thermal. Time averages of the conduction profile over the interval $(0, t_*)$ should therefore give estimates of the mean temperature profile and the heat flux which are equivalent to horizontal averages in the real flow. Howard's results, after averaging in this way, are

$$\overline{T} = \tfrac{1}{2}\Delta T[(1 + 2\xi^2)\operatorname{erfc}\xi - 2\pi^{-\frac{1}{2}}\xi\, e^{-\xi^2}] \qquad (7.3.2)$$

with $$Nu = d(\pi\kappa t_*)^{-\frac{1}{2}} = d/\delta = (Ra/Ra_\delta)^{\frac{1}{3}} \qquad (7.3.3)$$

and $$\xi = \tfrac{1}{2}z(\kappa t_*)^{-\frac{1}{2}} = \tfrac{1}{2}\pi^{\frac{1}{2}}z/\delta,$$

the other quantities being defined above. If t_* is determined by putting $Ra_\delta \approx 10^3$ then $Nu = 0.1\,Ra^{\frac{1}{3}}$, in reasonable agreement with experiment.

A comparison of this mean temperature profile with experiment is given in fig. 7.11. Townsend's (1959) measurements in air are shown and also some data in water obtained by Elder. Though only the shapes of the profiles are compared here, the results are consistent and the implied values of Ra_δ are reasonable. The r.m.s. temperature fluctuations at each level can also be calculated (considering only the conductive phase). This gives a profile having a peak in the boundary layer region of magnitude $(\overline{\theta^2})^{\frac{1}{2}} = 0.182\Delta T/2$, in remarkable agreement with Townsend's measured amplitude of 18 % of the temperature difference across the boundary layer, especially since nothing has been said about the contribution of the thermals themselves to $\overline{\theta^2}$.

This model has received strong support from the detailed laboratory and numerical experiments of Elder (1968). He has studied the flow both in a porous medium and a viscous fluid near a suddenly

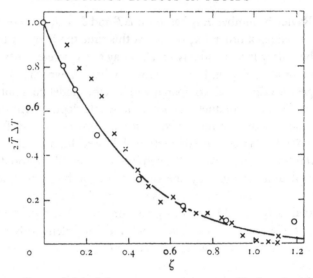

Fig. 7.11. Comparison of the mean temperature profile (7.3.2) with the experimental results of × Elder, ○ Townsend. (From Howard (1964), *Proc. 11th Congress Appl. Mech. Münich*, Springer-Verlag.)

heated horizontal boundary, and also (numerically) at an 'interface' or thin region of strong unstable temperature gradient separating two uniform layers. In all cases, when there is an initial source of temperature fluctuations, fairly regular cellular motions at first grow rapidly but remain confined within the boundary layer or interfacial region, with a horizontal scale which is of the same order as the interface thickness, and is unrelated to the total depth of the fluid. This simple but important result goes some way towards determining the scale and form of buoyant elements which was left entirely arbitrary in the calculations presented in chapter 6.

At later times, this layer becomes distorted and individual disturbances grow irregularly and break away from the boundary layer. These accelerate, even in the porous medium where inertial forces are by assumption negligible, because they can increase their buoyancy rapidly by entraining buoyant fluid from a surrounding volume. This happens before they lose their connection with the surface, and the process temporarily denudes a considerable area of the plate of its hot fluid, as shown clearly in fig. 7.12 where the stream function and isotherms are plotted at successive times for the special case of heating confined to a small area of the boundary.

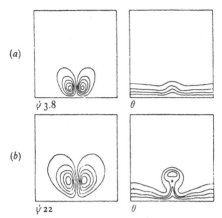

Fig. 7.12. Growth of a buoyant element near a heated section of the boundary, and its escape. Streamlines and isotherms calculated by Elder (1968).

Comparable (but of course less regular) patterns are obtained for the buoyant elements forming with random temperature fluctuations, and two photographs of a laboratory experiment of this kind are shown in fig. 7.13 pl. XVII. In similar experiments both Rossby (1969) and Sparrow, Husar and Goldstein (1970) observed a periodic formation and escape of thermals above preferred sites on the boundary, which they interpreted using Howard's ideas. (See fig. 7.14 pl. XVII.)

The time-dependent stability problem implied by this model has been studied by Foster (1965). He computed the time needed for an imposed disturbance to grow to specified multiples of its initial amplitude, and confirmed the importance of the conductive boundary layer thickness rather than the total depth. He showed that, except for very small heating or cooling rates, a quasi-steady calculation is misleading because the disturbances develop more slowly than the boundary layer itself. Foster has applied his theory to the interpretation of the onset of convection due to evaporative cooling of a free liquid surface. This occurs in the form of thin two-dimensional sheets plunging away from the surface, possibly because of the smaller constraint on the horizontal motion at a free surface compared to a solid wall. He has also shown (Foster 1971 b) that it is possible to reproduce the intermittent character of the observations using a numerical model.

The extension of such 'initial instability' studies to fully turbulent convection of course requires further justification. The numerical calculations of Elder (1968) for the turbulent case do show the same features of formation and acceleration of thermals out of the sublayer with a characteristic size, though the details are affected by larger scale mean and turbulent motions. Considerable regularity of convection patterns is often observed in the atmosphere, as remarked in §7.2.1, but the prediction of the form and spacing of the convection elements as they leave the ground remains a major problem.

7.3.2. *The environment as an ensemble of convection elements*

Priestley (1959) has shown how one can obtain information about the mean properties and the fluctuations in the fluid over a heated plane by considering it as the superposition of many closely spaced convection elements having one of the special forms considered in chapter 6. As attention is shifted to the whole population instead of an individual plume or thermal, the process of entrainment is assumed to include the mixing of one element into another. First, the idea of the 'open parcel' (§6.4.4) can be extended to analyse the relative contributions to the heat flux of elements of different sizes. In a stable environment both very small and very large parcels are inefficient transporters of heat, the former because they are rapidly mixed, and the latter because their motion approaches a simple harmonic oscillation with w and ρ' a quarter cycle of phase. Imposed fluctuations in the form of parcels of an intermediate size produce the maximum buoyancy flux B_m, which can be written in terms of the variance of buoyancy fluctuations σ_ρ^2 and the (potential) density gradient as

$$B_m \approx \sigma_\rho^2 G^{-\frac{1}{2}} = \sigma_\rho^2 N^{-1} \qquad (7.3.4)$$

using the notation of §§5.2 and 6.4. This form is obtained immediately from a dimensional argument, given that B_m is to depend on σ_ρ^2 and G alone; the detailed mechanistic calculation only adds a numerical factor.

In the same way, if one interprets the results for the free convection regime (§5.1.3) in terms of a population of plumes, one must inevitably get the same form of dependence of the various quantities on height, though the multiplying constants depend on the particular model chosen. We have in fact already written down some relevant solutions in (6.4.2), for axisymmetric plumes with top hat profiles; specializing to $p = -\frac{4}{3}$ (appropriate for free convection) gives

$$w = C^{\frac{1}{2}}(\tfrac{3}{16})^{\frac{1}{2}} z^{\frac{1}{3}}, \quad g' = \tfrac{1}{2} Cz^{-\frac{5}{3}}, \quad b = \tfrac{6}{7} \alpha z. \qquad (7.3.5)$$

The non-dimensional heat flux H_* for a single plume of this kind is by definition (5.1.14)

$$H_* = g'w/(Cz^{-\frac{5}{3}})^{\frac{3}{2}} z^2 = 0.22. \qquad (7.3.6)$$

The important point here is that the argument gives a purely numerical constant characteristic of the plume, though as it happens the point source with top hat profile is a poor model of the atmospheric process, and more realistic results are obtained with line sources and Gaussian profiles. In comparing with the observed value of H_* in the atmosphere allowance must also be made for the fraction of the area convered by ascending plumes, which observation suggests is about $\frac{1}{2}$. (Priestley 1959, p. 88.)

7.3.3. Convection from small sources in a confined region

In this section we consider another kind of model which illuminates certain unsteady aspects of the problem of convection from sources near boundaries of a confined region. The type and spacing of the buoyancy sources will be regarded as given; the difficulties of specifying these for a uniformly heated surface have already been pointed out, but even a broad pattern of non-uniform heating will tend to organize the unstable regions into narrow plumes. (See Stommel 1962.) The convection elements are assumed to be far enough apart not to interact directly, which in practice means a separation greater than about twice the total depth. The environment will be assumed to be non-turbulent (though a more elaborate model which takes account of turbulence inside and outside rising plumes has been proposed by Telford (1970)). Thus the methods of chapter 6 can be used, but attention will now be focused on the properties of the environment rather than exclusively on the plumes or thermals themselves.

It is often clear in the laboratory that experiments on plumes can only be continued for a limited time before the walls of the containing vessel begin to affect the flow and the buoyant fluid modifies the environment through which the plumes are moving. The same is true of the lower atmosphere; during the course of a day, the air below cloudbase heats up, due to the transfer of heat from the ground in the form of plumes or thermals. In this relatively deep layer the heat flux is decreasing with increasing height, not constant as was assumed in §5.1.3 for the layer very close to the ground. Eventually, in a region of limited extent, fluid which has been in the convection elements begins to 'fill up the box' and become part of the environment, and so changes in the latter become an integral part of the problem and cannot be ignored as they were in chapter 6. The effect is shown in fig. 7.15 pl. XVIII taken during an experiment in which a turbulent plume of salt solution was supplied at the top of a tank of fresh water for a long time. Dye put into the flow at an early stage spreads out along the floor, and this marked fluid thereafter becomes part of the environment, in which turbulence is quickly suppressed. A later marked patch of plume fluid passes through this slightly heavier layer and, as a result of mixing with it, becomes heavier still. The net result of this process is to produce a *stably stratified* environment, made up of fluid which has all passed through the plumes.

Baines and Turner (1969) have based a detailed theoretical model on this picture, using the entrainment equations (6.1.6) for plumes, but with $G = N^2$ now regarded as an unknown function of height. Assuming that the plume fluid spreads out instantaneously across the region at the top or bottom of the region (of height H, radius R) and becomes part of the non-turbulent environment at the same level, two extra equations can be written down linking G and the vertical velocity U in the environment. The first is just the continuity equation at any level which is

$$wb^2 = -UR^2 \qquad (7.3.7)$$

(provided $b \ll R$). The second expresses the fact that the density of a particle in the non-turbulent environment remains constant, and that the only density changes at a fixed height occur because of the vertical motion:

$$\frac{\partial g_0}{\partial t} = -U\frac{\partial g_0}{\partial z} = UG, \qquad (7.3.8)$$

where g_0 is the buoyancy parameter for the environment.

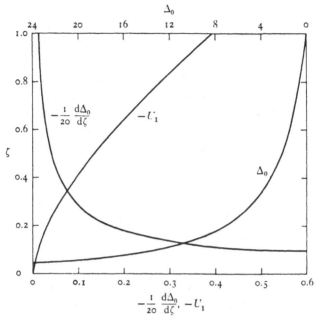

Fig. 7.16. The asymptotic properties of the environment, in non-dimensional form, produced by an axisymmetric plume in a closed region. U_1 is the vertical velocity of the environment, and $d\Delta_0/d\zeta$ is the density gradient.

The general problem is time dependent, but several special cases are more tractable. In the asymptotic state at large times t, only g_0 depends on t; the density of the environment changes linearly in time at all heights, while the *shape* of the density profile and all the other properties remain fixed. The equations can be reduced to non-dimensional form by scaling with H and the buoyancy flux F_0 at the source; for instance the environment is described by

$$\left.\begin{aligned} g_0 &= \tfrac{1}{4}(\pi)^{-\frac{2}{3}}\alpha^{-\frac{4}{3}}F_0^{\frac{2}{3}}H^{-\frac{5}{3}}[\Delta_0(\zeta)-\tau], \\ U &= 4\pi^{-\frac{1}{3}}\alpha^{\frac{2}{3}}F_0^{\frac{1}{3}}H^{-\frac{1}{3}}\left(\frac{H}{R}\right)^{2}U_1(\zeta) \end{aligned}\right\} \qquad (7.3.9)$$

using subscripts to denote non-dimensional functions of $\zeta = z/H$. The width R enters only into the scaling of U and t; the distributions of density and the plume properties depend on H alone, but the time to achieve these does depend on R (i.e. on the spacing of the plumes).

The solutions for several of these non-dimensional functions are shown in fig. 7.16 (and essentially the same qualitative behaviour was obtained theoretically for two-dimensional plumes and axisymmetric thermals). The form of the density distribution was checked experimentally in two ways, directly and by relating it to the motion of the dyed fluid layers, and these were both found to be in excellent quantitative agreement with the theory. The sharp stable density gradient in the environment near the source was clear in the experiments; it is not realistic for the atmospheric near the ground, but the theory can be modified to allow for a well mixed layer there. The slightly stable gradient at greater heights is, however, a feature of aircraft measurements below cloudbase (Warner and Telford 1967) even in conditions of strong convection, and it is a property which is preserved in the theoretical solutions even when horizontal averages are taken which include the plumes.

The relation of this calculation to the parallel plate convection problem now becomes clear. If buoyant elements emerging from one boundary layer can travel nearly to the opposite plate before losing their identity then this experiment can be thought of as a box filling up from both above and below with thermals (or unsteady plumes). (The picture is complicated, but not qualitatively changed, by the fact that the environment is turbulent.) This process can in time produce the slightly stable gradient in the interior which is observed, and also perhaps the more stable region at the edge of the unstable boundary layer which has been reported in the numerical experiments described in §7.2.3.

The model also explains in physical terms how there can be a 'counter-gradient' heat flux, which at one time caused a great deal of controversy (Priestley 1959, Deardorff 1966). Individual buoyant elements can clearly transport heat upwards in the atmosphere even though the mean temperature is greater at the top and the environment stably stratified. It is not helpful to interpret the transfer in terms of an eddy diffusivity when the motions are dominated by buoyancy, since the convective mechanism is not properly described by a relation between the flux and the local mean gradient.

7.3.4. *Penetrative convection*

The term 'penetrative convection' has come to mean the process whereby convective motions arising in an unstable region penetrate into an adjacent stable layer. It has also sometimes been used, logically enough, to distinguish the 'buoyant element' models treated in chapter 6 from the cellular motions observed at low Rayleigh number. When the convection is turbulent, this latter is no longer such a useful distinction to make, and so the first definition is used here. The problem is an important one for the lower atmosphere and upper ocean, where a layer of convecting fluid, driven by heat transfer across the surface, is bounded above (or below) by a stable gradient. As heating (cooling) is continued the depth of the convecting layer increases at the expense of the stable region, and so in general one must consider a time-dependent problem. Some of the properties of such systems, particularly the mixing across the fluid boundary, will be treated more fully in chapter 9.

An instructive laboratory model has been described by Deardorff, Willis and Lilly (1969). They set up a nearly linear stable temperature gradient in a tank of water, increased the temperature of the bottom to a new fixed value so that convection began, and measured vertical profiles of the horizontally averaged temperature as functions of time (fig. 7.17). (Note that the gradient becomes slightly stable below the top of the 'well-mixed' layer, suggesting that a 'filling box' model might be appropriate here too, with now a variable depth, instead of a fixed upper boundary.) Deardorff *et al.* deduced heat flux profiles, and showed that the upward flux decreases nearly linearly with depth above the bottom (as is to be expected for a layer of almost constant temperature). In a thin region at the top, however, the heat flux is downwards i.e. heat has been added to the convecting layer from the gradient region above. This latter effect is small, and the rate of rise of the interface can therefore be inferred pretty accurately by considering only the heat balance of the layer, equating the input to that necessary to heat a layer of increasing depth. A somewhat unrealistic feature of these experiments was that the temperature of the lower boundary was kept fixed, rather than the heat flux (a more natural boundary condition at the earth's surface). The flux can of course

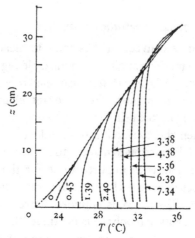

Fig. 7.17. Vertical profiles of temperature in a laboratory tank, set up initially with a linear stable temperature gradient and heated from below. The profile labels give the time in minutes. (From Deardorff, Willis and Lilly 1969.)

always be calculated from a relation of the form (7.1.10), but when it is assumed to be constant, especially simple relations can be derived for the depth and temperature of the convecting layer, which both increase like $t^{\frac{1}{2}}$. (See §8.2.1, where these are derived and compared with experiments, in the context of a *salinity* gradient heated uniformly from below.)

Another kind of experiment was reported by Townsend (1964) who made use of the non-linear density behaviour of water (which has its maximum density near 4 °C). He set up a tank of water with an ice-covered bottom and an upper surface at about 25 °C, so that the water at the bottom was convectively unstable while a stable gradient at the top was maintained by molecular conduction. After an establishment period (of about ten hours, starting with water at room temperature), during which the convecting layer grows in the manner described above, a steady state is attained which is very convenient for the detailed study of penetrative convection, both visually and by recording the fluctuations.

Townsend showed that vigorous overturning was confined to a thin region near the bottom, but that columns of fluid emerge from this, and penetrated through the whole 'convecting' layer, finally being deflected horizontally as they reached the region of very stable stratification. The temperature of the mixed layer was nearly constant at 3.2 °C (below that of maximum density) but there was evidence of a slightly stable gradient, with more pronounced overshoots near the edges of this region. This again suggests the 'filling box' model (but without the location and spacing of the sources being specified), and it also gives some experimental support for the reality of the peaked form of profiles obtained numerically by Herring (1964) and Elder (1969 a).

Temperature measurements in the ice–water system revealed the surprising feature that the fluctuations are a maximum just beyond the point of furthest penetration of the convective columns, in a thin region at the bottom of the stable layer. They are continuous rather than intermittent and so cannot be due to individual impacts of the 'columns', but they can be explained in terms of internal wave theory. In subsequent papers Townsend (1965, 1966, 1968) explored further the properties of waves produced by a convecting layer and also by a turbulent boundary layer generated by shear at the ground. He showed (as we should expect from §2.2) that when the duration of the impacts is long compared with N^{-1} the spread is predominantly horizontal, in agreement with his laboratory observations. He also suggested that the results can be applied to stratocumulus clouds, and that the 'clear air turbulence' observed just above and sometimes several kilometres to the side of convecting clouds is in fact a vigorous internal wave motion generated by the penetrative convection. The theory leads to numerical predictions which are consistent with this interpretation, provided the motions in clouds are persistent plumes rather than isolated thermals. (It is not yet clear from direct observation that this assumption is valid, and we mention in passing an alternative suggestion of Turner and Yang (1963). They regarded the region above a stratocumulus cloud as dynamically part of the turbulent cloud itself, which happens to be invisible because of evaporation following turbulent mixing with drier air above. A non-linear density behaviour was also included in their laboratory model of this process.)

Several numerical models of penetrative convection have been based directly on the ice–water convection experiments (though they have been applied to solar convection rather than geophysical problems). Veronis (1963) used a series expansion method (a modification of that used earlier by Malkus and Veronis (1958)) to investigate the stability problem for a fluid in which the equation of state is parabolic, i.e.

$$\rho = \bar{\rho}[1 - \alpha(T - T_0)^2]. \tag{7.3.10}$$

His infinitesimal amplitude analysis showed that the value of a critical Rayleigh number (based on the thickness of the unstable layer, between 0 and 4 °C) is affected by the presence of stable fluid above 4 °C; the minimum Ra_c with the relaxed upper boundary condition is about one third of that in the ordinary Bénard problem and occurs when the temperature of the upper plate is about 6.7 °C. The second order terms indicated further that a finite amplitude instability can occur at an even smaller Rayleigh number, because any finite mixing motion which mixes two parcels of water with temperatures above and below 4 °C will create a layer of fluid near the maximum density, but deeper and therefore more unstable than the corresponding layer in a purely conductive state.

Subsequent stability calculations by Whitehead and Chen (1970) have modified Veronis' conclusion. They treated a variety of temperature profiles through a thin layer of unstable fluid bounded by a stable layer, and showed that, whereas a weak stable stratification reduces Ra_c, a very stable layer can remove energy from the unstable region and thus lead to an increase in the critical Rayleigh number above that for a solid boundary. They also produced comparable temperature distributions using radiant heating, and studied the stability and the subsequent supercritical motion, in which plunging vertical plumes were prominent (cf. §7.3.1).

Musman (1968) obtained steady-state finite amplitude solutions for the parabolic initial profile (7.3.10). He used the 'mean field' approximation of Herring (1963) (with one mode in the horizontal but many in the vertical), and compared his results with Townsend's (1964) measurements. The agreement through the convecting region is excellent (including a mean temperature less than 4 °C, and the stable gradient), but not so good in the conducting region above, where side wall heat fluxes were certainly important in the experiment. Musman showed that when the temperature of the upper boundary is about 7 °C or higher, convection first takes place

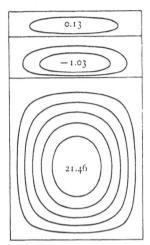 

Fig. 7.18. Streamlines and isotherms calculated for 'convection over ice' by Musman (1968). The maximum vertical velocities in each cell are marked, and isotherms are plotted for each degree C from 0 at the bottom to 24 at the top.

at finite amplitude (in agreement with Veronis) and that when.it is above 10 °C, the upper boundary is no longer dynamically important. The largest motions consist of a convecting cell extending from the lower boundary to the level where the temperature is 7 or 8 °C, with above this one or more counter cells, whose structure depends on the Rayleigh number. (See for example fig. 7.18.) The penetration of convection into the stable layer takes the form of nearly horizontal motions associated with the upper part of the main cell.

7.4. Convection with other shapes of boundary

We turn now more briefly to motions produced by buoyancy forces in different geometries, which have not been so intensively studied as the Bénard–Rayleigh problem, but which are nevertheless important in many applications.

Many of the flows of practical interest can be illustrated by considering a circular pipe of diameter d in various orientations. Suppose first that the axis of the pipe is vertical, that it is open above and below, and contains a fluid which is set into laminar motion because of a density difference between the fluid inside and outside the pipe.

Balancing the buoyancy force against the viscous stress at the walls shows that the velocity is $\bar{w} \sim g'd^2/\nu$ and the Reynolds number $\bar{w}d/\nu \sim g'd^3/\nu^2 \sim Gr$, justifying the statement made in §7.1.1. The Grashof number is the appropriate dimensionless parameter (and is equivalent to a Reynolds number) in convection problems where the density or temperature difference is *given*. The molecular diffusivity, and hence *Pr* or *Ra*, enters when the mechanism of heating the flow from the solid walls is of interest; when flows in a single fluid, and therefore at a fixed Prandtl number, are being compared the distinction is less important.

'Fully forced' convection, and heat transfer into a moving stream, will not be considered here (for an outline of the principles of this subject see, for example, Prandtl (1952), p. 397). One case of 'mixed' convection, the flow under a fixed pressure gradient in a long vertical heated pipe, will be described qualitatively. When the pressure gradient and buoyancy forces act together (i.e. the forced flow is upwards in a heated pipe), velocity and thermal boundary layers are formed, as they are when there is no pressure gradient. Provided the Rayleigh number (based on the pipe diameter and the change in temperature along the pipe in a length equal to the diameter) is sufficiently large, the upflow and heat flux are concentrated close to the wall, and there can be a slow reversed flow in the stably stratified fluid near the centre. On the other hand, if the pressure gradient and buoyancy forces are opposed, there will be a tendency for the whole motion to become unsteady and turbulent above a certain critical Rayleigh number, since the density distribution in the vertical will then be unstable. It will therefore be most effective, because of the convective motions which are induced, to heat fluid by passing it *down* a hot pipe. (cf. §6.2.3). Detailed similarity solutions exhibiting these features have been worked out by Morton (1960) for the special case of a pipe having a constant linear temperature gradient along it.

The phenomena observed in heated *horizontal* pipes, with or without an axial flow, are entirely different. Then there will be a circulation in the plane of the cross section, whose form depends on the distribution of heating and cooling and on the Rayleigh number; a problem of this kind is discussed in §7.4.3. The convective flow produced in the fluid *outside* a heated pipe (or horizontal cylinder) is also of some importance. Near the body this can be treated by much the same methods as a vertical surface (§7.4.1), to determine the form of the boundary layer and the total heat flux into it with a given temperature difference. Far from the body, the flow detaches from the cylinder and becomes a plume whose properties can be expressed in terms of the height and the buoyancy flux, using the similarity arguments of chapter 6.

7.4.1. *A heated vertical wall*

The essential properties of free convection near a solid body can be brought out by considering a two-dimensional flow near a vertical boundary. Typically, this is held at a temperature T_1, different from that of the fluid (T_0) in which it is immersed, but the problem will be discussed here (as in §3.3.4) in terms of the local density difference ρ' set up between the fluid and the surroundings. Let ρ_1' denote the (fixed) density difference between fluid at the surface and at large distances from it, and $\gamma = \rho'/\rho_1'$.

When the convecting fluid is confined to a thin layer near the wall, the vertical equation of motion can be simplified to a boundary layer form, neglecting higher derivatives in the verticial (z) direction. This and the diffusion equation (cf. (3.3.15)) may be written

$$\left. \begin{aligned} u\frac{\partial w}{\partial x}+w\frac{\partial w}{\partial z} &= \nu\frac{\partial^2 w}{\partial x^2}+\frac{g\rho_1'}{\rho_0}\gamma, \\ u\frac{\partial \gamma}{\partial x}+w\frac{\partial \gamma}{\partial z} &= \kappa\frac{\partial^2 \gamma}{\partial x^2}, \end{aligned} \right\} \qquad (7.4.1)$$

with boundary conditions $u = w = 0$, $\gamma = 1$ at $x = 0$, and $u = 0$, $\gamma = 0$ at $x = \infty$. Defining the stream function ψ in the usual way (§3.1.3) similarity solutions can be found for ψ and γ using the variable

$$\eta = \left(\frac{g\rho_1'}{\rho_0}\right)^{\frac{1}{4}}\nu^{-\frac{1}{2}}xz^{-\frac{1}{4}} = (Gr)^{\frac{1}{4}}\left(\frac{x}{H}\right)\left(\frac{z}{H}\right)^{-\frac{1}{4}}, \qquad (7.4.2)$$

where H is an overall lengthscale, say the total height of the wall, and Gr is the Grashof number based on this length and ρ_1'; they take the form

$$\left. \begin{aligned} \psi &= \nu Gr^{\frac{1}{4}}\left(\frac{z}{H}\right)^{\frac{3}{4}}f(\eta) \\ \gamma &= \rho'/\rho_1' = g(\eta). \end{aligned} \right\} \qquad (7.4.3)$$

and

It is clear from the form of η that the thickness of the buoyant layer increases like $z^{\frac{1}{4}}$. Further, (7.4.3) shows that both the local Nusselt number and the integrated value (which compares the actual heat flux over the whole depth H with the purely conductive flux) are proportional to $(Gr)^{\frac{1}{4}}$. This is a general result for laminar flows, arising from the structure of the boundary layer equations (7.4.1), and it is applicable to other shapes provided only that the heated layer remains thin compared to the overall dimensions of the body.

The resulting ordinary differential equations, from which $f(\eta)$ and $g(\eta)$ are obtained, still contain an explicit dependence on κ and thus on Pr. Ostrach (1964) has calculated their forms numerically for a range of values of Pr. The dependence is complicated, but for values near that of air and greater, it is approximately $Pr^{\frac{1}{4}}$, so that $Nu \propto Ra^{\frac{1}{4}}$. The measured values of the flux, and the profiles predicted by these calculations, are in excellent agreement with laboratory measurements made in air.

The stability of these flows against a vertical plate has also been studied theoretically and experimentally. It is observed that turbulence sets in at overall Grashof number of about 10^9, corresponding to a boundary layer Reynolds number of a few hundred (see §4.2.3). It was first assumed that the stability criteria (for both the inviscid and viscous problems) are identical to those obtained for the same velocity profile in classical stability theory, the buoyancy forces just acting to produce this mean profile. It has been shown, however (see for example Gill and Davey (1969) and the review by Gebhart (1969)), that the inclusion of the buoyancy source of disturbance energy makes a substantial difference to the theoretical stability curves, lowering the critical Gr or Re and the wavenumber at which this is attained. These predictions, for the case of a plate in air through which a constant flux is maintained (a variation of the more usual condition of constant temperature), were confirmed by Polymeropoulos and Gebhart (1967). They created artificial disturbances of known frequency and amplitude using an oscillating ribbon, and watched the growth or decay of the waves, using an interferometer which gives contours whose separation is inversely proportional to the temperature gradient; two of their beautiful photographs are reproduced in fig. 7.19 pl. XVIII. More will be said about this stability problem and its dependence on Prandtl number in §7.4.3.

7.4.2. *Buoyancy layers at vertical and sloping boundaries*

If the plane surface is not parallel to gravity but is at an angle θ to the horizontal, the equations describing the flow in an otherwise unconfined uniform environment remain the same except that $\sin\theta$ multiplies the body force term. The above solutions will therefore apply to the laminar case with $Gr\sin\theta$ replacing Gr in (7.4.2) and (7.4.3) (see also §4.2.3). This extension will of course be valid only when the flow is statically stable (for example when it is confined to the underside of a heated plate). On the upper surface of an inclined heated plate instability will set in much earlier. The observations of Lloyd and Sparrow (1970) show that for angles of inclination from the vertical greater than 14°, the wave-like instability illustrated in fig. 7.19 is replaced by longitudinal rolls (which is the preferred mode for Bénard convection in the presence of a shear). At higher Grashof numbers, the flow becomes more like the turbulent convection over a horizontal plate described in §7.2.2, with the addition of a mean flow up the plate (Tritton 1963).

When there is a stable density (temperature) gradient in the environment however, the heating of the upper surface of an inclined plate has quite a different effect, since the buoyant fluid now cannot rise indefinitely but remains confined to a layer near the boundary. A 'buoyancy layer' of this kind is important whenever the boundary conditions cannot match those imposed in the interior; it occurs at both vertical and sloping boundaries when there is a flux through the wall. Another, more surprising, result discussed below is that a sloping boundary through which there is *no* flux will also produce a flow near the wall in a stably stratified fluid. Gill (1966) and Veronis (1967) have pointed out that there is an exact analogy between this buoyancy layer, whose growth is limited by stratification, and the Ekman layer in a rotating fluid (see §9.2.4).

The laminar problem will first be considered, with the simple boundary conditions proposed by Prandtl (1952, p. 422) namely, a uniform gradient $G = N^2$ in the interior and a *fixed* density difference between the fluid near a sloping wall and that in the interior at the same horizontal level. No position along the boundary is then distinguishable from any other, so one can expect solutions which depend only on the coordinate n normal to the slope and not on s parallel to it. Writing the differences from a standard density ρ_0 in the form

$$g\frac{\Delta\rho}{\rho_0} = Gz - g' \qquad (7.4.4)$$

the equation of motion giving the velocity u parallel to the slope is

$$g' \sin\theta + \nu\frac{\partial^2 u}{\partial n^2} = 0. \qquad (7.4.5)$$

The diffusion equation must include derivatives in the s-direction, but using (7.4.4) and the relation $z = s\sin\theta + n\cos\theta$ it can be reduced to

$$uG\sin\theta = \kappa\frac{\partial^2 g'}{\partial n^2}. \qquad (7.4.6)$$

Either u or g' can be eliminated from (7.4.5) and (7.4.6) to give a single equation for the other variable. The solutions can be written as

$$\left.\begin{array}{l} \rho' = \rho_1' e^{-\xi}\cos\xi, \\[2mm] u = g_1'\left(\dfrac{\kappa}{\nu G}\right)^{\frac{1}{2}} e^{-\xi}\sin\xi \end{array}\right\} \qquad (7.4.7)$$

and these are shown in fig. 7.20. Here $\xi = n/l$, and the lengthscale l ('the thickness of the buoyancy layer') is given by

$$l^4 = 4\kappa\nu/G\sin^2\theta = 4\kappa\nu/N^2\sin^2\theta. \qquad (7.4.8)$$

Notice that with these boundary conditions the maximum velocity is independent of the angle θ, though the normal distance from the boundary at which this is achieved does depend on θ through l. Prandtl has suggested an application to the phenomenon of 'mountain and valley winds', which blow down or up a valley when the slopes are cooled or heated.

Phillips (1970) and Wunsch (1970) have given solutions to the same equations with a stable linear gradient but with a boundary condition corresponding to no flux of solute through a solid sloping wall. This condition implies that the surfaces of constant density will be distorted so that they are normal to the slope as the slope is

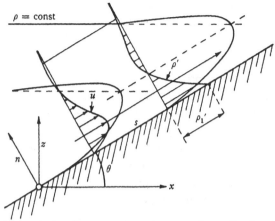

Fig. 7.20. The velocity, density and density anomaly profiles for a buoyancy layer in a linearly stratified environment, with the boundary condition of constant density difference. (From Prandtl (1952), reproduced by permission of Vieweg und Sohn.)

approached; the fluid against the wall will therefore be lighter than that in the interior at the same level and there will be an upslope current in a layer of depth l given by (7.4.8). Under laboratory conditions using dissolved salt this flow is very small (but detectable), though Phillips has shown that similar convective transports along narrow nearly horizontal slots could greatly exceed the purely diffusive vertical flux.

Prandtl has proposed a turbulent solution for the 'valley wind' problem, based on mixing length arguments which make κ and ν irrelevant. The velocity and depth are again constant, and by dimensional reasoning depend on g_1' and N through $u \propto g_1'/N$ and $l \propto g_1'/N^2$; the dependence on θ can be obtained by a more mechanistic argument. Such solutions can only remain valid when the entrainment is small. When turbulent mixing dominates (on steep slopes) then the depth will increase and be a function of slope (see §6.2); so too will the velocity when the buoyancy flux is changing. For example, a two-dimensional turbulent plume with a constant density difference between the plume and the environment just outside it will have $h \propto s$ and $u \propto s^{\frac{1}{2}}$.

7.4.3. *Convection in a slot*

An important example of a convective circulation in a totally enclosed region is that of fluid contained in a two-dimensional slot with vertical walls maintained at different temperatures. The flow is completely specified by three parameters, a Rayleigh number Ra_L based on the temperature difference and the width L, the Prandtl number, and the aspect ratio $h = H/L$, which can generally be taken as large so that the top and bottom boundaries have only a minor effect on the flow. We follow the description of Elder (1965*a*, *b*) who used a variety of laboratory techniques to obtain a clear picture of the flow at high Prandtl numbers in different ranges of *Ra*, and outline various theoretical analyses which have been suggested by the observations.

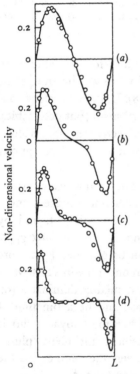

For small $Ra_L(< 10^3)$ the temperature field is very close to that due to conduction alone, with a linear horizontal variation and no vertical temperature gradient. This horizontal density contrast will inevitably generate a slow steady circulation, up the hot wall and down the cold, but this makes a negligible contribution to the flux (except right at the top and bottom). As Ra_L increases, the isotherms become progressively more distorted, as shown in fig. 7.21. In the range $10^3 < Ra_L < 10^5$ large temperature gradients are established near the walls, with a uniform *stable* vertical gradient in the interior. (Compare with the mechanism discussed in §7.3.3.) There is a strong convective flow concentrated near the walls that transfers a larger fraction of the heat as *Ra* is increased. Above

Fig. 7.21. Vertical velocity profiles in a slot at various Rayleigh numbers Ra_L; measurements compared with theory by Elder (1965*b*). (*a*) $Ra_L = 3.08 \times 10^4$, (*b*) 2.95×10^5, (*c*) 6.56×10^5, (*d*) 3.61×10^6.

$Ra_L = 10^5$ the interior region has superimposed on it a secondary flow consisting of a regular cellular pattern. The number of cells in the vertical increases with Ra, until at $Ra_L = 10^6$, when the amplitude of this secondary motion is large, counter-rotating cellular motions are generated in the weak shear regions between the larger cells.

The first theoretical discussions of the primary convective flow in the boundary layers (due to Pillow (1952) and Batchelor (1954 b)) assumed that the interior region should have a constant temperature and a constant non-zero vorticity, which is not in accord with the observations described above. Many other experiments, such as those of Eckert and Carlson (1961) in air, now also support the view that the interior region has a stable stratification. Gill (1966) derived the 'buoyancy layer' solutions described above, and showed how it is possible to maintain boundary layers which are not entraining, but have constant velocity (7.4.7) and thickness (7.4.8). The vertical gradient G in Elder's experiments was approximately related to the overall temperature difference ΔT and height H by $G = \frac{1}{2}(g\Delta T/T_0)/H$, so that (7.4.8) becomes

$$l^4 \approx 8\kappa\nu H \left/ \left| \frac{g\Delta T}{T_0} \right. \right. . \qquad (7.4.9)$$

Gill was able to go further, and include the boundary layer and interior in the same solution, which is shown in fig. 7.22. This allows for some entrainment in the lower part and outflow at the top of an upflowing layer, and its form is independent of Ra_L and H/L. The lengthscale l still determines the maximum thickness of the boundary layer, however, and the condition for this regime to be established is that l should be small compared with L. The solutions contain no disposable constant, the scaling being determined entirely by the boundary condition $w = 0$ at the top and bottom boundaries, and they are in good agreement with Elder's experimental results.

The stability of the laminar boundary layer regime on a vertical wall has been studied by Gill and Davey (1969), using numerical methods to cover the whole range of Prandtl numbers. They showed that the buoyancy-driven instability (as distinct from the purely mechanical instability arising from the shape of the velocity

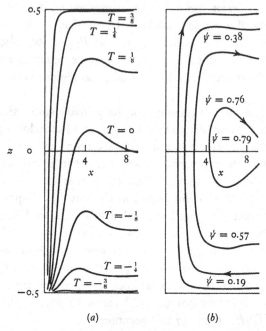

Fig. 7.22. (a) Isotherms and (b) streamlines for the boundary layer on the heated wall of a vertical slot. (From Gill 1966.)

profile) is increasingly important at large Pr, and that the Reynolds number for instability decreases with increasing Pr. Both diffusion and the stable gradient have an inhibiting effect on the growthrate of disturbances, which have the form of waves travelling faster than the flow. In the application to the slot, the critical condition is determined by a value of $Ra_L h^3$, i.e. a Rayleigh number Ra_H based of the vertical dimension (cf. §7.4.1), and varies strongly with Prandtl number.

In contrast, the breakdown of the conduction dominated flow (before strong boundary layers have formed) occurs in the form of stationary waves, and appears to be 'mechanically' driven. The critical value of Ra_L is independent of the aspect ratio h, but still depends on Pr. These results have an important application in the field of thermal insulation. The effectiveness of double walls or windows in minimizing the heat losses in air seems to depend on the stability of the conduction regime, rather than the change from conduction to the boundary layer form of motion. Both instabilities are suppressed in a porous medium, and this is probably the main effect of filling a cavity with insulating foam.

The stability of the flow in a vertical slot when there is a vertical temperature gradient in either sense as well as a horizontal gradient has been investigated by Birikh, Gershuni and Zhukhovitshii (1969). Space does not allow us to discuss this here, but the reference quoted also gives a good introduction to earlier Russian work in this field. Much more could be said too about observations and theories dealing with related geometries; for example, Hart (1971 a) has examined the convective motions (and their stability) in a tilted narrow slot, and Walin (1971) has considered the relation between the boundary layer flows and the stratification in a closed region of any shape. Convection in a region of variable depth will be considered in the more specialized context of the next chapter (§8.3.4), though some of the results are valid regardless of how the buoyancy flux is supplied.

Finally, let us consider the experimental results of Elder (1965 b) for unsteady and turbulent flows in a slot. Using water as the working fluid, he observed travelling wave-like motions at the edge of the boundary layers (cf. fig. 7.19), which appeared at $Ra_H \sim 10^9$, much as they do at a single wall. At higher values of Ra_H the waves grow and break, and there is an intense interaction between the wall layer and the interior, both of which become turbulent. This strongly turbulent flow, in which the mean temperature is nearly constant and the mean velocity zero across the interior region, can be used to test some of the predictions of the theories of thermal turbulence. For example the horizontal temperature profile has a conduction region with linear temperature gradient (presumably in a marginally stable state, similar to that discussed in §7.3.1), a substantial region where the temperature varies with $z^{-\frac{1}{3}}$ and finally the uniform interior. The existence of the $z^{-\frac{1}{3}}$ region, which did not appear in Townsend's (1959) experiments on a horizontal surface but which is characteristic of free convection profiles in the atmosphere (§5.1.3), is at first sight surprising. As Elder pointed out however, the dimensional arguments leading to (5.1.15) require only that there should be a constant heat flux normal to the boundary, that the motion be driven entirely by buoyancy forces, and that molecular processes be unimportant. All of these are satisfied in the intermediate mixing region, and the angle between the direction of gravity and the heat flux is irrelevant except for the evaluation of the numerical factor. The total heat transport between tall vertical plates in the turbulent range can be expressed (Jakob 1957) as $Nu = 0.053 \, Ra_L{}^{\frac{1}{3}}$, so the multiplying constant is rather smaller than it is for horizontal plates.

CHAPTER 8

DOUBLE-DIFFUSIVE CONVECTION

A comparatively recent development in the field of convection has been the study of fluids in which there are gradients of two (or more) properties with different molecular diffusivities. When the concentration gradients have opposing effects on the vertical density distribution, a number of surprising things can happen, and these are the subject of the present chapter. The phenomena were first studied with an application to the ocean in mind (see §8.2.4), and because heat and salt (or some other dissolved substance) are then the relevant properties, the process has been called 'thermohaline' or 'thermosolutal' convection. Related effects have now been observed in the laboratory using a pair of solutes, and in solidifying metal alloys, and the name 'double-diffusive convection' has been chosen to encompass this wider range of phenomena.

The stability problem will first be reviewed, somewhat more fully than was done in previous chapters because of the comparative novelty of the double-diffusive phenomena. It will then be shown that when two components contribute to the vertical density gradient, a series of steps tends to form, with well-mixed layers separated by sharper density interfaces. The detailed structure of these interfaces and measurements of the coupled fluxes across them will also be described.

8.1. The stability problem

8.1.1. *The mechanism of instability*

In such a system with opposing gradients, the existence of a net density distribution which decreases upwards is *not* a guarantee of stability. Diffusion, which was seen in §7.1.1 to be generally stabilizing in a fluid containing a single solute, can now act so as to allow the release of the potential energy in the component which is

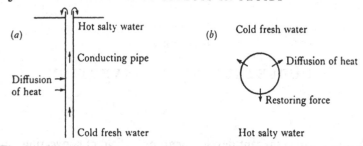

Fig. 8.1. Illustrating the two types of motion which are possible in a fluid containing opposing vertical gradients of heat and salt (*a*) salt fountain, (*b*) oscillating element.

heavy at the top. The form of the motions depends on whether the driving energy comes from the substance of higher or lower diffusivity, and our physical intuition based on observations of ordinary thermal convection is of little direct help here. Two simple conceptual experiments (the first of which was the starting point of the whole subject) do bring out the main features of the two possible kinds of motion.

Consider first, following Stommel, Arons and Blanchard (1956), a long narrow heat-conducting pipe inserted vertically through a region of the ocean where warm salty water overlies colder fresher (and of course denser) water (fig. 8.1*a*). Water which is pumped upwards, say, would quickly reach the same temperature as the surroundings at the same level (by conduction of heat through the wall of the pipe), while it remains fresher and therefore lighter. A 'salt fountain' started in this way (in either direction) will continue to flow so long as there is a vertical gradient of salinity to supply potential energy. Some confusion (mostly semantic) has arisen in describing the role of heat in this case. The temperature distribution is stabilizing in the sense that initially the warmer water is on top, but the diffusion of heat is the essential destabilizing process.

In the opposite case (of warm salty water underneath colder, fresher, lighter water) consider a parcel of fluid which is isolated from its surroundings by a thin conducting shell and then displaced upwards (fig. 8.1*b*). It will lose heat but not salt, and buoyancy forces must drive it back towards its initial position. In this way an oscillatory motion will be produced, with the buoyancy reversing each half cycle and the parcel overshooting its position of neutral equilibrium. Any lag in temperature between the moving fluid and its surroundings implies that there is a net buoyancy force in the direction of motion over much of the cycle, and so the oscillations can grow even when some energy is dissipated.

A most important step forward was taken by Stern (1960), who pointed out that solid boundaries are not essential to the above arguments. A slower transfer of salt relative to heat is assured by its smaller molecular diffusivity, and motions of both the kinds described above can be set up in the interior of a fluid. In the first case, long narrow convecting cells, called 'salt fingers', have been predicted and observed (see fig. 8.8 pl. xx), and in the second, the existence of oscillatory motions has been verified. For reasons which will become clearer in § 8.2, these two will be distinguished by calling them the 'finger' and 'diffusive' regimes respectively. Some of their properties can be derived using the stability analysis described in the following section. The extent to which these simple predictions need modification to describe the fully developed convection phenomena will be discussed in conjunction with laboratory observations.

8.1.2. *Linear stability analysis*

Only the simplest case will be described here, the stability to infinitesimal disturbances of a system containing linear opposing gradients of two properties, with free horizontal boundaries above and below held at fixed concentrations (temperatures and salinities). This problem was treated by Stern (1960), but we now add the restriction of two-dimensional motion and incorporate further results due to Veronis (1965) and Baines and Gill (1969). Stability criteria for a wide variety of conditions at the boundaries have been obtained by Nield (1967) and for the case of an unbounded fluid by Walin (1964). These make a small quantitative difference to the values of the critical parameters, but introduce no new qualitative effects.

The density distribution can be written in the linear approximations as

$$\rho = \rho_m(1 - \alpha T^* + \beta S^*), \qquad (8.1.1)$$

where ρ_m is the mean density at height z_* (which is assumed to vary linearly between $z_* = 0$ and $z_* = d$). T^* is the temperature, or more generally the concentration of the property with higher molecular diffusivity, and S^* is the salinity (concentration of the component with lower diffusivity): α and β are the corresponding 'coefficients of expansion' defined in such a way that they are both positive for temperature and salinity changes at constant pressure

$$\alpha = -\frac{1}{\rho}\left(\frac{\partial\rho}{\partial T^*}\right)_{S^*,p^*}, \quad \beta = \frac{1}{\rho}\left(\frac{\partial\rho}{\partial S^*}\right)_{T^*,p^*}. \quad (8.1.2)$$

The linearized two-dimensional Boussinesq equations (1.3.10) can be reduced to the dimensionless form (scaling with d and κ, and introducing dimensionless parameters to be defined below)

$$\left(\frac{1}{Pr}\frac{\partial}{\partial t} - \nabla^2\right)\nabla^2\psi = -Ra\frac{\partial T}{\partial x} + Rs\frac{\partial S}{\partial x}. \quad (8.1.3)$$

The diffusion equations for the two components, with similar scaling, are

$$\left.\begin{aligned}\left(\frac{\partial}{\partial t} - \nabla^2\right) T &= -\frac{\partial\psi}{\partial x}, \\ \left(\frac{\partial}{\partial t} - \tau\nabla^2\right) S &= -\frac{\partial\psi}{\partial x},\end{aligned}\right\} \quad (8.1.4)$$

where ψ is a stream function, defined as before, and the non-dimensional T and S have been defined relative to the mean linear gradients. The boundary conditions at $z = z_*/d = 0$, 1 become

$$\psi = 0, \quad \partial^2\psi/\partial z^2 = 0, \quad T = S = 0. \quad (8.1.5)$$

With the introduction of two independent concentration differences ΔT and ΔS between the bottom and top boundaries, and two different molecular diffusivities, four non-dimensional parameters are needed to specify the system completely. In addition to $Ra = g\alpha\Delta Td^3/\kappa\nu$ and $Pr = \nu/\kappa$ used earlier, the above equations contain $\tau = \kappa_S/\kappa < 1$, the ratio of diffusivities, and the density ratio $\beta\Delta S/\alpha\Delta T$. The latter can be replaced by an equivalent 'salinity Rayleigh number' Rs such that

$$Rs = \frac{\beta\Delta S}{\alpha\Delta T}Ra = \frac{g\beta\Delta Sd^3}{\kappa\nu} \quad (8.1.6)$$

(but note that κ, not κ_S, appears in the denominator here).

Rather than eliminating two of the variables to give a single equation in the third (the procedure implied by the form of (7.1.6) quoted earlier), the boundary conditions (8.1.5) are such that one can write down directly a set of functions satisfying (8.1.3) and (8.1.4):

$$\left.\begin{array}{l} \psi \sim e^{pt} \sin \pi a x \cdot \sin \pi n z, \\ T, S \sim e^{pt} \cos \pi a x \cdot \sin \pi n z, \end{array}\right\} \qquad (8.1.7)$$

where πa and πn are horizontal and vertical wavenumbers; n must be an integer but a is so far unrestricted. The characteristic equation derived by substitution is a cubic in p,

$$p^3 + (Pr + \tau + 1)k^2 p^2 + [(Pr + \tau Pr + \tau)k^4 - (Ra - Rs)Pr\pi^2 a^2/k^2]p$$
$$+ \tau Pr k^6 + (Rs - \tau Ra)Pr\pi^2 a^2 = 0, \quad (8.1.8)$$

where $k^2 = \pi^2(a^2 + n^2)$.

The subsequent infinitesimal motion is described by the solutions of (8.1.8) for various combinations of parameters. (See Baines and Gill (1969) for a detailed discussion, especially of the case $Pr = 10$, $\tau = 10^{-2}$ which corresponds roughly to salt and heat in water.) At least one value of p must be real, but $p = p_r + ip_i$ is in general a complex number whose real part represents the growthrate, and whose imaginary part allows for an oscillatory behaviour. For fixed Prandtl number Pr and fixed τ, the stability boundaries (minimum Ra with $p_r = 0$) correspond to two straight lines in the Ra, Rs plane whose equations are

$$\left.\begin{array}{ll} XZ: & Ra = \dfrac{Rs}{\tau} + \dfrac{27\pi^4}{4}, \\[4mm] XW: & Ra = \dfrac{Pr + \tau}{Pr + 1} Rs + (1 + \tau)\left(1 + \dfrac{\tau}{Pr}\right)\dfrac{27\pi^4}{4}. \end{array}\right\} \qquad (8.1.9)$$

These are drawn, diagrammatically and not to scale, in fig. 8.2.

In the quadrant where Ra is negative and Rs positive, both gradients are stabilizing and no growth is possible. In the opposite (upper left) quadrant, all points are unstable above XZ, so that $Ra_e \equiv Ra - Rs/\tau = \frac{27}{4}\pi^4$ represents an effective Rayleigh number having the same role as Ra_c in ordinary convection; Ra_e reduces to Ra_c when either component is present alone, and the numerical values also vary with the boundary conditions in the same way.

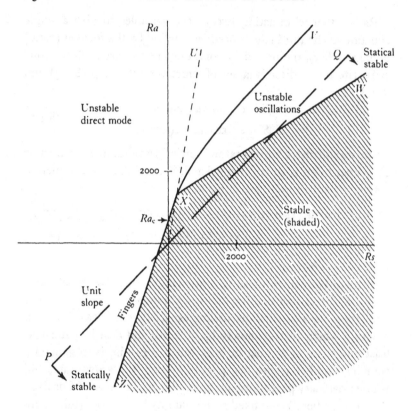

Fig. 8.2. Diagram of the various convection regimes described by (8.1.8). The sign convention is such that negative Ra and positive Rs represent stabilizing gradients of the respective components. (After Baines and Gill 1969.)

In the 'finger' regime (Ra and Rs both negative) instability sets in (when τ is small) at values of $|Rs|$ only slightly larger than the critical value for a salinity gradient alone, and certainly while the net density gradient is still statically stable (i.e. below the line PQ on which $Ra = Rs$). The motion just to the left of XZ is a direct, exponentially growing mode ($p_1 = 0$), drawing its potential energy from the component of lower diffusivity. For large negative values of both Ra and Rs, convection can occur only while $|Ra| < |Rs|/\tau$ or

$$\frac{\alpha \Delta T}{\beta \Delta S} < \frac{\kappa}{\kappa_s}, \qquad (8.1.10)$$

which becomes an important consideration when the diffusivities are comparable.

Instability in the 'diffusive' regime (Ra and Rs both positive) occurs, as suggested by the above qualitative discussion, in an 'overstable' or oscillatory mode, since $p_1 \neq 0$ just above XW. When $\tau \ll 1$ the effective Rayleigh number may now be defined as $Ra_e = Ra - Pr \cdot Rs/(Pr + 1)$ and the critical 'thermal' Rayleigh number is

$$Ra_c{}^0 = \frac{Pr}{1 + Pr} Rs + \tfrac{27}{4}\pi^4. \qquad (8.1.11)$$

For a given stable 'salt' gradient, $Ra_c{}^0$ must exceed the ordinary criterion by $Pr \cdot Rs/(1 + Pr)$; in some problems (see §8.2.1), this correction can be so small as to be negligible, though the oscillatory mode of breakdown is preserved. When $Rs \gg \tfrac{27}{4}\pi^4$, the instability occurs when $Ra_c{}^0 \approx (Pr + \tau)Rs/(1 + Pr)$, i.e. when the net density field is statically stable, since XW crosses PQ at some Rs. The frequency of the most unstable mode, when translated into the dimensional variables, is then

$$[(1 - \tau)/3(Pr + 1)]^{\frac{1}{2}}N; \qquad (8.1.12)$$

this is always less than the buoyancy frequency, the factor being about $\frac{1}{5}$ for the case of heat and salt. Above the line XV the motion has a direct instead of an oscillatory form, if the limiting case $p_1 \to 0$ is also interpreted as steady convection. As $Rs \to \infty$, XV is parallel to PQ and the direct mode becomes possible just after the net density gradient becomes statically unstable (not only at the much larger Ra, above XU, as had previously been suggested).

8.1.3. The form of the convection cells

The analysis outlined above has been extended to calculate the growthrate of disturbances with any specified wavenumber. For given Ra and Rs (not just the critical values) one can determine the aspect ratio of the cells which will grow most rapidly. In spite of the limitations imposed by the linearized analysis, which is strictly valid only very close to the critical state, it appears that this gives some guidance as to what is in fact observed under highly supercritical conditions. (See §8.3.3.)

For all Ra, Rs the most unstable mode has $n = 1$, i.e. the cells extend from top to bottom of the unstable region. On the 'marginal' lines XZ and XW, $n = 1$ and $a^2 = \frac{1}{2}$ are the appropriate values (this most unstable horizontal wavenumber has already been taken into account in the numerical values quoted above). The behaviour of the maximum growthrate, and the value of a^2 at which this occurs for general Ra, Rs, were again given by Baines and Gill for the case $Pr = 10$, $\tau = 10^{-2}$. The line above which the most unstable mode is direct then lies just above XV in fig. 8.2, and is so close to it that where both modes are possible it is virtually certain that the motion will be direct. Through the whole top part of the diagram, where the difference in temperature (the property with the higher diffusivity) is driving the motion, the most unstable wavenumber is near unity; the cells are about as wide as they are high whether the motion is direct or oscillatory.

In the lower left-hand quadrant, however, maximum growth at supercritical $|Rs|$ (to the left of XZ) implies a relatively large a^2 so that cells tend to be tall and thin. Asymptotically, as $|Ra| \to \infty$ with $\tau \ll 1$, the wavenumber a_* of maximum growth is (in dimensional units)

$$\pi a_* \approx \left(\frac{g\alpha\Delta T/d}{\nu\kappa} \right)^{\frac{1}{4}}. \qquad (8.1.13)$$

This result was obtained originally by Stern (1960) as the scale of the fastest growing disturbances of square planform but, as discussed in §7.1.2, linearized analysis can give no information about the actual structure in the horizontal. Although cells with $a \approx n$ are preferred just above the critical point, at large $|Ra|$ thin columns can evidently release the potential energy of the salt field more efficiently. This justifies the name given to the regime where Ra and Rs are both negative, and it is also in agreement with the observed form of 'salt fingers' in their fully developed state.

Note also that the form of (8.1.13) is the same as that derived in §7.4.2 for the thickness of a 'buoyancy layer' at the boundary of a fluid stratified with a single component (7.4.8). In both cases the basic gradient is statically stable, and it is the horizontal diffusion of this component which allows the fluid to react to an extra source of potential energy.

When τ is no longer very small (i.e. the diffusivities are comparable), another limiting case is of interest. As $|Ra| \to \infty$ with $Ra \sim Rs$ (nearly compensating density gradients), the predicted horizontal lengthscale in the finger regime is of the form (8.1.13), but with the net density gradient $\Delta\rho/\rho d = (\alpha\Delta T - \beta\Delta S)/d$ replacing $\alpha\Delta T/d$.

8.1.4. *Finite amplitude calculations*

Detailed calculations of the motions produced by finite amplitude disturbances of a double-diffusive system (starting with the same initial state considered in §8.1.2) were first made by Veronis (1965, 1968) and Sani (1965). Both these authors concentrated on the 'diffusive' regime (e.g. a stable salinity gradient, heated from below); and the description given here follows the most recent of these papers.

With the non-linear (product) terms added to (8.1.3) and (8.1.4) (cf. (7.2.2) and (7.2.3)), Veronis expanded ψ, T and S as double Fourier series in x and z (an extension of his method for a single component). Substituting in the governing equations and truncating after a finite number of terms (M in the horizontal and N in the vertical) gives a set of equations for the (time-dependent) coefficients. These can be solved numerically, taking into account all components and interactions such that $M + N \leqslant K$. The complexity increases rapidly with K, and Veronis treated $K = 4, 6, 8$ and 10. He regarded the answers as reliable when they changed little as K was increased, and showed that his earlier (1965) approximation with $K = 2$ was inadequate. The horizontal wavenumber was kept fixed throughout at $a^2 = \frac{1}{2}$ which is the most unstable value in the linear theory (cf. §7.1.2).

These calculations show that with $\tau = 10^{-\frac{1}{2}}$ and $Pr = 1$ and above, the onset of instability appears as an oscillatory mode at the $Ra = Ra_c{}^0$ predicted by linear theory (8.1.11). As Ra is further increased, a finite amplitude instability, leading to steady convection and an increased flux of T, sets in below the line XV (fig. 8.2). As $Rs \to \infty$, the criterion for this direct mode is again $Ra \to Rs$ (a marginally stable density gradient). When Pr is small, on the other hand, the first instability appears in the finite amplitude steady mode.

The main results of Veronis' analysis are the values of horizon-

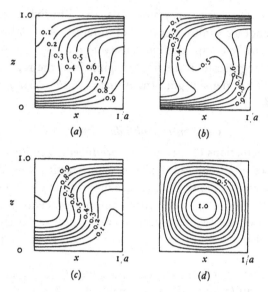

Fig. 8.3. Contours of (a) T, (b) S, (c) ρ and (d) ψ in a half cell, calculated by Veronis (1968) for the case $\tau = 10^{-\frac{1}{2}}$, $Pr = 1$, $Rs = 10^3$ and $Ra = 2500$. The lines $x = 0$ and $x = 1/a$ are lines of symmetry for T, S and ρ and lines of antisymmetry for ψ.

tally averaged T and S fluxes (F_T and F_S in density units), made non-dimensional with the molecular fluxes, i.e. the 'Nusselt numbers' Nu and Nu^s (see §7.1.1), which are functions of Ra, Rs, τ and Pr. A stabilizing gradient of solute inhibits the onset of convection and reduces the calculated F_T at moderate values of Ra. When Ra is sufficiently large, the strong finite amplitude (steady) motions which develop will tend to mix the solute so that the interior region is more nearly neutrally stratified; the fluid layer then transports nearly as much of the destabilizing T component as it does when this is present alone. Contours of T, S, ρ and ψ in a half cell show this tendency even at $Ra = 2500$ (fig. 8.3) and horizontal averages exhibit a slight reversal of the gradients in the central region.

Both F_T and F_S are increased by convection, and the ratio of these two fluxes is of special interest. In the limit $R_S \to \infty$, $Rs/Ra \to 1$, Veronis has shown that $F_T/F_S \to -\tau^{-\frac{1}{2}}$ (a result which will be derived in a more mechanistic way in §8.3.1). It also follows that

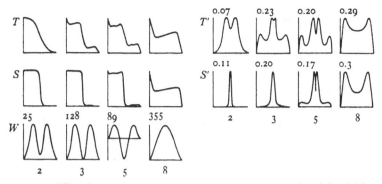

Fig. 8.4. The development of a T, S interface, as calculated by Elder (1969b). Mean values of T and S, the velocity W and fluctuations T' and S' are plotted at four times, with the height axis horizontal and the origin taken at the lower boundary. The parameters are $Ra = 10^6$, $Rs = 4 \times 10^6$, $\tau = 10^{-2}$ and the relative magnitudes of the peaks are marked.

the mean 'coefficient of diffusion for density' κ_ρ (defined as the ratio of the flux to the gradient) is in this limit.

$$\kappa_\rho \to -(\kappa\kappa_S)^{\frac{1}{2}}. \qquad (8.1.14)$$

The negative sign reflects the fact that for energetic reasons F_T must always exceed F_S, and so the density flux is against the gradient.

In the case of the periodic oscillatory mode, calculations at finite amplitude show that the horizontally averaged density gradient changes sign at a particular height. For most of the cycle, and in the temporal mean, there is a stably stratified region near the boundaries and an unstable interior; but note that the overall density gradient is unstable, in contrast to the observed flows to be described later.

Elder (1969b) has used a 'mean field' approximation (cf. Herring 1964, and §7.2.3) to treat a double-diffusive problem with different and very revealing initial conditions. He started with a step (or sharp interface) between two well-mixed layers, each having the properties of the adjoining boundaries, with higher T and S below and a net stable stratification (again the 'diffusive' regime). The sequence of events is shown in fig. 8.4, where both mean and fluctuating properties are plotted. At first the interface thickens by diffusion, the T interface more rapidly than the S interface. The outer part thus becomes unstable, and fluctuations which were imposed initially begin to grow, with at first a weak oscillation appearing. Later there is a rapid exponential growth and breakdown, and the interface is sharpened by the increased mixing at its edges. During this period there is a balance between diffusion which tends to spread the interface, and mixing on either side which sharpens it – features which are also prominent in the laboratory experiments. The transfers across the interface produce two layers which for a time behave like isolated thermal layers, and the fluxes from the boundaries also become important. Eventually the interface is disrupted (by a mechanism which is not clear from the numerical experiment), and a single convecting layer is produced, with reversed gradients across the central layer.

8.2. The formation of layers; experiments and observations

We now turn to a discussion of laboratory experiments on double-diffusive systems, commenting at the same time on their relation to the predictions set out in §8.1, and adding some more theoretical ideas where appropriate.

8.2.1. *The 'diffusive' regime*

The simplest example of a 'diffusive' system (defined in §8.1.1 as one in which the energy to drive convective motions comes from the component T having the larger diffusivity) is a linear stable salinity gradient, initially at constant temperature and heated strongly from below. Convection is first observed in a layer close to the bottom, as the diffusion of heat makes the density gradient unstable there (cf. §7.3.1). The motions do not immediately extend through the whole depth of the fluid, however, as in ordinary thermal convection, and as predicted by the theories based on linear gradients of both properties. Instead, a series of convecting layers forms in succession, from the bottom up, as shown in fig. 8.5 pl. XIX. The salinity distribution maintains a stable density jump across each of the interfaces while the more rapid diffusion of heat relative to salt provides an unstable buoyancy flux which drives the convection. Each stage of this process will now be examined briefly in turn.

Shirtcliffe (1967, 1969a) has shown that the initial instability does occur as a growing oscillation, as predicted by theory. (See fig. 8.6.) He obtained fair quantitative agreement with both the

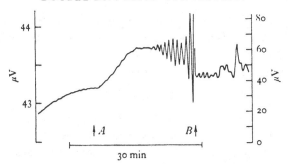

Fig. 8.6. Section of a differential thermocouple record obtained by Shirtcliffe (1967), *Nature*, **213**, 489–90, showing the overstable breakdown of a stratified sugar solution heated from below. The heating rate was changed at '*A* and the right-hand voltage scale applies after *B*.

critical Rayleigh number criterion (8.1.11) and the frequency (8.1.12), provided the depth of the layer is defined as that which makes the effective Rayleigh number a maximum. Allowance for the non-ideal boundary conditions (Nield 1967) improves the agreement between theory and experiment. Only a very small stabilizing gradient is needed to produce the overstable mode of breakdown (see fig. 8.2). For example, Hurle and Jakeman (1971) have shown that the Soret effect (i.e. mass diffusion caused by a temperature gradient) can produce concentration gradients near a boundary which are sufficiently large for this purpose, even in extremely dilute solutions. They have documented this phenomenon with experiments in water–methanol mixtures, and have also suggested it as an explanation of temperature oscillation and impurity striations observed during the growth of crystals from liquid metals.

As heating is continued beyond the point of instability, a well-mixed layer develops; this behaviour too is consistent with the finite amplitude calculations leading to fig. 8.3. The top of the layer is rising, however, as fluid from the gradient region above is incorporated into it, and its growth can be calculated as follows. Assuming a constant heat flux H and a linear initial gradient of solute denoted by $N_S{}^2 = -g\beta\,dS/dz$, the heat and solute balances in a uniformly mixed layer with depth d at time t give

$$Ht = -\rho C\,d\Delta T,$$

and
$$g\beta\Delta S = \tfrac{1}{2}dN_S{}^2, \qquad (8.2.1)$$

where T and S are the steps of 'temperature' and 'salinity' at the top of the layer, C is the specific heat, and the rest of the notation is that used in §8.1.2.

Experimentally it is observed that

$$\alpha\Delta T = -\beta\Delta S, \qquad (8.2.2)$$

i.e. the steps in T and S are compensating to within the experimental error, and the net density structure is as shown in fig. 8.7b (compare with §7.3.4, where an experiment using heat alone is described). Combining (8.2.1) and (8.2.2), and using the buoyancy flux $B=g\alpha H/\rho C$ defined in (5.1.7), we obtain

$$d = 2^{\frac{1}{2}}B^{\frac{1}{2}}N_S^{-1}t^{\frac{1}{2}} \qquad (8.2.3)$$

and $$g\alpha\Delta T = -g\beta\Delta S = 2^{-\frac{1}{2}}B^{\frac{1}{2}}N_S t^{\frac{1}{2}}. \qquad (8.2.4)$$

If, on the other hand, a fixed fraction of the kinetic energy generated by buoyancy in the unstable layer were used to mix fluid across the interface, d and ΔS would be increased above the values just calculated, though the form of the relations remains the same. (This possibility could become relevant on a larger scale, and will be referred to again in chapter 9.) In the limit where the energy is conserved between the initial and final states, (8.2.2) is replaced by $\alpha\Delta T = -\frac{1}{3}\beta\Delta S$ and the distribution of properties is as shown in fig. 8.7a. There is now a definite density step at the top of the layer; d is increased and ΔT is decreased by a factor of $\sqrt{3}$ compared to (8.2.3) and (8.2.4).

The relations (8.2.3) and (8.2.4) have been verified directly over a range of strong heating rates by measuring the depth and temperature of the growing convecting layer (Turner 1968a). (Rather better agreement with experiment is obtained by taking into account the diffusive boundary layer ahead of the convecting zone; this will be invoked explicitly below.) At some point the growth stops abruptly; a second layer forms on top of the first and behaves in the same way, until the many layers illustrated in fig. 8.5 have built up. (Note that sidewall heating can lead to a similar effect, which will be discussed separately in §8.2.3.) A purely vertical heat flux produces layer depths which depend on the initial gradient and heating rate, and the measured thickness of the first layer at the time a second forms above it can be explained as follows.

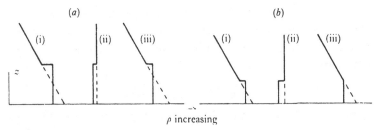

Fig. 8.7. The distributions of (i) density due to salinity, (ii) density due to temperature, and (iii) the net density produced by heating a linear salt gradient from below. It has been assumed that (a) all the potential energy released by heating is used to redistribute salt, and (b) the top of the mixed layer is marginally stable. (From Turner 1968 a.)

The diffusive thermal boundary layer growing ahead of the well-mixed layer has thickness $\delta \propt t^{\frac{1}{2}}$, and is thus a constant fraction of the layer depth d which can be written as

$$\delta/d = 2\kappa N_S^2/B. \qquad (8.2.5)$$

When δ/d is small (i.e. B is sufficiently large), Rs based on δ and N_S^2 is much less than Ra based on δ and ΔT, so the criterion for instability (8.1.11) is effectively a condition on Ra alone. Instability occurs when the depth d is

$$d_c = \left(\frac{\nu Ra_c}{4\kappa^2}\right)^{\frac{1}{2}} B^{\frac{3}{4}} N_S^{-2} \qquad (8.2.6)$$

or when ΔT (proportional to dN_S^2) reaches a corresponding value determined only by B. The form of (8.2.6) has been verified experimentally, though the implied numerical value of Ra_c is larger than the application of the simplest theory would suggest. The corresponding calculations for later layers would have to allow for the dependence of the fluxes through the first interface on the changing S and T differences across it (see §8.3.1).

Shirtcliffe (1969 b) has used another approach to explain his laboratory observations of successive layering in the limit where δ/d is not small. He modelled the process numerically, using diffusion coefficients which changed abruptly from molecular to much larger 'eddy' values when a Rayleigh number criterion for instability was exceeded locally. The observed features of successive layer formation at the top, and the merging of previously formed layers at the bottom when their densities become equal (see fig. 8.5) are reproduced well, but of course many detailed questions about the mechanism of transfer through the various regions are left unanswered.

8.2.2. The 'finger' regime

In the opposite case to that described above, when a little hot salty water is poured on top of a stable temperature gradient, 'salt fingers' very quickly form, as predicted by the first of the stability criteria (8.1.9). A cross section of these long narrow convection cells, made visible by adding fluorescent dye at the bottom, is reproduced in fig. 8.8 pl. xx. (Later it will be shown that the planform in the idealized case is square, with upward and downward motions alternating in a close packed array. Practically, this means that each 'finger' tends to have four near neighbours, though their orientation may gradually change. See fig. 8.18 pl. xxi.)

This process looks very different from that illustrated in fig. 8.5, for example, but consider the results of another related series of experiments. If a layer of lighter sugar solution (the more slowly diffusing component, denoted by S in the notation of §8.1.2) is placed above a stable gradient of salt (T), the configuration again gives rise to 'fingers', since the substance of lower diffusivity is again unstably stratified, though τ is about $\frac{1}{3}$ instead of 10^{-2} for heat and salt. These solutes are very convenient to use in the laboratory, since they allow large gradients to be set up without requiring any special precautions to prevent losses at the boundaries. When the salinity gradient is large in relation to the concentration step, fingers develop and eventually extend right to the bottom of the container (as in the experiment with heat and salt shown in fig. 8.8). With a rather smaller T gradient and the same step of S at the top, however, a convecting layer is formed; the fingers become unstable and overturn over a finite depth. The layer deepens, bounded by a sharp interface below which fingers persist. These in turn can become unstable, leading to a second convecting layer separated from the first by an interface containing fingers. In time, a series of such convecting layers can be formed, as shown in fig. 8.9 pl. xx.

Looked at in the large, therefore, and ignoring for the present the details of interface structure and the differences in the mechanism of transport which will be taken up in §8.3, there is a complete correspondence between the 'finger' and 'diffusive' systems. The inequality of molecular diffusivities results in an unstable buoyancy flux across a statically stable interface in both cases, and this maintains the convection in the layers above and below. In the 'finger' system this buoyancy flux is dominated by the S component, and hence such interfaces must reverse the relative rates of transport due to molecular processes alone (e.g. salt is transported faster than heat across an interface containing fingers).

A theory of layer formation in this case has been proposed by Stern and Turner (1969) along similar lines to that outlined previously, but now taking more explicitly into account the 'finger' structure of the interface. The criterion for instability of a deep layer containing fingers must also be taken into account here (see §8.3.4).

8.2.3. Side boundaries and horizontal gradients

It was remarked in §8.2.1 that laboratory experiments on the formation of layers by heating a salt gradient from below can be complicated by heating or cooling from the side walls. Though vertical transfer processes are always important once layers have formed, their depth can be set by these side wall effects, rather than by the criterion (8.2.6). The mechanism is readily understood (qualitatively at least) by extending the arguments of §7.4.2. There we saw that a distortion of the interior distribution of a single component can produce a steady flow along a boundary, in which a constant difference of (say) temperature is maintained between the boundary layer and interior. If a stabilizing salt gradient is distorted by heating, however, such a balance cannot be maintained; two different boundary layer length scales become relevant and there is no velocity profile which can balance both T and S.

The formation and growth of layers in a two-component system is an unsteady process, which is related to the instability of the sidewall thermal boundary layer as it grows by conduction (cf. §7.3.1). When instability sets in, salt is lifted by the heated wall layer, but only to a level where the net density is close to that in the interior. Fluid then flows out away from the wall, producing a series of layers which form simultaneously at all levels and grow inwards from the boundaries. This is shown beautifully in the experiments of Thorpe, Hutt and Soulsby (1969); it is clear from fig. 8.10 pl. xx, that each layer (marked by dye) then slopes downwards as it extends and cools. Chen, Briggs and Wirtz (1971) have shown that the layer thickness (when τ is small) is close to the natural lengthscale

$$l \approx \frac{\alpha \Delta T}{\beta \, dS/dz} \qquad (8.2.7)$$

(where ΔT is the imposed horizontal temperature difference), which is the height to which a heated fluid element would rise in the initial density gradient.

Analogous effects can be obtained using two solutes instead of salt and heat. If opposing gradients are set up, the surfaces of constant concentration are horizontal and so the no-flux boundary conditions are satisfied for both properties on vertical walls. Such distributions can be very stable in the laboratory, a considerable finite amplitude disturbance being required to initiate layered convection in the interior. At an inclined boundary, however, density anomalies are produced by diffusion which tend to drive a flow along the wall. Again the second component prevents this continuing indefinitely, and fluid spreads into the interior as a series of layers.

The observed motions are much stronger than with a single solute and have some surprising features. Fig. 8.11 pl. xix shows an experiment started with linear gradients of salt overlying sugar. Just above the boundary is a downflow (as one might expect from the distortion of the thicker salinity boundary layer) but above this is an upslope counterflow. As the fluid moves outwards into the layers the vertical gradients are reversed locally and fingers form, though the original distribution was in the 'diffusive' sense. At a much later stage, a series of sharp diffusive interfaces remained, with flow down the boundary dominating. This produced a mean circulation which caused the layers to spread far from the boundary at the bottom and suppressed their extension near the top of the slope. (Under a sloping lid, the motions are reversed.)

The theoretical stability problem corresponding to these observations has not been solved (it is analogous to the Bénard problem with large gap widths) but Thorpe, Hutt and Soulsby (1969) have analysed the simpler case of a fluid contained in a narrow slot. They assumed linear gradients of S and T initially in both the vertical and horizontal directions, with a statically stable (or zero) density gradient in the vertical and exactly compensating horizontal gradients. The stability condition can be expressed as a relation between two effective Rayleigh numbers defined as in §8.1.2., Ra_z being based on the vertical differences and Ra_x on horizontal differences of the two properties over a distance d, the separation of the plates. Asymptotically, at large Ra_z, the stability criterion for vertical walls is

$$|Ra_x|_{\min} = 2.76|Ra_z|^{\frac{1}{4}}, \qquad (8.2.8)$$

so a minimum horizontal gradient is necessary for instability. Linear theory predicts that a steady motion will then be set up, with cells reaching right across the gap and having a height $d(2/|Ra_z|)^{\frac{1}{4}}$. A definite slope of the cell boundaries can also be calculated.

These relations were shown to be consistent with laboratory measurements carried out using a salinity gradient contained in a vertical slot 6 mm wide, with one wall heated slowly. Rather better agreement with these same experiments was obtained in a subsequent theory by Hart (1971 b), who took into account the mean flow due to the sidewall heating at finite values of Ra_z, and also used more realistic boundary conditions. The observed motions in all cells do, however, differ from the predictions in one respect; they have the same sense of rotation (dominated by the flow up the hot wall and down the cold) instead of being counter-rotating in alternate cells as the theory suggested. More detailed observations of the motion and the temperature and salinity structure in a wider tank showed reversals of the gradients which are consistent with the same circulation pattern.

Horizontal gradients in the *interior* of a fluid can also lead to instabilities and the release of energy which produces a layered structure, without the need for disturbances propagating in from boundaries. Some of the essential physical features are illustrated by experiments which are a variation on the 'lock exchange' problem considered in §3.2.4. When one half of a channel is filled with hot salty water and the other with cold fresh water of the same density, and a vertical barrier separating them is withdrawn, horizontal diffusion gives rise to vertical motions near the interface. A small initial tilt in either direction, introduced by removing the barrier, will grow and lead to the formation of two layers, stratified either in the 'diffusive' or 'finger' sense. In both cases vertical transports will increase the density difference between the layers (see §8.3), so reinforcing the original (fortuitous) direction of motion. In a deep fluid it seems likely that a preferred vertical scale which depends on both the vertical and horizontal gradients of the two properties should emerge from a stability argument.

8.2.4. *Related observations in the ocean*

Various observations in nature, and especially in the ocean, can be interpreted as examples of the phenomena discussed above. With this background we can now describe, more concisely and with better physical understanding, some of the most striking examples. Of course not all layered structures are the result of double-diffusive processes, and other mechanisms of layer formation will be discussed in chapter 10. The examples given here are distinguished fundamentally by the existence of opposing vertical gradients of temperature and salinity. In the observations so far ascribed to double-diffusive convection there is also a notable regularity and persistence of the steps, though the process could also be important under transient conditions which are not so easily recognized.

There are many fresh water lakes which have become stably stratified with salt, for example by an intrusion of sea water at some time in the past. Some of these are observed to be hotter at the bottom than the top, and well-mixed layers separated by 'diffusive' interfaces are formed. An especially well-documented example is Lake Vanda in Antarctica, studied by Hoare (1966), among others. Well-mixed layers of several scales are recorded, those nearest the surface being about $1\frac{1}{2}$ m deep, with the sharpest temperature transitions about 1 °C in 1 cm.

Steps maintained by cooling a salinity gradient from above, underneath an ice island in the Arctic ocean, are shown in the record of fig. 8.12, obtained by Neal, Neshyba and Denner (1969). Here the temperature steps are only a few hundredths of a °C, but they are

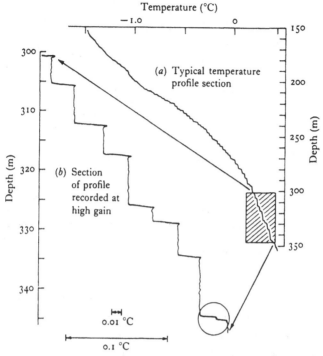

Fig. 8.12. Temperature profile under Arctic Ice Island T3, showing steps formed by the double-diffusive mechanism. (From Neal, Neshyba and Denner (1969), *Science*, **166**, 373–4. © 1969 by the American Association for the Advancement of Science.)

extremely regular, indicating (according to the results of §8.3.1) that the heat and salt fluxes must be uniform with depth. At the other extreme is the hot salty water found at the bottom of the Red Sea (Degens and Ross 1969). Two well-mixed layers have been observed, the bottom one nearly saturated and at about 56 °C, and the other at 44 °C and intermediate in salinity between this and the overlying water. The interface between them is so sharp that no sample has been obtained from it, and this too can be interpreted as a 'diffusive' interface separating two strongly convecting layers.

Good examples of layering consistent with the 'finger' process have also been found by many oceanographers, using continuously recording temperature–salinity–depth devices. Fig. 8.13 shows a series of layers about 20 m thick, recorded by Tait and Howe (1968,

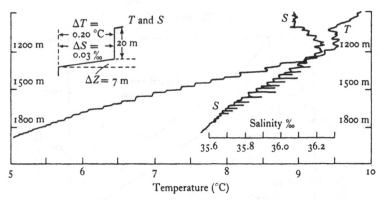

Fig. 8.13. A series of layers in T and S observed under the Mediterranean ouflow into the Atlantic, and attributed to the 'finger' mechanism. The sharp spikes to the right of the salinity trace are instrumental and can be ignored. The inset summarizes the mean properties of the layers and interfaces. (From Tait and Howe (1971), *Nature*, **231**, 178–9.)

1971) deep in the Atlantic underneath the Mediterranean outflow, where warm salty water overlies colder fresher water. Such layers have been observed relatively rarely, though the mean distributions of temperature and salinity appear to favour their formation elsewhere as well. The results discussed in §8.3.4 suggest that externally generated turbulence may be responsible for inhibiting the formation of fingers. No observations of the 'fingers' themselves have been made in the ocean, though of course such a direct measurement would greatly increase our confidence in the double-diffusive explanation of this regular layering.

It is worth pointing out that conditions are particularly favourable for double-diffusive convection when a layer with compensating temperature and salinity differences intrudes into an environment with different properties (as the Mediterranean outflow does). This guarantees that the net density difference will be small, and it is shown in the following section that the fluxes will then have their maximum values. Another example of this effect has arisen in the context of sewage disposal in the sea off California (Fischer 1971). Effluent, which can be regarded as nearly fresh water, flows out of a pipe laid along the bottom and rises as a line plume. The sea in this area is strongly stratified in temperature, and so the effluent, diluted with cold sea water, spreads out in a layer below the thermocline (see §6.4.2). This layer, however, remains colder and fresher than the water above it, and the salt finger mechanism can cause it to thicken and even extend to the surface.

The importance of boundary processes is brought out by observations of layers formed in the main thermocline near Bermuda (where the stratification is in the 'finger' sense) (see Wunsch 1970). The stepped structure is much clearer near the slope than few miles from it, but the validity of applying the laboratory results of §8.2.3 here is rather doubtful. It seems likely that layers form not as a result of a diffusive instability but directly because of turbulent mixing near the slope (as described in §4.3.4). However they are formed initially, such layers will be mixed and the interfaces kept sharp by the buoyancy flux carried by the salt fingers.

Layering and fine structure associated with horizontal gradients of T and S has also been reported. For example, Stommel and Fedorov (1967) were able to trace layers several kilometres horizontally, and to show that the density in them remained constant (to the accuracy of measurement) while T and S both varied in a compensating manner. Pingree (1971) has demonstrated the importance of horizontal variations, in his discussion of the less regular small scale structures above and below the Mediterranean outflow. The T and S fluctuations are highly correlated and compensating in density (whatever the signs of the mean gradients); this observation cannot be explained in terms of vertical mixing alone, but requires some horizontal advection or vertical shear. The regular layering described previously may, on the other hand, be favoured in more uniform regions where the horizontal motions are weak (see Zenk 1970).

8.3. The fluxes across an interface

Once layers and interfaces have formed in one of the ways described, the most important property is the flux of S and T across them. In this section we discuss laboratory measurements of these fluxes, and more detailed observations aimed at relating them to the structure of the interface. Ideally one would like to examine a number of identical layers in a steady state, but so far only quasi-steady two-layer systems have been used, with a single interface and solid boundaries (or a free surface) above and below. The 'diffusive' systems will be discussed first, but many of the ideas carry over to the 'finger' case.

8.3.1. Measurements in the 'diffusive' regime

The overall behaviour of a two-layer convecting system, with a hot salty layer below cold fresh water for example, can be interpreted using an extension of the earlier dimensional arguments, which led to (7.1.9) and (7.1.10) in the case of high Rayleigh number single component convection. The flux of the driving component F_T across a convecting layer of depth d, and therefore through the bounding interfaces in a steady state, must now be of the form

$$Nu = f(Ra, Rs, Pr, \tau), \qquad (8.3.1)$$

where $\quad Nu = \dfrac{F_T}{\kappa \Delta T/d}, \quad Ra = \dfrac{g\alpha\Delta T d^3}{\nu\kappa}, \quad Rs = \dfrac{g\beta\Delta S d^3}{\nu\kappa}$

as defined in §8.1.2, and $Pr = \nu/\kappa$, $\tau = \kappa_S/\kappa$. ΔT and ΔS are the T and S differences 'across a layer', or by symmetry in the ideal situation, between the centre of one layer and the centre of the next.

In any experiment $\beta\Delta S/\alpha\Delta T = Rs/Ra$ is given and Pr and τ are constants, and a plausible form of (8.3.1) which removes the dependence on d is

$$Nu = f_1\left(\frac{\beta\Delta S}{\alpha\Delta T}\right) Ra^{\frac{1}{3}}. \qquad (8.3.2)$$

Note that the choice of Ra as the major parameter in (8.3.2) is to some extent arbitrary; a more general combination of Ra and Rs (such as the effective Rayleigh number used in §8.1.2) would also be consistent with (8.3.1). Equation (8.3.2) is equivalent to (cf. (7.1.10))

$$\alpha F_T = A_1(\alpha\Delta T)^{\frac{4}{3}}, \qquad (8.3.3)$$

where A_1 has the dimensions of velocity. It is to be understood that f_1 and A_1 can depend on molecular properties but once the two components are specified, they should be functions only of the density ratio $R_\rho = \beta\Delta S/\alpha\Delta T$.

The deviations of A_1 from the value A obtained for solid boundaries with only a difference ΔT between them are a measure of the effect of an increasing ΔS on F_T. Measurements by Turner (1965) in the heat–salt system, which were compared with the 'solid plane'

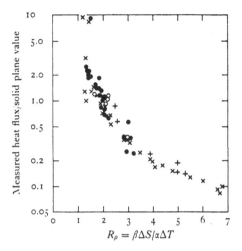

Fig. 8.14. The measured heat flux across the interface between a hot salty layer below a cooler fresher layer, compared with the value calculated for solid plane boundaries. This ratio has been plotted against the density ratio R_ρ; the different symbols refer to experiments with different heating rates at the lower boundary. The three points at the top of the figure are now judged to be unreliable. (From Turner 1965.)

values in this way, are shown in fig. 8.14. They confirm that the ratio A_1/A can indeed be expressed as a function of R_ρ alone. For $R_\rho < 2$, the heat flux is greater than it would be above a solid plane (i.e. if a thin conducting foil were inserted at the centre of the interface). This increase can be associated with two effects: the interface behaves more like a free surface, so that there is a weaker constraint on the horizontal motion, and it also supports waves which can break, so increasing the effective surface area. When $R_\rho > 2$ the heat flux falls progressively below the solid plane value as R_ρ is increased, reaching about a tenth of that value when $R_\rho = 7$. Huppert (1971) has suggested that over the whole of the measured range, the empirical functional form

$$A_1/A = 3.8(\beta\Delta S/\alpha\Delta T)^{-2} \qquad (8.3.4)$$

fits the observations to the experimental accuracy.

Similar arguments can be used to express the salt flux as a function of Ra and $R_\rho = \beta\Delta S/\alpha\Delta T$, which will have a form like (8.3.2) independent of d. It follows that the ratio of the fluxes, expressed in

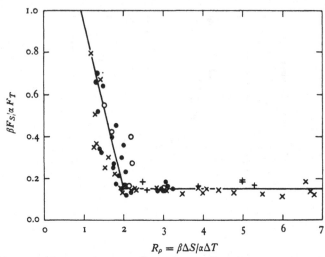

Fig. 8.15. The ratio of the potential energy changes due to the transfer of salt and heat across an interface between a hot salty layer below a cooler fresher layer, plotted as a function of the density ratio R_ρ. (From Turner 1965.)

terms of their respective contributions to the density (or potential energy) flux will be

$$\beta F_S/\alpha F_T = f_*(\beta\Delta S/\alpha\Delta T). \qquad (8.3.5)$$

For given molecular properties, the flux ratio should be another systematic function of R_ρ, and this is again borne out experimentally as shown in fig. 8.15. As $R_\rho \to 1$, $\beta F_S/\alpha F_T \to 1$ also, suggesting (in agreement with our interpretation of the increased heat flux in this range) that heat and salt are here being transported by the same turbulent motions at a breaking interface. As R_ρ increases to 2, the flux ratio falls rapidly, since molecular processes become increasingly important. The most striking feature of the results is that for $R_\rho > 2$ the flux ratio remains constant to within the experimental error, the mean value for salt and heat being $\beta F_S/\alpha F_T = 0.15$. Note that when these results are combined with (8.3.4) they imply that the maximum net density flux for a given ΔT occurs at $R_\rho = 2$.

Shirtcliffe (personal communication) has carried out the corresponding measurements for a diffusive interface between two concentrated solutions at the same temperature (salt above sugar). These confirm the constancy of $\beta F_S/\alpha F_T$ over a range of R_ρ, his measured value for NaCl and sucrose being 0.60 ± 0.02. Using an elegant optical technique, he also deduced profiles of the two components through the interface. He showed that the fluxes of each of them in his experiment could be explained completely by ordinary molecular diffusion through the measured gradients (using diffusion coefficients corrected for the presence of the other property). This deduction makes the name 'diffusive interface' a particularly suitable and descriptive one.

There is as yet no completely satisfactory theory which reconciles these last two results, though it is clear from both the laboratory and numerical experiments that the adjustment between the stable, diffusing interface and convecting layers must take place in a thin boundary layer separating them. A simple model of this layer has been proposed by Rooth (quoted by Veronis (1968)), which has features in common with that of Howard (1964) for a single component (see §7.3.1). Starting with sharp steps of both properties $((\Delta T)_*$ and $(\Delta S)_*$ say), T and S boundary layers will grow by diffusion to thicknesses proportional to $\kappa^{\frac{1}{2}}$ and $\kappa_S^{\frac{1}{2}}$. If these are swept away intermittently by the large scale convecting motions, then the ratio of the amounts of buoyancy removed (i.e. the ratio of the fluxes) will be

$$\frac{\beta F_S}{\alpha F_T} = -\frac{\beta(\Delta S)_*}{\alpha(\Delta T)_*}\tau^{\frac{1}{2}}. \qquad (8.3.6)$$

Under the conditions envisaged in Veronis' calculation (§8.1.4), $\alpha(\Delta T)_*$ and $\beta(\Delta S)_*$ are the whole differences $\alpha\Delta T$ and $\beta\Delta S$ imposed at the boundaries and are equal when $Rs/Ra \to 1$; it follows that $\beta F_S/\alpha F_T \to -\tau^{\frac{1}{2}}$ in that case. The same argument will apply when the differences are unequal, provided one adds the assumption that only the statically unstable edge of a diffusive interface will be swept away, i.e. down to the level where $\alpha(\Delta T)_* = \beta(\Delta S)_*$, even when $\beta\Delta S > \alpha\Delta T$. The constancy of the flux ratio certainly implies that there is some kind of self-limiting mechanism which makes the convecting layers insensitive to the total differences.

The measured flux ratios for salt–heat and sugar–salt are roughly in agreement with this prediction, though the correspondence is not exact and the details of the balance mechanism are probably more complicated. Corresponding measurements have also been made using layers containing several solutes S_x with different molecular diffusivities κ_x, which are driven across an interface by convection induced by heating from below. Turner, Shirtcliffe and Brewer (1970) have shown that the individual 'eddy transport coefficients' (defined as the flux divided by the mean gradient) are different, and approximately proportional to $\kappa_x^{\frac{1}{2}}$. This is consistent with (8.3.6) only if $(\Delta S)_*$ for each stabilizing component is proportional to the *whole* difference of each property across the interface; this result too awaits a detailed theoretical explanation.

Linden (1971 b) has given a mechanistic argument which puts some numbers into the explanation of the increase in the flux ratio at lower values of R_ρ. As the density ratio is reduced, the direct entrainment of both properties across the interface becomes increasingly important relative to the intermittent breakdown mechanism. The buoyancy flux through the interface can be used to estimate the turbulent velocities in the layers (cf. §5.1.3), which are then related to the entrainment through the results of the mechanical stirring experiments described in §9.1.

8.3.2. *The time-history of several convecting layers*

Using experimental data for the two fluxes across an interface (such as those summarized in figs. 8.14 and 8.15 for a 'diffusive' salt–heat system), it is straightforward to calculate the changes with time of T and S in the individual layers. Consider first the simplest case of two layers, the upper depth d_1 and the lower of depth d_2, convecting at high Rayleigh number because of a double-diffusive flux across the bounding interface and with impermeable top and bottom boundaries. The fluxes are given in terms of either the rate of change of properties or the differences between the layers by

$$\left. \begin{array}{l} \alpha F_T = d_1 \alpha \dfrac{dT_1}{dt} = -d_2 \alpha \dfrac{dT_2}{dt} = A_1 (\alpha \Delta T)^{\frac{4}{3}}, \\[2mm] \beta F_S = d_1 \beta \dfrac{dS_1}{dt} = -d_2 \beta \dfrac{dS_2}{dt} = f_* \alpha F_T, \end{array} \right\} \qquad (8.3.7)$$

and

where A_1 and f_* are functions of $(\beta \Delta S / \alpha \Delta T)$, which are in principle known. Rearrangement gives two equations for ΔT and ΔS, from which T_1, S_1 and T_2, S_2 can be deduced as functions of time.

Because the driving energy comes from the T distribution, f_* must be less than one, and the net density difference $\Delta\rho/\rho = \beta\Delta S - \alpha\Delta T$ will *always* be increasing with time. When f_* is small the density difference increases most rapidly, since $\beta\Delta S$ remains little changed while the destabilizing $\alpha\Delta T$ is decreasing. As f_* approaches unity, the two fluxes are more nearly compensating, and there will be only a small change in $\Delta\rho$, but always in the stable sense. Both transports will fall to zero as $\Delta T \to 0$, i.e. when the driving component is nearly uniformly distributed in the vertical, but in this 'final state' $\Delta\rho/\rho = \beta\Delta S$ is non-zero and stable.

These points may seem obvious, but confusion can arise in defining the 'state of no motion' if they are not properly appreciated, and a closely related problem has led to a controversy in the astrophysical literature. Some large stars have a helium-rich core, which is heated from below and therefore convecting. Outside this core, lighter hydrogen predominates, but helium can be transported upwards by the convective stirring. Spiegel (1969) has pointed out that this is also a double-diffusive system, with the abundance of helium replacing salinity as the property of lower κ. The basic assumption used in computing some stellar models is that the convection outside the core can regulate itself to remain in a state of near convective neutrality, a process which has been called 'semiconvection'. From the arguments given above, it is clear that the criterion for convective neutrality must be that of uniform (potential) temperature, i.e. uniformity of the *driving* component, as assumed by Schwarzschild and Härm (1958). Uniform potential density is not the right assumption, since motions can persist when this increases with depth and appears hydrostatically stable.

The behaviour of a three-layer system, consisting of two effectively infinite layers between which ΔT and ΔS are specified (and unchanging) and which are separated by a finite layer with intermediate properties, has been investigated by Huppert (1971). He has proved that when $f_*(\alpha\Delta T/\beta\Delta S)$ is constant (e.g. when $R_\rho > 2$ in the heat–salt experiments), there is a range of combinations of

$$T = \tfrac{1}{2}\Delta T(1 + \theta) \quad \text{and} \quad S = \tfrac{1}{2}\Delta S(1 + \sigma)$$

of the middle layer which are in stable equilibrium. That is, under these conditions the fluxes of both properties across the two inter-

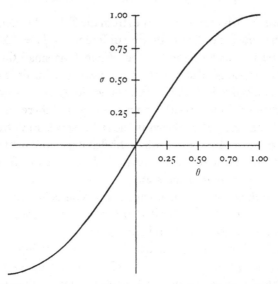

Fig. 8.16. The possible equilibrium states of an intermediate layer of temperature $T = \frac{1}{2}\Delta T(1 + \theta)$ and salinity $S = \frac{1}{2}\Delta S(1 + \sigma)$ lying between two semi-infinite layers whose properties differ by ΔT and ΔS. Note that the same steady fluxes of heat and salt can be maintained across the upper and lower interfaces even when the temperature and salinity steps are unequal. (From Huppert 1971.)

faces are equal, and layers started with other values of T and S will tend towards some point on the equilibrium curve. When f_* is not constant, the only equilibrium point occurs when T and S are at the mean values of the deep layers; even this is unstable, and the central layer will in time approach the density of one of the others and merge with it. Detailed results have been obtained for heat–salt diffusive interfaces in this geometry, and fig. 8.16 shows the 'equilibrium' curve calculated using the data from figs. 8.14 and 8.15. These results can be immediately generalized to apply to any number of interfaces and layers.

8.3.3. Fluxes and structure at a 'finger' interface

Because of the symmetry between the two systems remarked on in §8.2.2, many of the overall properties of a 'finger' interface can be described using a minor modification of the dimensional argument outlined in §8.3.1 for the diffusive interface. It seems logical (but not essential) to emphasize still the difference in the driving component across the interface (now ΔS), so that (8.3.3) is replaced by

$$\beta F_S = A_2(\beta\Delta S)^{\frac{4}{3}}. \qquad (8.3.8)$$

Again the factor A_2 depends on molecular properties, but when these are given, it is a function of the density ratio $R_\rho{}^* = \alpha\Delta T/\beta\Delta S$ (which is written here in the inverse sense so that again $R_\rho{}^* > 1$ denotes static stability). This form again receives support from laboratory experiments (Turner 1967). In the heat–salt system, the factor A_2 ($\approx 10^{-1}$ cm/s) is, however, about fifty times as large at $\alpha\Delta T \approx \beta\Delta S$ as it would be if the same constant salinity difference were maintained between solid boundaries; A_2 falls with increasing $R_\rho{}^*$ but is still about fifteen times this value at $R_\rho{}^* = 10$. A 'finger' interface is a much more efficient means of injecting a salt flux into a convecting layer than is pure diffusion.

The ratio of the fluxes can be expressed in the same way (cf. (8.3.5)) as

$$\alpha F_T/\beta F_S = f_*(\alpha\Delta T/\beta\Delta S). \qquad (8.3.9)$$

Experimentally, f_* has again been shown to be constant over a wide range of $R_\rho{}^*$. Turner (1967) obtained the value $\alpha F_T/\beta F_S = 0.56$ for heat–salt fingers over the range $2 < R_\rho{}^* < 10$, but Linden (1971 b) has since made direct measurements of the velocity in the fingers which can be used to estimate the fluxes (see (8.3.15)). The results suggest that the conditions in the earlier experiments were unsteady, and that the steady-state flux ratio may be much smaller. Preliminary experiments with layers of sugar over salt solution (Stern and Turner 1969) give a flux ratio of 0.91 in the range $1.05 < R_\rho{}^* < 1.6$, and fingers are clearly a very effective means of transporting both components in the vertical. The value of A_2 in this second case ($\tau \approx \frac{1}{3}$) was $A_2 = 10^{-2}$ cm/s, about a tenth of the value for heat–salt fingers.

As yet there is no detailed quantitative theory for the dependence of F_S and F_T on τ. Quite a lot is known, however, about the structure of the fingers through the interface, and this information can be used to support mechanistic arguments which describe the fluxes in an alternative self-consistent way. The 'shadowgraph' picture of fig. 8.17 pl. XXI obtained by shining a light beam horizontally through an interface containing sugar–salt fingers, shows them to be a regular array of long convecting cells, sharply limited above and below at the edges of the interface, and the same structure has been observed for heat–salt fingers. This is even more graphically illustrated using a colour Schlieren system devised by Shirtcliffe (1972), in which the colour of the image depends on the direction in which light is bent on passing through the tank. This technique shows that the mean density gradient is stable through most of the interface containing fingers, but there is a thin unstable region just at the edges where the fingers break down and feed an unstable buoyancy flux into the convecting layer. A vertical beam of light shone through a thick interface produces a plan view of the fingers (fig. 8.18 pl. XXI) which shows that they tend to be square in cross section, each heavy downward-moving element being surrounded by four in which the motion is upwards (Shirtcliffe and Turner 1970).

With these experimental facts in mind, a region containing steady fingers can be described by the following momentum and diffusion equations:

$$\left.\begin{aligned} \nu\nabla_H{}^2 w &= g(\beta S' - \alpha T'), \\ w\,\partial\bar{T}/\partial z &= \kappa\nabla_H{}^2 T', \\ w\,\partial\bar{S}/\partial z &= \kappa_S\nabla_H{}^2 S', \end{aligned}\right\} \qquad (8.3.10)$$

where $\nabla_H{}^2$ denotes $(\partial^2/\partial x^2 + \partial^2/\partial y^2)$, T' and S' are deviations from the horizontal mean and $\partial\bar{T}/\partial z$ is the mean vertical gradient of T. (The close relationship between the physics of salt fingers and buoyancy layers, already remarked on in §8.2.3, becomes apparent when (8.3.10) are compared with (7.4.5) and (7.4.6).) These equations have solutions of the form

$$(w, T, S) = (0, \bar{T}(z), \bar{S}(z)) + (-w_*, T_*, S_*)\sin\frac{x}{L}\sin\frac{y}{L} \quad (8.3.11)$$

provided the amplitudes satisfy relations which are conveniently expressed in terms of the flux ratio

$$F_R = \alpha F_T/\beta F_S = \alpha T_*/\beta S_* = \frac{\kappa_S\alpha\partial\bar{T}/\partial z}{\kappa\beta\partial\bar{S}/\partial z} \qquad (8.3.12)$$

as
$$\frac{g\alpha(\partial\bar{T}/\partial z)L^4}{4\kappa\nu} = \frac{F_R}{1 - F_R} \qquad (8.3.13)$$

and
$$w_* = gL^2\beta S_*(1 - F_R)/2\nu. \qquad (8.3.14)$$

Notice that (8.3.13) has the form of a Rayleigh number, which is of order one when $F_R \approx \frac{1}{2}$, the value suggested by the experiments on heat–salt fingers. In this case the temperature gradient is observed to be nearly linear through the depth of the interface h; and when the density ratio R_ρ^* is large, there is only a small step at the edge, so $\partial \bar{T}/\partial z \approx \Delta T/h$. Thus the width L of the fingers, determined by this large amplitude steady model, is just the same as that found earlier (8.1.13) to be the scale of the fastest growing initial disturbances in a layer of depth h. A constant flux ratio corresponds to a particular value of this Rayleigh number (which again must depend on τ).

In making the above comparison, it has been assumed that the results (8.3.11) to (8.3.14) can be applied within an interface of limited depth, though they are strictly valid only in an unbounded fluid. Now suppose further that S_* is some (specified) fraction of the total step ΔS between the layers and that the system is in a steady state, with the same flux passing through the fingers and the layers. Comparing (8.3.8) and (8.3.14) shows that

$$L \propto (\beta \Delta S)^{-\frac{1}{3}}, \quad w_* \propto (\beta \Delta S)^{\frac{1}{3}}. \tag{8.3.15}$$

We have so far ignored the boundary layer between the finger interface and the convecting layers, where the transition between the two kinds of motion must take place. This is again the least well understood region, though a qualitative description of the processes occurring there is given below.

8.3.4. *The thickness of a 'finger' interface*

The relations (8.3.15) are consequences of two assumptions: that the fluxes should be independent of any external lengthscale (which led originally to the $\frac{4}{3}$ power law), and that all salinity differences are proportional to ΔS (a 'similarity' assumption). They also imply that the interface thickness must be an internally determined quantity, not one which can be arbitrarily imposed: h must in fact be proportional to L, which is clearly consistent with (8.3.13) and (8.3.15) when $\alpha \Delta T \sim \beta \Delta S$. Laboratory measurements in the heat–salt (small τ) case do give values of L consistent with (8.3.13), though there is little evidence to support a simple relation between L and h. Another limit is relevant in the sugar–salt system ($\tau \approx \frac{1}{3}$): as

$\kappa \rightarrow \kappa_S$, the appropriate gradient in (8.3.13) becomes the net density gradient or $\Delta\rho/h$ (cf. the remarks at the end of §8.1.3). During the course of an experiment, $\Delta\rho$ changes little because of the nearly compensating fluxes, so $L \propto h^{\frac{1}{2}}$, a result which is in accord with the observations of Shirtcliffe and Turner (1970).

An equivalent prediction that there should be a 'self-determined' interface thickness has been made by Stern (1969) in an entirely different way. He considered the stability of an established field of salt fingers to disturbances in the form of long internal gravity waves, and showed that when

$$\frac{\beta F_S}{\nu\alpha\,\partial\overline{T}/\partial z} \geqslant \text{const (of order 1)} \qquad (8.3.16)$$

the disturbances will grow. Thus a given S flux can only be carried by finger convection provided the stabilizing T gradient is large enough. On this view, if an interface is too thick for the imposed $\beta\Delta S \sim \alpha\Delta T$, the fingers will be unstable over a considerable fraction of their length, and the interface must thin. It adjusts itself so that the centre of the interface is near the critical state given by (8.3.16), while at the edges, where $\partial\overline{T}/\partial z$ must initially be smaller, the fingers will still be unstable. Large scale convection should rather abruptly take over the transports here and sharpen the edges, as is indeed observed. The identity of this result with that obtained by the similarity argument is apparent if (8.3.16) is interpreted as a critical Reynolds number of the fingers. With $\beta S_* \sim \beta\Delta S \sim \alpha\Delta T$ and $h \propto L$, it follows that $w_* L/\nu$ = constant (or more generally, a function of molecular properties alone), which is also implied by (8.3.15).

Given that the flux is proportional to the gradient $\Delta T/h$ across an interface (which holds, with different multiplying constants, in both the finger and diffusive cases), and that the flux ratio is constant, the equilibrium thickness of an interface separating two layers of thickness d is given by

$$F_T = \tfrac{1}{2}d\frac{d(\Delta T)}{dt} = a\frac{\Delta T}{h}.$$

Thus
$$h = h_0 + \frac{2a}{d}t, \qquad (8.3.17)$$

where a has absorbed several constants of proportionality. This has been obtained without explicitly invoking the flux law (8.3.8), but when that is added (with A_2 constant), it follows that $h \propto (\Delta S)^{-\frac{1}{3}}$, in agreement with (8.3.15). An accurately linear spread of interface thickness in time has been observed in sugar–salt experiments in both the finger and diffusive configurations, so it seems clear that (8.3.17) is not restricted to the similarity (small τ) regime.

When extra mechanical stirring is imposed on each side of a 'finger' interface, the equilibrium is upset and h can even decrease with time. Linden (1971 a) has shown also that the fluxes in the heat–salt case change systematically with increasing stirring, between rates corresponding to undisturbed fingers and a purely mechanical transport (see §9.1). Using the difference in turbulent velocity u' across the interface as a measure of the stirring, the criterion for the disruption of the salt finger regime can be expressed as a particular value of the ratio of u' to finger velocity w_*. Using (8.3.14) this can be written alternatively as an internal Froude number based on u', the density difference and the undisturbed interface thickness h. With a steady shear imposed across an interface, on the other hand, the laboratory evidence suggests that the fluxes are changed rather little. Both a stability calculation and direct observations show that, while the cross-stream modes are heavily damped, the down-stream modes are practically unaltered (cf. §7.1.2). That is, the fingers are changed into sheets aligned down shear which can still transport heat and salt efficiently. (See fig. 8.19 pl. XXI.)

8.3.5. *Convection in a region of variable depth*

All the processes considered so far in §8.3 have been essentially one-dimensional, i.e. they have treated as if they were taking place in a container with vertical walls and constant depth. It will now be shown that the introduction of a sloping boundary can have as important effect: large scale circulations can then be driven by a uniform flux imposed over a horizontal surface. The idea is applicable however this flux is supplied (for example, it could be due to a uniform cooling at the surface), but it arises in the context of double-diffusive convection and is conveniently described here. It is quite different from the weak local effects of a sloping boundary

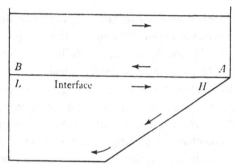

Fig. 8.20. Sketch of the motions produced in a container with a sloping
boundary, by a flux of buoyancy across an interface.

described in §8.2.3; the effect to be considered now is a purely
geometrical one.

The essential point is contained in the equations (8.3.7). The rate
of change of density of a fluid layer, for a given flux, depends inversely on the depth. In the configuration sketched in fig. 8.20 the
fluid at H will become heavier faster than that at L. A horizontal
pressure gradient is set up, which drives a circulation in the sense
shown, including a flow right down the slope to the bottom. At the
same time the flux across the interface at the corner will be reduced
because of the decreased concentration difference there, and the
fluid at A will get lighter more slowly than at B. This sets up a
circulation in the same sense (even with an upper layer of uniform
depth) and there is a strong shear across the interface.

These circulations have been observed in the laboratory with
sugar–salt layers set up in both the diffusive and finger senses. With
a finger interface, the flow down the slope is quite thin and rapid.
The growing pool of fluid at the bottom is stably stratified (cf.
§7.3.3), but shows a notable absence of fingers; in fact a series of
sharp 'diffusive' interfaces is produced, since the relative concentrations are reversed in this region (see fig. 8.21 pl. xxII). With
salt above sugar, on the other hand (a diffusive interface), the bottom
layer and the slope current show a secondary finger effect. A tentative application of the idea has been made by Gill and Turner
(1969) to the formation of bottom water in the Antarctic ocean, where
warm salty intermediate water underlies colder fresher water.

The Antarctic situation is ambiguous, because cooling and

freezing at the surface can also produce a downward flux of buoyancy. But in this case too the depth variation will be important, with the densest water forming on the shallow shelf and tending to flow down the slope (subject to the constraint of the earth's rotation which must be important in the oceanic case). Similarly, any process which increases the density, such as uniform cooling at the surface, could have an important effect on currents at other coastlines, just because of the change in depth through which mixing can occur.

The depth changes need not be smooth for the same argument to apply; a step is equally effective. If a constant buoyancy flux is imposed at the surface of a fluid in a region consisting of a long shallow arm, connected to a deeper reservoir, then a circulation is set up with an inflow at the surface and an outflow along the bottom of the shallow region. One can in this way realize in the laboratory a flow which resembles the similarity solution described in §5.3.4 (see fig. 5.14).

CHAPTER 9

MIXING ACROSS DENSITY INTERFACES

In this chapter we will consider the processes of formation of mixed layers by externally driven turbulence, and entrainment of fluid across the interfaces bounding such layers. Mixing across a density interface has so far been treated only in the special context of double-diffusive convection (§8.3), but more general stirring mechanisms must now be discussed. The problems of interest here may be identified with cases (d) or (e) of fig. 4.19, in which stirring at one level is used to produce mixing across an interface located some distance below or above the source of turbulent energy.

Various laboratory experiments which have shed light on the mechanism of entrainment at a density interface will be described first. Mixing can be generated by mechanical stirring, or by the production of a mean turbulent flow in the surface layer, and both methods have been used. The observed structure of the surface layers of the ocean and atmosphere discussed in §10.1 suggest that mixing in those regions is dominated by the boundary processes. The laboratory results can immediately be applied to these geophysical examples, and, with the addition of results from the earlier chapters on convection, they can be extended to take account of convective as well as mechanical mixing. It is assumed throughout this chapter that the mixing processes under consideration can be treated as one-dimensional in depth, implying that mixing is uniform (in the mean) over a large horizontal area, with no significant contribution from large scale lateral convection.

9.1. Laboratory experiments
9.1.1. *Stirring with oscillating grids*
A convenient way to study one-dimensional mechanical mixing in a stably stratified fluid is to use a vertically oscillated grid of solid

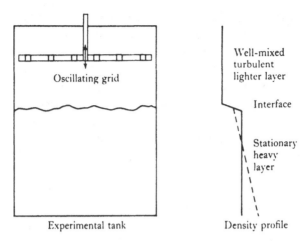

Oscillating grid

Experimental tank

Well-mixed
turbulent
lighter layer

Interface

Stationary
heavy
layer

Density profile

Fig. 9.1. Sketch of the experimental tank and stirring grid, and the density distribution produced by stirring.

bars (see fig. 9.1). Rouse and Dodu (1955) first applied this method in a two-layer experiment, and Cromwell (1960) stirred at the top of an initially stable density gradient. In both cases, a well-mixed layer is formed, bounded by a *sharp* interface which moves away from the stirrer as fluid is entrained across it from the quiescent fluid into the stirred layer. Thus the external stirring produces and maintains the well-mixed layer, and also sharpens the interface and causes the mixing across it. (This geometry corresponds to case (*e*) of fig. 4.19.)

The nature of the entrainment process is shown in the two photographs of fig. 9.2 pl. xxiii. When the stirred layer is dyed, one can see that large eddies of the imposed motion are thrusting into the unstirred fluid and trapping some of it between them, whereas smaller scale motions are rapidly damped. (In a uniform density gradient the larger eddies would be damped first – refer to the discussion of the 'buoyancy subrange' in §5.2.2, and the experiments of Grigg and Stewart (1963). The new point here is that the eddies which dominate the entrainment process are larger than the thickness of the interface and are outside it, so they are substantially unaffected by the interfacial density gradients.) On the other hand, when the unstirred fluid is dyed (fig. 9.2*b*) thin sheets of entrained fluid are at first visible, before being rapidly mixed through the whole of the stirred layer. There is too little evidence to judge yet whether this entrainment process is a slightly modified version of that occurring at the edge of a turbulent plume or jet (§6.1.1) or whether the density step introduces an essentially different mechanism.

There have been few detailed measurements of interfacial structure in these experiments, though visual observations give some important clues. The instantaneous thickness of the interface decreases as the stirring rate is increased, but this thickness depends only very weakly on the property being transferred. Thus the transfer cannot be described as a simple balance between molecular diffusion and large scale stirring, and indeed it can be shown that under typical laboratory conditions an interface sharp enough to transfer even the measured heat fluxes by pure diffusion will be unstable to the shear instability mechanism described in §4.1. The mixing takes place largely through a process which looks like the intermittent breaking of steep forced internal waves, which tends to thicken the interface, followed by the sweeping away of this fluid by the stirring in the layers, which sharpens the interface again.

Most quantitative measurements of mixing rates have been interpreted using arbitrary overall measures of length and velocity scales, derived from the geometry and stirring frequency. Thompson (1969) has gone a step further, and has shown how one can relate the entrainment rates to the properties of the turbulence. Using a hot-film technique he measured an integral lengthscale l_1, and the r.m.s. value u_1 of the horizontal component of the turbulent velocity, as functions of the distance z from the grid in the same apparatus which had been used earlier (Turner 1968b) to measure entrainment rates. Over a comparable range of conditions, and using the same grid of square bars, Thompson showed that u_1 remains proportional to the stirring frequency while the turbulence decays rapidly with increasing z, and that l_1 increases linearly with z. (With other geometries of stirring grid, however, the relations were not so simple, and this can lead to ambiguities in the interpretation of mixing rates in other experiments. See §9.1.3.)

Thus one can now be confident that the entrainment rates measured in Turner's experiments have been related to scales which are not arbitrary, but are characteristic of the fluid motion near the interface. By a similar argument to that used in §6.2.1 for a

Limit at
zero Ri_0 ←

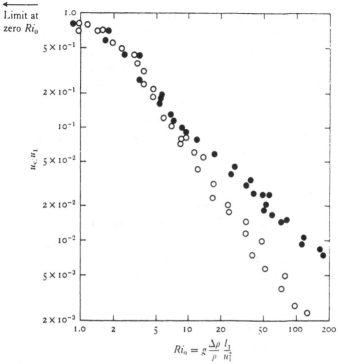

Fig. 9.3. A comparison between entrainment velocities produced by a stirring grid, across density interfaces formed with ● temperature differences and ○ salinity differences. The dimensionless mixing rates obtained by Turner (1968 b) are plotted logarithmically against an overall Richardson number, using the length and velocity scales l_1 and u_1 suggested by Thompson (1969). These are genuine flow parameters, though their numerical values are subject to revision.

turbulent gravity current, we can suppose initially that the mixing process represents a balance between buoyancy and inertia forces alone. The entrainment velocity u_e will then be a function of l_1, u_1 and the density difference $\Delta\rho$ between the layers, which can be expressed as

$$u_e/u_1 = f(Ri_0),\qquad (9.1.1)$$

where $$Ri_0 = g\Delta\rho l_1/\rho u_1{}^2$$

is another overall Richardson number. The experimental results (with stirring in one layer) shown in fig. 9.3 support the choice of Ri_0 as the major governing parameter. As $Ri_0 \rightarrow 0$ and the buoyancy effects become negligible, u_e again approaches a constant fraction of

u_1, and u_e falls as Ri_0 increases. Notice, however, that the functional forms shown on fig. 9.3 for the rates of mixing with temperature or salinity differences across the interface are *different*, so there is also a molecular effect, which will be discussed further in §9.1.3.

When *both* layers are stirred in similar experiments, it is found that the mixing rate (in one direction) is not substantially changed (though now, of course, an indirect measure of u_e must be used, based on the rate of change of properties in the layers). This implies that not only is the turbulence very strongly damped at an interface, but the events which cause the removal of fluid are so rare that the two sides can be regarded as statistically independent. When the stirring is applied at the same rate in the two layers, the interface remains sharp and central. If stirring is unsymmetrical, the interface moves away from the region of more vigorous stirring until the entrainment rates on the two sides balance. Superficially, the two-layer system with stirring on both sides of an interface has much in common with the double-diffusive interfaces treated in §8.3, since in both cases the interface is kept sharp by turbulence in the layers on each side. The latter are complicated, however, by the double boundary layer structure and by the fact that the stirring is through convection driven by the interfacial flux; no theoretical framework which can include both of these has yet been suggested.

9.1.2. *Mixing driven by a surface stress*

Grid stirring is experimentally convenient, and it will also be relevant for modelling natural processes in which turbulent energy is put in on a scale much smaller than the layer depth (for example, by the breaking of waves at the sea surface; see §9.2.1). It does impose a rather special structure characteristic of the grid, however, and various experimenters have suggested other techniques for producing a turbulence structure which is related to the geometry of the flow alone. As we have seen earlier, turbulence is often produced by a boundary stress, either at the bottom (as in §6.2 – see fig. 6.9), or by the action of a wind on the surface of the sea (§5.3.2), and in both these cases there will be a mean flow as well as turbulence.

Kato and Phillips (1969) have reported a laboratory experiment designed to model this latter process directly. To avoid end effects

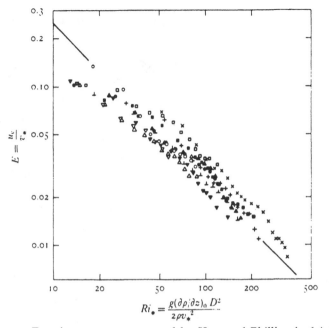

Fig. 9.4. Entrainment rates measured by Kato and Phillips (1969) in a turbulent stratified flow produced by a surface stress. The overall Richardson number is defined using the friction velocity and the depth of the mixed layer, and different symbols are used for experimental runs with various initial density gradients and surface stresses.

they used an annular tank containing salt solution with a linear density gradient, and applied a controlled constant stress τ at the top by rotating a plastic screen immersed just below the surface. During the first few seconds of motion they observed a regular train of waves, aligned across the flow, arising from a shear instability of the kind discussed in §4.1. These quickly broke down, and a turbulent layer formed, bounded below by a sharp interface which advanced at a decreasing rate and across which the density difference was increasing with time. (Compare with the layer growing by convection, fig. 8.8.)

Their measurements of the layer depth D as a function of time, together with the known initial gradient and the imposed stress, can be plotted non-dimensionally in the form shown in fig. 9.4. The dimensional argument which led to (9.1.1) now suggests that

$$E = u_e/v_* = f(Ri_*), \qquad (9.1.2)$$

where the friction velocity in the water (defined by $\tau = \rho_0 v_*{}^2$) is used as the velocity scale and

$$Ri_* = \frac{g\Delta\rho D}{\rho_0 v_*{}^2} = \frac{g(\partial\rho/\partial z)_0 D^2}{2\rho_0 v_*{}^2}. \qquad (9.1.3)$$

The observations are not inconsistent with this prediction, but the scatter is considerable and the range of E rather small, so there is some uncertainty in fitting a curve to these points.

A particular form of this function is suggested by a rather more explicit dimensional argument than the one used above, which also has an attractive physical interpretation. If $\Delta\rho$ and u_e are supposed to be separately unimportant, and to occur only as the product $(g\Delta\rho/\rho)u_e$ (i.e. the buoyancy flux across the interface) then dimensional reasoning gives

$$\left.\begin{array}{l} g\Delta\rho u_e D \propto \rho v_*{}^3, \\[4pt] E \propto Ri_*{}^{-1}. \end{array}\right\} \qquad (9.1.4)$$

or rearranging,

The first form of this relation is equivalent to the statement that the rate of change of potential energy V, say, of the well-mixed layer is proportional to the rate of working of the surface stress. It also implies, incidentally, that the rate of change of density of the layer is independent of the density difference, for a fixed rate of working.

The numerical constant obtained by fitting this form to Kato and Phillips' (1969) experiments over the range $30 < Ri_* < 300$ (see fig. 9.4) implies that

$$\frac{u_e}{v_*} = 2.5 Ri_*{}^{-1} = 2.5\frac{\rho_0 v_*{}^2}{g\Delta\rho D} \qquad (9.1.5)$$

and also that

$$\frac{\mathrm{d}V}{\mathrm{d}t} = \tfrac{1}{2}g\Delta\rho D u_e = 1.25\rho v_*{}^3. \qquad (9.1.6)$$

J. B. Hinwood (personal communication) has found that these (and various other experimental results) fit this simple assumption rather better if the change in potential energy of the whole profile, including the detailed structure of the interface, is used instead of that based on the layer alone (with the interface regarded as a simple step).

Qualitative observations in Kato and Phillips' experiment at much later times (and greater mixed layer depths) showed that viscosity can eventually become important. The rate of increase of depth, and the amplitude of wave motions at the interface, became very small, but a drift velocity was observed below the interface. A layer of increasing depth was set into motion by the viscous stress across it, and at this stage viscosity will also be damping the waves (cf. §2.4.2) and affecting the entrainment rate. The possibility that some momentum is extracted in the form of waves propagating into the stable gradient will be considered in §9.2.

A more elaborate experiment along the same lines has been described by Moore and Long (1971). They produced a steady turbulent stratified flow in a cyclic tank by injecting and withdrawing fluid through the floor and ceiling. As is to be expected in a flow which is driven at the boundaries, two homogeneous layers formed, separated by a thin interface. Experiments with one or both layers flowing were interpreted in a similar overall way to that already described. Detailed profile measurements were also made, and these indicate that the gradient Richardson number through the interface has a value of order one.

9.1.3. *The influence of molecular processes*

An explicit dependence of mixing rate on Reynolds number was proposed by Rouse and Dodu (1955) on the basis of their grid stirring experiments. The more recent measurements suggest, however, that this interpretation may be inappropriate, and a consequence of their use of the grid frequency as a measure of the velocity near the interface, under conditions where this assumption was not justified. In other words, their results could be an indication of a viscous effect occurring near the generating grid, rather than during the decay process or near the interface. Viscosity can certainly affect motions across very stable interfaces at low Reynolds number, but there is a substantial range of conditions where it can be ignored even in the laboratory.

Whatever view one takes about the importance of viscosity, it cannot be used to explain the results shown in fig. 9.3. Since these experiments were conducted with essentially constant viscosity, and the same range of density differences and stirring rates, the large differences in u_e using salt and heat can only be explained by

invoking the molecular diffusivity. A second dimensionless parameter must enter the problem; this can be taken as a Péclet number,
say $Pe = u_1 l_1/\kappa$, so (9.1.1) is replaced by

$$u_e/u_1 = f_1(Ri_0, Pe). \tag{9.1.7}$$

In these experiments u_1 and l_1 were varied little, so the major change
of Pe was due to the change in κ. Effectively Pe had two different
constant values, one for salt and another for heat, and the experimental curves are just two sections of the surface $f_1(Ri_0, Pe)$. These
tend to the same form at low Ri_0 (where neither buoyancy nor diffusion is important) but diverge at larger Ri_0, becoming approximately $u_e/u_1 \propto Ri_0^{-1}$ (heat) and $u_e/u_1 \propto Ri_0^{-\frac{3}{2}}$ (salt).

The apparent correspondence between the results with temperature differences and the 'energy' argument (implied by (9.1.4)) led
Turner (1968 b) to suppose that the Ri_0^{-1} law is in some way fundamental, and that the salt flux is reduced relative to this because of a
slower incorporation of an entrained element into its surroundings
by diffusion. More recent evidence supports the contrary view.
C. G. H. Rooth (personal communication) has used a wider range of
stirring rates, in experiments with temperature differences across
an interface, to show that the basic rate of entrainment in the buoyancy controlled turbulent regime at large Pe (u_1 large) is close to

$$u_e/u_1 \propto Ri_0^{-\frac{3}{2}}$$

in the range studied. As Pe is reduced, the curves break away from
the $Ri_0^{-\frac{3}{2}}$ line at successively lower values of Ri_0, rising above it with
decreased slope as molecular diffusion increases the transfer rate.
(The tendency can also be detected in fig. 9.3.) Baines (1975) has
investigated the mixing across an interface caused by a turbulent
plume or jet directed normal to it, and his results also support the
$Ri_0^{-\frac{3}{2}}$ form.

A mechanistic explanation of the $Ri_0^{-\frac{3}{2}}$ mixing law in terms of
inertial and buoyancy forces alone has recently been proposed by
Linden (1971 b). He has projected vortex rings against an interface
and has observed the characteristics of the depression produced by
the ring over a range of approach velocities and density differences.
His experiments suggest that the important transport mechanism is
the storage of energy in the deflected interface, followed by ejection

of a jet up the centre on a timescale which is determined by the density difference and the scale of the eddies. The increased distortion of the interface at low Richardson numbers is the important factor which had previously been neglected.

At very large values of Ri_0, all the mixing curves flatten out and ultimately become independent of Ri_0. This last limit was studied by Fortescue and Pearson (1967) in the context of gas absorption into a water surface. They showed experimentally that u_e/u_1 is proportional to $Pe^{-\frac{1}{2}}$, and explained this using a theory based on molecular diffusion into the large eddies of turbulence in the water. Their arguments also seem appropriate for a very stable liquid–liquid interface.

9.1.4. *Comparison of various methods of stirring*

Several different mechanisms for the production of the turbulence which is responsible for mixing at density interfaces have now been discussed – surface jets and bottom currents in chapter 6, oscillating grids and a surface stress in this chapter – and it seems useful now to try to compare them. Absolute comparisons are uncertain because of the variety of length and velocity scales used, and it is not possible to judge yet, for example, whether a given level of turbulence is more effective in a mean shear flow than it is in a stirred box. Eventually one must decide how to relate the mean, turbulent and friction velocities in different geometries, but provided all the scales used are genuine flow parameters, much can be learned just from the shape of the curves. On a logarithmic plot of u_e/u_1 against Ri_0, the choice of l_1 and u_1 affects the magnitudes but not the slope.

In fig. 9.5 are plotted all the available results. The scales are appropriate for the grid stirring experiments transferred from fig. 9.3 (full lines, labelled I) but they must be regarded as arbitrary for the other data. The dotted outline II shows the range of entrainment rates obtained by Kato and Phillips and previously plotted in fig. 9.4; this has been placed at the value of Ri_0 where the curves are flattening out on the left. The upper dashed line III is taken from the experiments of Ellison and Turner (previously plotted as the full line in fig. 6.8), which has been fitted so that u_e/u_1 is the same as $Ri_0 \rightarrow 0$ and the curves match at small Ri_0. (Note that this differs

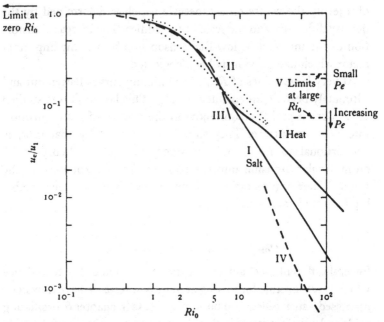

Fig. 9.5. Comparison of entrainment rates as a function of overall Richard-son number for various methods of production of the turbulence. The scales are transferred from the earlier plot of I, but are arbitrary for the other curves. I Grid stirring (fig. 9.3). II Surface stress (fig. 9.4). III Bottom current on steep slope (fig. 6.8). IV Bottom current on low slope (fig. 6.8). v Limits at large Ri_0.

from all the other experiments shown, in that *interfacial* stresses dominate in the production of turbulence.) The lower dashed line IV (due to Lofquist, and also shown in fig. 6.8) is scaled in exactly the same way as III. The limits for large Ri_0 and various values of Pe are shown as the dotted lines on the right, marked v.

When compared in this way, the measurements suggest the following generalizations. In the limit of high Re and high Pe, u_e/u_1 is a function of Ri_0 alone, approaching a constant value as $Ri_0 \to 0$ and having the approximate form $Ri_0^{-\frac{3}{2}}$ at intermediate Ri_0. As Pe decreases, the mixing rate is increased and the curves break away from the high Pe limit at successively smaller values of Ri_0; at large Ri_0, u_e/u_1 approaches a constant value proportional to $Pe^{-\frac{1}{2}}$. The curves III and IV suggest that a similar statement can be made about the effect of Reynolds number. When Re is reduced the mixing rate is also reduced, the break from the high Re curve coming at lower values of Ri_0 the smaller the scale of the experiments. The mechanism whereby viscosity affects the mixing, whether by increasing the rate of decay in the turbulent layer or (more probably) by damping the interfacial waves, awaits a detailed study.

9.2. Geophysical applications

9.2.1. The wind-mixed surface layer

The above results will now be applied to the problems of mixing across oceanic thermoclines and atmospheric inversions. We first consider the case where the stirring is entirely mechanical, and driven by the wind stress at the surface of the sea. The effects of geometry and a stabilizing buoyancy flux must also be taken into account here, so that cases (e) and (f) of fig. 4.19 become relevant, as well as (d). A later step will be to add convective stirring, using some of the results introduced in chapters 7 and 8.

Let us suppose that the mixing at the bottom of the surface layer can be described by (9.1.4), derived above using dimensional reasoning (but also shown to be equivalent to an energy argument). This is a convenient (but not essential) assumption which simplifies the subsequent analysis while retaining all the important qualitative features. If the upper ocean can be described as a two-layer system, with a surface layer of constant total buoyancy and depth D lying above a deep column of uniform density, the rate of change of potential energy V as the layer deepens due to mixing across the interface is

$$\frac{dV}{dt} = \tfrac{1}{2}(gD\Delta\rho)\frac{dD}{dt} = \tfrac{1}{2}bu_e. \qquad (9.2.1)$$

Thus with a fixed rate of input of mechanical energy and no buoyancy flux across the sea surface, the interface will descend at a constant rate, which is inversely proportional to the total buoyancy b. (Other mixing laws, with given v_*, will still give a constant rate of descent depending on b alone, though the functional form will be different.)

With an arbitrary initial density distribution in the water column $\rho(z)$, a relation of the form (9.1.4) (or more generally, (9.1.2)) will still hold, and this can be integrated to give D as a function of t using

$$\bar{\rho} = \frac{1}{D}\int_0^D \rho(z)\,dz \qquad (9.2.2)$$

and $$\Delta\rho = \rho(D) - \bar{\rho}$$

to evaluate the density jump at each stage. For example, with a linear density gradient the energy argument gives

$$D(t) = v_*\left(\frac{15t}{N_0^2}\right)^{\frac{1}{3}}, \qquad (9.2.3)$$

N_0 being the initial buoyancy frequency. Though the stress in this case is applied by a wind blowing over the surface, v_* is still the friction velocity in the water (i.e. defined using the stress and the density of water).

The numerical factor in (9.2.3) has been taken from (9.1.6), but it is most unlikely that the laboratory experiments in question adequately reproduce all the relevant processes. It is observed, for example, that the motion of an oceanic mixed layer can become organized into long rolls aligned downwind (the Langmuir cells), which will result in a stronger stirring on the scale of the whole layer depth. The presence of surface waves (and their suppression in the laboratory) could also have a significant effect on the energy transfer process. The total rate of working of the wind on the water is much larger than $v_* \tau$ (it is more nearly $c\tau$, where c is the phase velocity of the waves) and even if only a small fraction of this energy were eventually used for mixing, it could increase the estimate based on the stress alone. The analysis of some direct observations of the transient deepening of shallow surface layers (Turner 1969b) does suggest that the factor in (9.2.3) might be as much as twice as great.

For very deep wind mixed layers, v_* is not likely to remain the relevant velocity parameter. Turbulent energy put in through the breaking of waves near the surface will have a smaller scale than the layer thickness (cf. fig. 4.19e). It will decay with depth and be less effective for mixing, as does the grid-generated turbulence discussed in §9.1.1. More will be said about deep layers in the next section, in the context of convectively driven mixing.

9.2.2. *Seasonal changes of a thermocline*

An extension of these arguments can be applied to the description of the longer term behaviour of the thermocline, but it is essential then to take into account the surface heating and cooling as well. During periods when the surface is being heated, and both the heating rate and the rate of input of mechanical energy are constant, the depth can remain fixed at a value D_1. This is the result of a delicate balance between two opposing effects (since surface heating tends to make the layer shallower and the stirring to deepen it), and is given by

$$D_1 = av_*^3/(\overline{g'w'}) = av_*^3/B, \qquad (9.2.4)$$

where B is the buoyancy flux (which in general should include the effects of salinity variations cf. (5.3.6)). This form was first suggested by Kitaigorodskii (1960) using dimensional reasoning similar to that which leads to the Monin–Obukov length L (§5.1.2). It is based on the same concepts as (9.1.4), the only difference being the source of the buoyancy flux. Equation (9.2.4) implies that the kinetic energy input, proportional to v_*^3, is being used to change the density of the upper layer by mixing light fluid from the surface through the whole depth D_1 (case (f) of fig. 4.19). The mean value of $a = 2$ which Kitaigorodskii obtained using selected ocean data agrees approximately with Kato and Phillips' measurements, but the choice of a numerical value here and in the extension of this model described below is subject to all the uncertainties mentioned earlier.

When the heating rate is increasing, the thermocline will rise and the upper layer will warm, leaving previously affected layers of fluid below it; as shown by (9.2.4) a given stirring energy can now mix a larger buoyancy input only to a shallower depth. When the rate of heating is decreasing (but still positive) the upper layer will continue to warm, but will begin to deepen slowly as fluid from the gradient set up earlier is entrained across the sharp interface below it. Finally, when the surface is being cooled, there will be a more rapid cooling and deepening of the upper layer. During this stage another assumption must be made, regarding the proportion of the convective energy which is available for entrainment (cf. §§7.3.4, 8.2.1); more will be said about this in the following section.

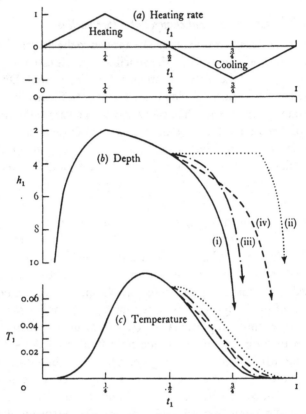

Fig. 9.6. Non-dimensional plots of the mixed layer depth h_1 and surface temperature T_1 as functions of time t_1, calculated for a saw-tooth heating and cooling function and a fixed rate of mechanical stirring. Curves (i) are based on the original energy-conserving assumption of Kraus and Turner (1967); (ii) assume no penetration at all during the cooling phase; (iii) take account of the entrainment due to convection alone, and (iv) include only the entrainment produced by the mechanical stirring. (After Kraus and Turner 1967.)

These ideas have been applied to situations where the surface buoyancy flux varies cyclically, as it does during the course of a year. A completely steady state (with the heat input at the surface exactly balancing the cooling over the year) is unlikely in the ocean without some advection, but this is nevertheless a useful case to study. Kraus and Turner (1967) treated an idealized 'saw tooth' heating and cooling cycle, with the rate of mechanical working

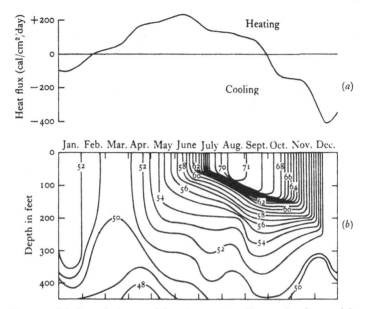

Fig. 9.7. (a) A typical annual heating and cooling cycle observed in an area of small advection (the subarctic Pacific) compared with (b) the variation of temperature with depth in the same region. Contours are marked in °F. (See Tully and Giovando 1963 for a more detailed discussion of this structure.)

assumed fixed. All the features described above can be picked out in fig. 9.6, where the full lines show the annual variations of layer depth and surface temperature calculated with these boundary conditions, using the principles outlined in the next paragraph. A laboratory experiment which modelled the buoyancy fluxes using fresh and salt water also supported the main conclusions of this theory.

The whole of the deepening process in this model can be described using the time-integrated buoyancy and mechanical energy equations

$$\rho_s D - \int_0^D \rho(z)\,\mathrm{d}z = \int_{t_0}^t \frac{\rho_0 B}{g}\,\mathrm{d}t, \qquad (9.2.5)$$

$$\tfrac{1}{2}\rho_s D^2 - \int_0^D \rho(z)\,z\,\mathrm{d}z = \int_{t_0}^t \frac{G}{g}\,\mathrm{d}t. \qquad (9.2.6)$$

These simply express the assumptions that in passing from an arbitrary density distribution $\rho(z)$ to a step of depth D and density ρ_s, the changes in buoyancy and potential energy are equal to the integrated inputs of buoyancy B and effective mechanical energy G during the deepening process. Equation (9.2.6) implies that there is no dissipation, a point which will be taken up again in the following section. Scaling the equations with $G/\rho = G_*$, a characteristic value of B (say B_{max}) and the period P of the heating cycle (a year in the case treated) gives the functional dependence on the external parameters:

$$g\alpha T_{max} = c_1 B_{max}{}^2 P/G_* \qquad (9.2.7)$$

and $$h_{min} = c_2 G_*/B_{max} \qquad (9.2.8)$$

(cf. (9.2.4)). The multiplying constants c_1 and c_2 depend on the exact shape of the heating function; for example, with this scaling, c_1 is 0.12 for a sinusoidal heating function compared with 0.08 for the saw-tooth form.

The predicted behaviour should be compared with the measurements shown in fig. 9.7, which were made in an area where advection is small and the one-dimensional model should be appropriate. Most of the qualitative features are well reproduced, especially the phase relationship between the time of maximum heating and minimum depth (which occur together), and maximum surface temperature (which comes later). The physical assumptions underlying this model are also well supported by the more detailed observations summarized by Tully and Giovando (1963), who have explained them qualitatively in very similar terms. One feature is not predicted by the model described above, namely, the change in depth with time of the isotherms below the well-mixed layer. This implies continuing mixing in this deeper region, which must be due to energy propagating beyond the interface, in the form of internal gravity waves. Direct evidence for the process is also to be found in the observations, and it will be discussed further in § 10.3.3.

Also drawn on fig. 9.6 are the results of varying the energy conservation assumption in various ways during the cooling cycle only. The dotted lines (ii) show what happens when there is *no* penetration of the interface, driven either mechanically or convectively; the surface layer just cools without deepening until the density step is removed, then it deepens in a way which can be calculated using the heat balance alone (see §9.2.3). (This last assumption is supported by the work of Anati (1971), who showed

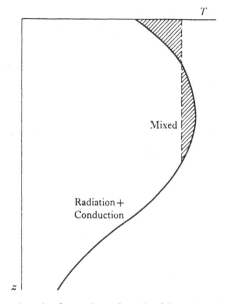

Fig. 9.8. Illustrating the formation of a mixed layer by the absorption of radiant heat in the interior combined with cooling due to evaporation at the surface. A heat balance alone has been assumed here, with no extra entrainment at the bottom of the mixed layer.

that the growth of a very deep well-mixed layer in the Mediterranean is described well using a non-penetrative model based on the density distribution and the net buoyancy flux. In that area in winter, vigorous mixing is produced by rapid surface cooling resulting from strong winds.) The dashed lines on fig. 9.6 show the penetrative effects of the two components, supposing that they can be considered separately. It is clear that whatever assumption is used, surface cooling and not mechanical mixing eventually has a dominant influence on the deepening process, though the time at which the deepening occurs does depend on the other factors.

In some cases it is essential to allow for stirring due to the absorption of radiated heat in the interior. Kraus and Rooth (1961) showed that the depth of a convective layer depends largely on the balance between the interior heating and the surface cooling due to evaporation; this effect too can be combined with mechanical stirring. Again one has to make an assumption about the fraction of the kinetic energy generated by the convective overturning which is used in entraining heavier fluid from below. The evidence in the laboratory and in lakes (Dake and Harleman 1966) and in the ocean (Defant 1961, vol. 1, ch. 3) suggests that this fraction is small. Thus the mixed layer depths in the absence of wind can be calculated using the heat balance alone, and replacing the hydrostatically unstable parts of the temperature profiles (which would otherwise result from the combination of radiation and conduction) by a well mixed layer with the same heat content and no temperature step below. (See fig. 9.8.) A model which develops this same idea for the diurnal temperature cycle has recently been proposed by Foster (1971 a).

9.2.3. Mixing at an atmospheric inversion

Theories of the development of the surface layer in the atmosphere have emphasized the stirring due to convection, though shear-driven turbulence can be important in this case too. The effect of a destabilizing buoyancy flux through a bounding surface is appropriately discussed in more detail in this context, and we can now add to the ideas about penetrative convection which were introduced in earlier chapters.

The results discussed towards the end of the previous section are also very relevant here. The rate of growth of a convecting layer into a region of stable temperature gradient will depend not only on the heat flux, but also on the rate of entrainment across the edge of the layer. The laboratory experiments of Deardorff, Willis and Lilly (1969) described in §7.3.4, and the corresponding experiments using a salinity gradient rather than a temperature gradient (§8.2.1), suggested that this effect is small. If viscous damping were significant in these experiments, then they might underestimate the importance of entrainment on the atmospheric scale, but the oceanic examples already given suggest that one will not go far wrong if the extra entrainment is ignored, and the convective layer depth is calculated using heat balance alone.

The opposite view was taken by Ball (1960), who calculated the variations in the height of an inversion on the assumption that *all* the energy produced by convection is available to entrain fluid across the top of the well-mixed layer. Though the energy-conserving case is still a useful limit to consider, this assumption is certainly unrealistic in its extreme form, and not consistent with what is now known about the convection and mixing processes. If convection occurs in the form of 'thermals', for instance, it was shown at the end of §6.3.1 that only a small fraction of the work done by buoyancy appears as kinetic energy of mean motion, and similar results hold for other convection elements. Thus even if all of this energy on the larger scales which are most effective for entrainment were used for this purpose (and this is by no means certain) there would already be a substantial reduction in mixing rate.

When the individual convection elements are taken into account, we should also bear in mind the 'filling box' mechanism described in §7.3.3. This can produce a weak stable stratification even in conditions of strong convection, an effect which has been observed in the atmosphere by Warner and Telford (1967). (The experiments of Baines (1975) which were mentioned in §9.1.3 include both this effect and the entrainment by a plume at the interface.)

It is again useful to write down some special results, derived from (9.2.5) and (9.2.6) for the case of a linear initial density gradient specified by N_0^2, with various buoyancy fluxes and the two extreme mixing assumptions. The results given in §8.2.1 corresponding to a constant flux B_0 can be rewritten for the present purpose as

$$D = (6B_0)^{\frac{1}{4}}N_0^{-1}t^{\frac{1}{2}}, \quad g\frac{\rho_0-\rho_s}{\rho_0} = (\tfrac{8}{3}B_0)^{\frac{1}{4}}N_0 t^{\frac{1}{2}}, \qquad (9.2.9)$$

where ρ_0 is the initial density at the ground and ρ_s that in the well-mixed layer. The constants given are upper limits corresponding to the assumption of no energy loss; as explained previously, they are reduced, though the power law dependence remains the same, when the heat balance equation alone is used and there is no penetration. Note that this change in depth with constant heating is more rapid than it is with constant mechanical stirring (the time dependence is $t^{\frac{1}{2}}$ instead of the $t^{\frac{1}{4}}$ predicted by (9.2.3)). For a linear density gradient and a linearly increasing flux $B_0 = Ct$ the corresponding solutions are also linear in time, the assumption of no energy loss giving

$$D = (3C)^{\frac{1}{4}}N_0^{-1}t, \quad g\frac{\rho_0-\rho_s}{\rho_0} = \left(\frac{4C}{3}\right)^{\frac{1}{4}} N_0 t. \qquad (9.2.10)$$

The multiplying constants with no penetration are both $C^{\frac{1}{4}}$.

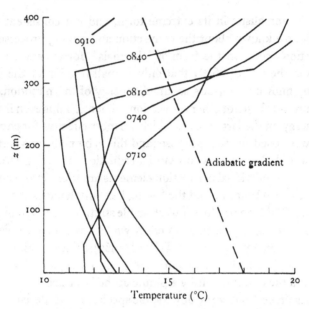

Fig. 9.9. Temperature profiles recorded by Izumi (1964) in the lower atmosphere as a mixed layer grew during the course of a morning. (The numbers on the curves are local times.)

Whatever the form of heating, there is a clear qualitative difference between the predictions in the two limiting cases. With penetration, the temperature at a given level will first fall abruptly as the step produced by entrainment passes a given level, and then gradually rise again. If there is no penetration, and just a change in gradient, the temperature at a fixed height can never decrease. Direct evidence for such a decrease in temperature (and therefore for extra mixing at the top of an atmospheric surface layer) was obtained by Izumi (1964), who followed the growth of a layer to about 400 m as the heat flux increased during two hours in the early morning (see fig. 9.9). Using these measurements, which were made when there was a strong wind, it is not easy to separate convective and mechanical energy inputs, however. It would be worth seeking examples of the convective growth of a well-mixed layer under conditions of no wind, to compare with those described in §9.2.1 for which the mechanical effect alone was believed to be relevant. There also seem to be some discrepancies between results obtained in the atmosphere and ocean, which may result from different mechanisms and scales of turbulent energy generation, but which can only be clarified by further observations of carefully documented natural phenomena.

The same kind of description can be extended to moist convection in the atmosphere, and to much larger scales. Riehl *et al.* (1951), for example, found evidence for entrainment through the trade wind inversion, the mixing elements in that case being cumulus clouds. The inversion acts as a lid for humidity, since the transport of properties by the entrainment process is always downwards into the turbulent layer; the consequent tendency of the inversion to rise is counteracted by large scale subsidence. Lilly (1968) added radiative cooling at the top of a convective layer to the heating below, and studied theoretically both the dry and the moist cases. He, too, used the two extreme assumptions about the conversion of energy, but a detailed discussion of his results would bring in the new effect of heat radiation, which it is not appropriate to pursue here. Note, however, that if losses of heat at the top are allowed, the existence of a step is *not* an indication that penetrative mixing is occurring.

We have emphasized the convectively unstable atmospheric surface layers, but mention should also be made here of some results obtained in stable conditions. Hanna (1969), following a suggestion of Laikhtman (1961), has shown that the observed depth D of the boundary layer can be described in terms of the velocity U_g at its top and the (potential) density gradient through it by

$$D = 0.75 U_g \bigg/ \left(\frac{g}{\rho}\frac{\Delta\rho}{\Delta z}\right)^{\frac{1}{2}}. \qquad (9.2.11)$$

This is equivalent to the statement that the overall Richardson number of all deep stable atmospheric boundary layers adjusts itself to a value of $\frac{1}{2}$ (cf. §§5.3.2 and 10.2.1). It is not as practically useful as (9.2.4) (since it involves a derived density gradient which is not easily observed), but (9.2.4) and (9.2.11) can be regarded as alternative 'flux' and 'gradient' forms resulting from related physical arguments.

9.2.4. *Other factors limiting the depth of a mixed layer*

So long as the turbulence near an interface is maintained in some way, the inviscid entrainment processes described above would continue to cause a well-mixed surface layer to deepen indefinitely. One way of counteracting this tendency and producing a steady state has been mentioned – large scale upwelling (in the ocean) or subsidence (in the atmosphere), with an associated divergent flow in the layer itself. For these geophysical flows, however, a steady state also becomes possible because of two factors not yet considered. It would be wrong to leave the impression that a simple entrainment model is always the relevant one to apply, so a few qualitative remarks should be made about each of these processes.

For the first time in this book, we must introduce explicitly the effect of the earth's rotation. In a rotating system, the influence of a boundary stress does not penetrate indefinitely, but is limited to the Ekman layer, in which it is entirely balanced by the Coriolis forces. (See, for example, Phillips (1966a, p. 228) for a fuller discussion.) Both the direction and magnitude of the mean velocity vary with depth, in a manner which depends in detail on the mixing, but in a rotating fluid of constant density (5.1.1) can be replaced by

$$\frac{d\mathbf{u}}{dz} = \frac{u_*}{z}\mathbf{h}\left(\frac{fz}{u_*}\right), \qquad (9.2.12)$$

where f is the component of angular velocity about a vertical axis, u_* is the friction velocity, and $\mathbf{h}$ is some function of the non-dimensional parameter in brackets. In the rotating case, therefore, with a boundary stress but negligible buoyancy flux, the lengthscale $L_e = u_*/f$ ('the depth of the Ekman layer') becomes relevant instead of the Monin–Obukov length L_b defined in (5.1.8) and implied by (9.2.4).

There is some confusion of these lengthscales, especially in the oceanographic literature where the 'upper mixed layer depth' and 'Ekman layer depth' often seem to be regarded as synonymous. This identification is valid only when $L_b \gg L_e$. Then the forced convection region (in the absence of rotation) would include the whole of the Ekman layer, and so rotation determines the extent to which the surface effects can penetrate. The opposite case, where

$L_{\rm b} < L_{\rm e}$, is the one already considered in arriving at (9.2.4). The penetration of a wind-stirred layer is then limited by buoyancy, before rotation can have an appreciable effect on the momentum flux.

The second factor which should be mentioned is the effect of a stable density gradient outside a well-mixed layer. Under these conditions momentum and energy can be lost from the layer (and therefore become unavailable for local mixing) in another way. They can be radiated away in the form of internal gravity waves, and the most favourable circumstances for this to happen can be found by examining the relations given in §2.2.2. If the forcing frequency ω is too high, waves cannot be generated, and no energy is radiated at all. When $\omega < N$, wave propagation is possible, and Townsend (1965, 1966) has shown that both convective and shear generated turbulence in the atmospheric boundary layer can produce disturbances of the interface which permit radiation. In the latter case most of the wave energy is in components having a phase velocity close to the convection velocity V of the dominant disturbances in the boundary layer, i.e. those which satisfy a 'resonance' condition, and these have a wavelength near $L_1 = 2\pi V N^{-1}$, where N is the buoyancy frequency of the region above the interface (not of the much more stable interface itself). Inserting appropriate numerical values for the atmospheric boundary layer, Townsend showed that wave amplitudes of order 100 m could develop in a few hours.

Detailed calculations for more general environmental conditions (including a shear outside the boundary layer) led Townsend (1968) to conclude that the rate of radiative energy loss from the boundary layer in both the atmosphere and the ocean can be large enough to affect the motion of the layer itself. He showed that the fraction of the energy produced in the outer part of the layer which is radiated away in gravity waves is a strong function of the ratio L_0/L_1, where L_0 is the scale of the turbulence (comparable with the boundary layer thickness), and L_1 has been defined above. The radiated energy is negligible if L_0 is small compared with V/N, but if $L_0 \sim V/N$, the thickness of the turbulent layer is near the upper limit set by the removal of energy in gravity waves. (Compare with (9.2.11).) The numerical values estimated for the atmosphere and ocean are such that the limits set by this wave radiation process and by the earth's rotation are approximately equal in each case.

Wave energy supplied to the interior in this (and other) ways also plays an important part in the transport and mixing processes outside the boundary layer. These will form the subject of chapter 10.

CHAPTER 10

INTERNAL MIXING PROCESSES

The mechanisms responsible for mixing in the interior of a stratified fluid are even less well understood than those described in chapter 9, since the sources of energy are not so obvious, and several different processes must be taken into account simultaneously. The important ideas have already been introduced in earlier sections, but it seems appropriate in this final chapter to take a broader and less detailed view, and to consider together the whole array of mixing phenomena which can be relevant in large natural bodies of stratified fluid. Geophysical examples have figured prominently in this book, and indeed the range of subject matter has been chosen with this final synthesis in mind. The basic facts requiring explanation are somewhat scattered, however, and in order to collect them together and to define the problems to be treated here, a brief summary will now be given of the observed structure of the ocean and atmosphere.

This field is developing rapidly, and the interpretation given here must necessarily be a somewhat tentative and personal one. Nevertheless, it seems useful to sketch how our present knowledge of the separate components can be fitted into a self-consistent picture, at the same time extending some of the earlier arguments so that they can be applied in this wider context.

10.1. The observational data

Routine density profiles made using reversing bottles and thermometers show that the ocean is everywhere stably stratified, except for limited regions where bottom water forms intermittently as the water column becomes convectively unstable. Near the surface, there is characteristically a layer of more nearly uniform density, bounded below by a thermocline, or region of rapid variation of temperature and density (see §9.2). This structure varies on a daily

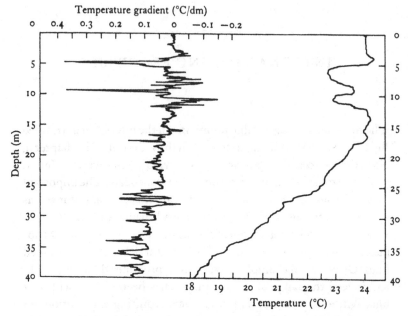

Fig. 10.1. Temperature and temperature gradient profiles recorded in the upper part of the summer thermocline near Malta by J. D. Woods (private communication). Over the depths where there are large temperature inversions (between 3 and 12 m) there must be compensating gradients of salinity since the density must increase with increasing depth, except in localized regions of active overturning.

and seasonal timescale, in a manner which depends on the local wind and heat flux, and on the vertical and horizontal gradients of temperature and salinity. (Heat is both added and removed at the surface of the ocean, which is a very different distribution of sources and sinks from that in the atmosphere.) At a depth of some hundreds of metres, there is a weaker permanent thermocline. Below this again is a region extending to the bottom in which the density gradient becomes progressively weaker with increasing depth. Nevertheless, the 'overall Richardson number' for the deep ocean (based on the geostrophic velocities produced by the measured density differences on the rotating earth) are typically very large, at least of order 10^2.

Recent detailed measurements of temperature and salinity profiles, made with continuously recording instruments in various parts

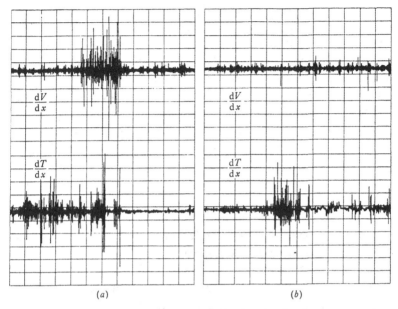

$\dfrac{\mathrm{d}V}{\mathrm{d}x}$

$\dfrac{\mathrm{d}V}{\mathrm{d}x}$

$\dfrac{\mathrm{d}T}{\mathrm{d}x}$

$\dfrac{\mathrm{d}T}{\mathrm{d}x}$

(a)   (b)

Fig. 10.2. Velocity and temperature derivatives in the deep ocean, obtained by towing sensors at about 200 m in the permanent halo-thermocline. The record length represents one minute or a distance of about 100 m. (a) Correlated velocity and temperature microstructure, indicative of active small scale mixing. (b) 'Fossil turbulence', or temperature microstructure unaccompanied by any visible velocity structure. The velocity signals shown are all caused by vibrations in the towing system. (Unpublished data from P. W. Nasmyth.)

of the ocean, have shown that the density distribution is not smooth, but contains significant variations on very small lengthscales even at great depths. Particularly when vertical gradients (or differences over a few centimetres) are recorded, rather than properties at a point, it becomes clear that the vertical density profile is often step-like, with layers of more uniform density separated by 'sheets' or interfaces where the gradients are large. (See fig. 10.1.) This is true not only when double-diffusive processes are important (§8.3.4), but small scale layering can occur even in situations where variations of one property alone determines the density. (Similar records of temperature variations have been obtained in the ocean and in fresh water lakes; see Simpson and Woods 1970.) Measurements of the vertical shear near the surface and in the deeper ocean also have a very non-uniform character, and give further evidence of the step-like structure. It has been shown, for example by Grant, Moilliet and Vogel (1968) in a tidal estuary, and by Woods (1969) in the summer thermocline near Malta, that turbulence in the interior is confined to thin, elongated patches which occur intermittently in space and time. Records of velocity and temperature fluctuations, obtained by towing a sensor through the upper ocean, reveal regions of vigorous small scale turbulence separated by others which are essentially laminar (fig. 10.2a). Patches of 'fossil turbulence', i.e. temperature microstructure remaining after the turbulence has decayed, have also been observed. (See Stewart 1969, and fig. 10.2b.)

Temperature profiles measured using radiosondes in the atmosphere also reveal a mainly stable stratification, but because heat is on average added to the atmosphere at the ground, and removed at higher levels by long wave radiation, the lowest layers are often convectively unstable. Dry convection occurs immediately above the ground, while higher up clouds are the more important mechanism for transporting heat and water vapour into the upper atmosphere. At the top of the surface layer there can be a sharp temperature inversion, implying a very rapid potential density change, or there may be a smoother transition to a stable gradient. The mean gradient increases with increasing height above this level. Again there is much fine structure, and a tendency towards layering, with turbulence occurring intermittently and in patches rather than uniformly through the atmosphere.

It is worth remarking that some of the observed non-uniformities of density gradient may be transient, and associated with internal waves rather than mixing. Even if a region is smoothly stratified, the propagation of a group of waves through it will cause a local squeezing and separation of the surfaces of constant density, which would be interpreted as variations in the vertical gradient. Repeated profiles made through regular layered structures also give evidence of spatial and temporal variations which can be attributed to waves.

It has been suggested (see, for example, Iselin 1939 and Munk 1966) that a major part of oceanic mixing might be due to boundary effects, that is, to turbulence produced at the margin of continents or at the surface, followed by a density-driven flow into the interior.

Large scale advection of this kind could indeed lead to vertical transports and the production of some of the layering which is observed, and both in the atmosphere and in the ocean the motion of fronts will have similar results. The effectiveness of these processes is limited by the horizontal constraint of rotation, however, and the sharpness of the observed gradients and the speed with which the microstructure can change show that local vertical mixing is often important. Only these smaller scale, more rapid processes will in fact be considered in this chapter, and again we use the assumption (already introduced in chapter 9) that they can for the most part be treated as one-dimensional. It would take us too far afield to give a proper discussion of horizontal diffusion and mixing, and the 'shear effect' referred to in §5.3, whereby the horizontal spread is strongly influenced by the interaction between vertical mixing and shear, but there is no doubt that these processes play a significant role. (See Pingree (1971), already mentioned in §8.2.3; Woods and Wiley (1971); and the papers by Okubo (1968, 1970).)

10.2. Critical Richardson number criteria

The central difficulty of the internal mixing problem is associated with the apparently very great stability of the geophysical flows. As was pointed out in §10.1, the overall Richardson numbers (based on the mean density and velocity differences $\Delta\rho$ and ΔU over the depth of the ocean, for instance) are typically very large. How, then, in view of the criterion for the breakdown of a shear flow discussed in §4.1.3, and that for the equilibrium state proposed in §5.1.4, can turbulence ever be produced and sustained in the interior of such a region?

A qualitative answer to these questions can be found by developing the ideas expressed in §§5.1.4 and 6.2.4 and paying special attention to the sources of turbulent energy. Long (1970) has gone further and proposed a detailed theory which combines several of the features discussed below, but it seems preferable here just to describe the essential ingredients of such a solution, rather than to give a particular recipe for their combination.

10.2.1. *Examples of equilibrium conditions*

It is less difficult to understand what is happening when localized processes can be identified which impose the differences $\Delta\rho$ and ΔU over a much smaller lengthscale than the whole depth of the ocean or atmosphere. When a sufficiently large shear is applied across a density interface and is such that the gradient Richardson number falls below a critical value of about $\frac{1}{4}$, Kelvin–Helmholtz waves will grow and overturn to produce patches of turbulent mixing. (The results given in §4.1.3 will be followed up in the geophysical context in §10.3.1.) If the shear is not sustained, this turbulence will decay as the mixed layer spreads out (§5.3.2). If further kinetic energy is provided, however, the arguments of §5.1.4 have shown that there should be an 'equilibrium', self-adjusting value of the gradient Richardson number Ri_e at which turbulence can be maintained. In this state no external lengthscale is relevant, and the mixing region adjusts its thickness to accommodate the imposed $\Delta\rho$ and ΔU.

The earlier applications to wakes (§5.3) and especially to the outer edge of inclined plumes (§6.2.4) were in accord with these ideas, and suggested that Ri_e is of order 0.1 (somewhat smaller than the critical value for the original breakdown). Various observers (see, for example, Browning 1971) have been accumulating evidence in the atmosphere which shows that the structure of frontal zones can be described in a similar way (cf. case (*e*) of fig. 4.19). They have combined Doppler radar measurements of winds with frequent temperature profiles made using radiosondes to deduce values of minimum Ri lying consistently in the range 0.15–0.3 in these regions (fig. 10.3), though Ri is poorly defined elsewhere.

An equilibrium value of Ri can be sustained in this case, sometimes for several hours, because the motion of a front tends to sharpen the gradients and decrease Ri, while mixing acts to increase it again. The numerical values quoted for the atmosphere can only be upper limits, since they were obtained by averaging vertically over 200 m and horizontally over several km diameter (see the discussion relating to (10.2.1)), but the principle of a limiting structure controlled by a Richardson number criterion seems to be confirmed. The regions above and below jet streams in the high atmosphere, and intruding layers in the ocean, should also be

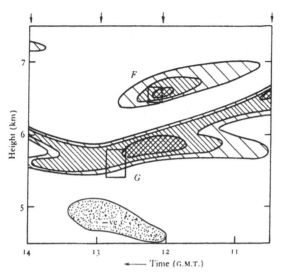

Fig. 10.3. Time–height pattern of Richardson number (defined over layers 200 m deep), which was obtained by Browning (1971) in the vicinity of two 'billow' events. These events were observed by radar and are indicated in duration and vertical extent by the rectangular frames marked F and G; the times of radiosonde ascents are marked by arrows. The increasingly dense cross hatching represents values of Ri of 0.3–0.5, 0.2–0.3 and 0.1–0.2 respectively, and in the stippled region $Ri < 0$.

governed by this criterion. In other situations, such as the atmosphere near the ground where the wind typically 'dies away' at night (as described in §5.3.2), energy cannot be supplied rapidly enough to the shear layer and so the turbulence must decay.

10.2.2. Non-equilibrium conditions: step formation

When $\Delta\rho$ and ΔU are imposed over a given depth Δz, it is most unlikely that the external boundary conditions will exactly match the equilibrium 'internal' criterion, i.e. that

$$Ri_0 \equiv g\Delta\rho\,\Delta z/\rho(\Delta U)^2 = Ri_e.$$

Only in this case, however, could the equilibrium state be maintained, with linear gradients of density and velocity extending through the whole depth, and turbulent transports of buoyancy and momentum independent of position. If $Ri_0 < Ri_e$, the shear will dominate over the density differences, and the mixing will be an 'external' process, controlled directly by the boundary turbulence (cf. case (a) of fig. 4.19). The distance from the boundary z then is a relevant parameter, and one must use the z/L scaling described in §5.1.2. Of most interest here is the opposite situation where $Ri_0 > Ri_e$, which is strongly satisfied in deep layers of the ocean or atmosphere.

In this latter case, the stratification (and therefore internal processes) must certainly dominate, but if 'equilibrium Richardson number' ideas are still to apply, another smaller lengthscale must become relevant. Stewart (1969) has pointed to a way out of this dilemma. If the gradients $\partial \rho/\partial z$ and $\partial u/\partial z$ are very non-uniform, then a constant value of the gradient Richardson number (say Ri_e everywhere) is compatible with *any* value of Ri_0 – essentially because it is always true that

$$(\Delta U/\Delta z)^2 \leqslant \overline{(\partial u/\partial z)^2}. \qquad (10.2.1)$$

If $Ri = Ri_e$ is the preferred state of a turbulent stratified fluid, then this can be achieved by a rearrangement of the density and velocity structure, for any given overall differences.

There are, of course, infinitely many ways of splitting up a given $\Delta\rho$ and ΔU to produce a constant $Ri = Ri_e$ (even if one ignores for the present the patchy nature of the observed distributions, and assumes horizontal uniformity), but a particular example will help to fix our ideas. Let us suppose that initially linear distributions of density and velocity, corresponding to $Ri_0 = \frac{1}{3}$, are to be transformed into 'well-mixed layers' and 'interfaces', subject to the conservation of mass and momentum. In both of these there is to be the same constant $Ri = Ri_e$, and the gradients are assumed to remain linear (but necessarily of very different magnitudes). Adopting for illustration the value $Ri_e = \frac{1}{16}$ suggested by the laboratory experiments described in §6.2.4, fig. 10.4 shows the various forms which can be taken by the velocity and density profiles for different ratios r of interface to total (layer plus interface) thickness H; the absolute value of H is arbitrary. Note that the contrast between the density gradients through the interface and in the layers is much greater than it is for the velocity gradients.

This idea resolves a basic difficulty, and shows that turbulence

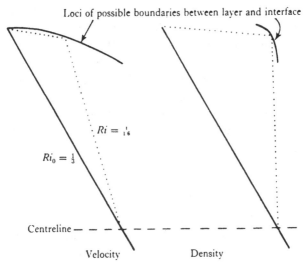

Fig. 10.4. Showing (half) the velocity and density profiles which are consistent with an initial layer Richardson number of $Ri_0 = \frac{1}{3}$, and have a linear 'layer' and 'interface' structure with $Ri = \frac{1}{16}$ everywhere.

need not be inhibited in a stratified fluid simply because the Richardson number is too high; a redistribution of properties can always reduce its value. It does, however, raise other questions which can be answered in principle but not yet in detail. The arguments of §5.1.4 imply that in the 'equilibrium' state, the ratio of the buoyancy flux to the momentum flux is proportional to the velocity gradient (5.1.20). If this gradient is non-uniform, as it must be according to the arguments given above, then the transport of momentum is relatively less efficient at the interfaces where the velocity gradient is larger, and it will be impossible to have steady turbulent fluxes of both properties through the whole depth. One must either abandon the one-dimensional and steady state assumptions (and it is conceivable that the transport processes might be essentially unsteady, with moving interfaces and horizontal spreading playing the dominant roles) or take into account other processes, besides the *turbulent* stresses implied in (5.1.20), which can transport momentum.

For this, and other reasons to be outlined below, one is led to consider the transports due to internal waves (as foreshadowed by the earlier discussion of case (g), fig. 4.19). The generation of interfacial waves provides just the right kind of complementary mechanism to transport momentum across the steep interfacial gradients without correspondingly increasing the buoyancy flux, and a steady state can plausibly be achieved by a combination of these two mechanisms. Thus a stratified fluid with large overall Richardson number can be turbulent at most intermittently and in patches, while a large fraction of its volume will be in a state of laminar, wave-like motion.

10.2.3. *Energetics of a layered system*

There are other physical constraints on the possible non-uniform profiles which it is again convenient to illustrate using the horizontally uniform 'layer and interface' model pictured in fig. 10.4. The condition for the layers to have constant velocity and density, while $Ri = Ri_e$ in the interfaces, is just

$$r = r_0 = Ri_e/Ri_0, \qquad (10.2.2)$$

using the earlier notation. Thus if Ri_e is regarded as given, and reversals of velocity gradients are excluded as implausible in a flow which has developed from a unidirectional shear, the interfaces when they first form will be relatively thinner the larger Ri_0 becomes. (Any horizontal advection which follows the local formation of a layer will of course tend to thicken the interfaces.)

Now consider the energy changes implied by the transition from linear gradients $u = \alpha z$, $\rho = -\beta z + \rho_0$ to a well mixed layer of depth H, bounded by sharp steps above and below. The change in kinetic energy is $T = \frac{1}{24}\alpha^2 H^3$ and in potential energy $V = -\frac{1}{12}g\beta H^3$; the decrease in T just balances the increase in V when

$$Ri_0 = g\beta/\alpha^2 = \tfrac{1}{2}. \qquad (10.2.3)$$

This result may readily be extended to the case where the transition between the well mixed layers takes place over an 'interface' of finite thickness (in which $r = r_0$ and $Ri = Ri_e$, as in (10.2.2) above) to give

$$Ri_0 = \tfrac{1}{2}(Ri_e + 1) \qquad (10.2.4)$$

as the condition for equality of T and V. At smaller values of Ri_0 than this, $T > V$ at $r = r_0$, and $T = V$ for $r < r_0$. Thus for each Ri_e there is an upper limit to Ri_0 such that all the energy for mixing could in principle have come from the *local* kinetic energy. The important point to emphasize is that the *overall* Richardson number Ri_0 is the parameter which determines the stability of the flow in the absence of an external energy supply, and in this sense flows with $Ri_0 \gg 1$ are stable. Note that the 'critical' Ri_0 values implied by (10.2.4) are somewhat larger than the limit $Ri_0 = \frac{1}{4}$ for infinitesimal disturbances found in §4.1.3, since a particular finite change in the distributions has been imposed. (See also Businger (1969) and Hines (1971) for discussions of this point. The latter has also shown that the energy condition for generation of turbulence is less stringent when the velocity vector is changing direction with height.)

A similar consideration of the overall energy balance also sets an upper limit on the thickening of an interface due to local turbulent mixing following the breakdown of an interfacial wave at $Ri < \frac{1}{4}$ (cf. §4.3.3). The profiles can at first become very non-uniform, with turbulence sustained at $Ri = Ri_e$, but when the thickness h of the mixing interface increases to a value such that $Ri_0 = g\Delta\rho h/\rho(\Delta U)^2$ is comparable with unity, the kinetic energy made available by the fluid entrained into the interfacial region will no longer be sufficient to change the potential energy (even if one neglects the reduction in turbulent energy due to viscous decay).

It seems appropriate to comment again here on the common tendency to confuse the several possible meanings of the 'Richardson number' (see §1.4). Not only is the term used without proper definition, but the numerical differences between various measured values are often disregarded once these have fallen well below some presumed critical value. The experiments described in §4.1.4, and the arguments given above show, however, that the growth rate of an instability and the amount of kinetic energy available for subsequent mixing are both strong functions of the magnitude of the relevant Richardson number, so it certainly is desirable to measure and quote this accurately. This question would probably not arise if an inverse parameter were in common use (10 and 100 are clearly different while 0.1 and 0.01 are lumped together as small!), and this is perhaps another argument in favour of using the 'gradient Froude number' F_g.

It is also of interest to calculate the Reynolds numbers in the layer and interface under various circumstances. The interface Reynolds number Re_1 (based on its thickness and the velocity difference across it) is a maximum at $r = r_0$, when the corresponding layer Reynolds number Re_1 is zero. For $r < r_0$, Ri_1 decreases slowly and Re_1 increases more rapidly, and at a particular $r = r_1$ it can be shown that

$$Re_1 = Re_1 = r_1(1 - r_1)H\Delta U/\nu$$
$$= \frac{Ri_e}{(3Ri_e + Ri_0)}\frac{H\Delta U}{\nu}. \qquad (10.2.5)$$

This can be much less than the overall Reynolds number

$$Re = H\Delta U/\nu$$

when $Ri_e \ll Ri_0$. The above result makes more explicit another suggestion due to Stewart (1969), that turbulence in a stably stratified fluid may be inhibited by viscosity, essentially because Re_1 is too low. The density gradient enters indirectly, because large Ri_0 implies a small value of r and a restricted scale of motion. (At very low Reynolds numbers, turbulence cannot even arise, since the shear instability is inhibited – see §4.1.4.)

The results of this section, though based on a particular example, are more generally valid. Most important is the conclusion that when $Ri_0 \gg 1$, sufficient kinetic energy is just *not* available in the mean motion to change a smooth shear flow, having roughly similar profiles of velocity and density, to a step-like structure. To produce the non-uniform profiles, which are commonly observed in nature and which seem to be necessary if turbulence is to exist in the interior of large bodies of stratified fluid, some other supply of energy is needed. In chapter 8 it was shown how layers can be formed by double-diffusive processes which draw on the potential energy of an unstably distributed component. In the absence of such a local source of convective energy (and thus in a fluid stratified with temperature or salinity alone), one is forced to take account of energy supplied at the boundaries, or at a discontinuity of properties such as a front, which propagates into the interior in the form of internal gravity waves. No lengthscale for the layering due to mechanical mixing has emerged from the arguments so far given, and this too must be a consequence of the particular mechanism which supplies the energy to the interior.

10.3. Wave-induced mixing

Now that the role of wave energy in the interior mixing process has been more firmly established, we must retrace our steps and develop in this geophysical context some of the ideas introduced earlier, especially in chapter 4. The most important application will be to the phenomenon of 'clear air turbulence', an all-embracing name applied to various processes which also occur in the ocean. We will review in turn the mechanisms leading to the intermittent formation of turbulent patches, to the production of a non-uniform structure from smooth gradients, and to the possibility of 'saturation' of the wave amplitude. The questions of the likely sources of the wave energy and the distinction between waves and turbulence will also be raised briefly.

10.3.1. *Mixing at existing interfaces*

Atmospheric examples of the production of turbulence due to the growth of interfacial waves have already been described in chapter 4. A sufficiently large steady shear leads to an instability of the Kelvin–Helmholtz form, which can be observed using radar in the clear air (fig. 4.15 pl. IX) or is made visible because cloud is present (fig. 4.14 pl. X). Of more general importance is the fact that similar instabilities can be produced in regions of relatively large density gradient by long internal waves propagating through the fluid (§4.3.3). The shear is a maximum and the gradient Richardson number Ri a minimum at such interfaces, and modest wave amplitudes, acting in combination with mean shears due to larger scale motions, can reduce Ri to the critical value for instability. The wave-induced growth of K–H billows, and their subsequent breakdown to give patches of 'billow turbulence', has been beautifully documented by Woods (1968 a, b) in the upper ocean (fig. 4.21 pl. X).

Woods and Wiley (1971) have gone further, and proposed that the whole of the vertical mixing in the ocean might be described in terms of this process, following the formation of a few sharp interfaces in some other way, for example by wind mixing or by larger scale intrusion of one water mass into another. The local breakdown of a single 'sheet' will produce a patch of relatively well-mixed fluid which is elongated horizontally by the vertical shear and by spreading along surfaces of constant density. At first this patch will be turbulent, but for energetic reasons (§ 10.2.3) the turbulence must decay as fluid is entrained from above and below. The net result, they argue, is to replace a single sharp interface by two interfaces, separated by a mixed layer (cf. fig. 4.11 p. 105). (The process is probably not as clear-cut as this, since a single breaking event may produce multiple interfaces, as suggested by fig. 4.12 pl. VIII; but this complication does not affect the substance of the argument.) The passage of subsequent long waves will produce further 'splitting' of the sheets, until in the mature state the whole water column is filled with microstructure which has developed from the original sharp interfaces, by a combination of splitting and superposition of horizontally non-uniform structures. Molecular diffusion must eventually act to limit and smooth out the microstructure. At any one time, only a small fraction of the volume contains turbulent patches, but the net effect of many such events is to give mean transports of heat, salt and momentum which are much greater than the molecular rates. The fluxes will of course depend on the large scale processes producing the shear, and will be largest in the neighbourhood of fronts.

This picture is consistent both with observations such as those of Woods and Wiley and with the theoretical constraints outlined in § 10.2.2. A major part of the vertical flux of horizontal momentum in the ocean (and also the atmosphere) must be carried by internal waves. The fraction is largest near density interfaces, where interfacial waves can grow, and as these intermittently break, they also produce a smaller turbulent transport of the scalar quantities. Plausible estimates of the various contributions to the total transports can be obtained from the existing data, but in the absence of a fully quantitative theory these will not be pursued here.

10.3.2. *Formation of layers from a smooth gradient*

When interfaces exist, they are the first regions to become unstable to internal waves, but waves can also lead to the formation of turbulent patches in a smoothly stratified fluid, with non-uniformities of the vertical gradients developing only as a result of the subsequent horizontal spreading of the mixed regions. The overturning of the streamlines in lee waves and the production of rotors is the most obvious example of such a process.

When the mean horizontal velocity u varies in the vertical, the 'critical layer' mechanism (§2.3.2) can lead to the growth of internal gravity waves and the absorption of energy in limited regions. At levels where u equals the horizontal phase velocity of the waves, the group velocity relative to the fluid also tends to zero and wave energy accumulates. The amplitude grows, dissipation is increased either because of viscosity or by wave breaking, and eventually wave energy and momentum are transferred to the mean flow. The distortion of the velocity profile produced in this way was remarked on earlier in connection with the laboratory experiment illustrated in fig. 2.15 pl. IV. Small scale jets, which have been interpreted by Woods (private communication) in terms of the local addition of momentum to the mean flow when a wave breaks, are also a common feature of the oceanic thermocline.

Bretherton (1969 *a*, *b*) has shown that a substantial fraction of the drag exerted by an obstacle can be effective well away from the boundary: for example the drag exerted by the Welsh mountains on the airflow over them often acts at great heights. Most of the momentum is added at the critical layer, but it can also be absorbed elsewhere in the flow, in any region where friction dissipates the waves. Thus patches of turbulence (i.e. locally well-mixed layers) could be self-sustaining so long as wave energy is propagating into them, even if they do not remain near a critical level. There is also the possibility that a more gradual absorption of wave momentum will so distort the velocity profile that a critical layer develops where there was none before. An internal gravity wave could in this way produce the optimum conditions to hasten its own decay.

Another mechanism for the concentration of energy in a smoothly stratified flow involves weak resonant interactions, which are most important in just the conditions of interest here (when the overall Richardson number is large). If is often difficult to separate this from the result of superposition of waves from several sources, so both effects will be considered together. A striking illustration of wave breakdown of this kind has been provided by the experiments of McEwan (1971), the earlier stages of which were described in §2.4.3, and it will be useful to enlarge on them here. He generated a standing internal wave in a laboratory tank, and observed that for sufficiently large amplitudes, the waveform became modulated by other waves having a frequency different from the forcing frequency (fig. 10.5 a pl. xxiv). In most cases the higher mode waves could be identified with an unstable pair which formed a resonant triplet with the forced wave. They grew by extracting energy from the forced mode, until the superposition of the several motions produced visible local disturbances of the smooth gradient. These were of two kinds, either an overturning or a violent horizontal convergence followed by vertical divergence. In both cases there was a subsequent collapse to form sharp density discontinuity layers, which seemed to be 'contagious', i.e. they tended to spread in consecutive cycles of oscillation in groups round the original discontinuity, as the first disturbances contributed to the kinematic distortion elsewhere. Such disturbances were often self-stabilizing, because the small scale structure so produced gave enough viscous dissipation to limit the growth and suppress further breaking. With stronger forcing, however, patches of turbulent fluid could be formed and sustained, as shown by the fine structure in shadowgraph pictures (fig. 10.5 b).

McEwan also demonstrated that, although the conditions leading to permanent distortion are realized more easily when growing resonant modes are produced, similar effects can be observed without such interactions provided the forcing amplitude is sufficiently large. The photographs of fig. 10.5 c and d pl. xxiv were taken in an experiment where the lowest 1/1 mode was generated (a case for which there was evidently no unstable interaction), and they show the growth of disturbances in time, and the final breakdown. Even in this simpler case, for which the amplitudes of the fundamental and its harmonics could be calculated, it was not possible to relate the kinematic conditions for breakdown to any of

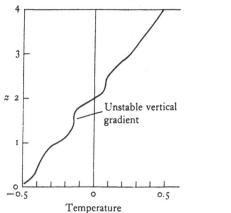
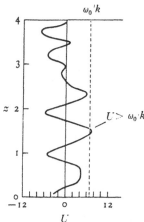

Fig. 10.6. Vertical profiles of temperature and horizontal velocity in the vicinity of an overturning event, according to the numerical calculation of Orlanski and Bryan (1969). Internal waves were generated by resonant forcing in the top quarter of the region, but at the instant shown the amplitude was largest in the lower part. (From Orlanski and Bryan (1969), *J. Geophys. Res.* **74**, 6975–83.)

those which apply in other better understood situations. The maximum slope of the constant density surfaces was always small, and the calculated minimum gradient Richardson number remained large (of order 5), so that neither an overturning nor a shear instability would have been predicted on the basis of earlier work.

A numerical model leading to an overturning instability has been described by Orlanski and Bryan (1969). They too considered a fluid with constant N^2 contained in a two-dimensional rectangular region, and assumed a disturbance consisting of an oscillating body force applied at $t = 0$ to the upper quarter of the fluid. Using the inviscid equations of motion in the Boussinesq approximation, they followed the growth from rest of a wave resonating with the forcing function, with a dominant vertical scale smaller than the total depth. Ultimately, non-linear effects dominated, and higher wavenumbers appeared, which by superposition led to overturning, at points half way between crests and troughs and between horizontal nodal lines. As shown in fig. 10.6, at particular times the points of instability can be remote from the generating region. The condition for overturning corresponds to a state where the disturbance velocity just exceeds the phase velocity locally (which is consistent with the 'critical layer' arguments of §2.3.3).

The energy input near the boundary, which is required to produce overturning in this numerical model, decreases as the vertical scale of motion is reduced. When viscosity is taken into account, however, the dissipation also increases rapidly because of the larger shears associated with the smaller scales. (See Johns and Cross 1970, McEwan 1971.) These two effects combine to give a sharp minimum, thus defining a scale for which the energy needed is a minimum, and Orlanski and Bryan have suggested that this will be of the order of tens of meters in the main thermocline. Though the numerical values are subject to revision, this does seem to be a promising approach to the question of a preferred layer depth.

10.3.3. *Statistical aspects of wave generation and breaking*

So far, only relatively simple generation and interaction processes have been considered, involving particular forced modes and resonant triads whose behaviour could in principle be calculated explicitly. Now we must recognize the possibility of interaction between many waves generated in different places, and abandon any attempt to follow each of them in detail. Broadly one can say that non-linear interactions will tend to transfer energy from large scale internal waves into shorter and shorter wavelengths, and eventually there will be a whole spectrum of resonant modes, which can only be treated by statistical methods. Associated with the smaller vertical lengthscale of the higher modes will be an increased shear $\partial u/\partial z$, and so the gradient Richardson number Ri will be decreased in localized regions, spaced a small distance apart compared with the total depth. Eventually there will be a breakdown to turbulence in randomly distributed patches, where the superposition of many contributions to the local shear and density gradient has led to Ri falling to some specified value, say below a critical value of about $\frac{1}{4}$.

Bretherton (1969c) has proposed a model of this process which is based on an analogy between weak resonant interactions and the theory of colliding molecules in a dilute gas. Using a linear superposition and assuming that the magnitudes of velocity and density gradients are normally distributed, he showed that the probability that $Ri < \frac{1}{4}$ depends only on the overall N^2 and on $\overline{(\partial u/\partial z)^2}$ through the parameter

$$\sigma^2 = \overline{(\partial u/\partial z)^2}/N^2, \qquad (10.3.1)$$

a kind of inverse Richardson number for the wavefield as a whole. For small σ the probability of breakdown is

$$P \sim \frac{\sigma}{\sqrt{\pi}} e^{-(1/2\sigma^2)}. \qquad (10.3.2)$$

Though this must certainly be an underestimate (since non-linear effects can only increase the distortion and the number of turbulent spots, and McEwan (1971) has demonstrated that instabilities can occur for much smaller shears) the conclusion that a substantial change in P can be associated with a small variation in σ will probably remain valid. Thus it seems likely that a statistically steady state could be achieved, in which the energy input to large waves is balanced by the dissipation in turbulent patches, with a minor change in the motions on intermediate scales. That is, (10.3.2) suggests that large differences of energy input can be accommodated by increasing the volume of the dissipating regions, while (for a given N^2) $\overline{(\partial u/\partial z)^2}$ changes little.

A similar argument due to Garrett and Munk (private communication) has gone a step further, and relates the input of wave energy to the rate of vertical mixing. They have derived a universal energy spectrum describing internal waves in the deep ocean, which is based on observations extending over the inertial range as well as gravity wave scales and frequencies. Using this to estimate the probability of breakdown, together with mechanistic models of the mixing produced by each breaking wave event, they arrive at estimates of the vertical eddy diffusivity K which are very sensitive functions of the shear. This again supports the view that the suggested universal spectrum could be a consequence of a saturation effect. The rather narrow range of K (of order 1 cm²/s) deduced from observations in the deep ocean (Munk 1966) implies that the mean rate of input of wave energy also varies little, at least over long time scales.

Some kind of saturation of the quasi-horizontal motions due to waves of large horizontal and small vertical scale is also indicated by certain more detailed observations in the ocean, which it therefore seems relevant to mention here. The form in which these are presented is most easily understood by quoting first a result for two-dimensional wave motions which follows from (2.2.5). The vertical velocity through a region in which N^2 is a slowly varying function of z, rather than constant, is

$$\hat{w}(z) \sim A N^{-\frac{1}{2}} \exp\left\{i\int (Nk/\omega)\,dz\right\} \qquad (10.3.3)$$

(see Phillips 1966a, p. 174). Here A is a dimensional constant determined by the boundary conditions, but the whole of the vertical dependence of the amplitudes of the velocity fluctuations for any high order mode is contained in the function $N^{-\frac{1}{2}}(z)$. Using the continuity equation with (10.3.3) shows that the corresponding horizontal velocity is

$$u(z) \propto N^{\frac{1}{2}}(z); \qquad (10.3.4)$$

that is, in regions of large density gradient, horizontal motions are increased while vertical motions are decreased.

Fofonoff and Webster (1971) have analysed current records at various depths in the deep ocean and have found that below the surface layer the horizontal kinetic energy in the band of frequencies associated with gravity waves is proportional to N, that is

$$\tfrac{1}{2}\rho(\overline{u^2}+\overline{v^2}) \propto N(z), \qquad (10.3.5)$$

in agreement with (10.3.4). Even the spectra can be brought to an identical form by scaling with the local value of N, as is shown in fig. 10.7. Moreover – and this is a point of special interest here – Webster (1971) has shown that the constant of proportionality varies surprisingly little with geographical position or with the strength of the local winds. This suggests again that the intensity is internally limited, and also that surface disturbances have little direct effect on the deeper motions.

Very little is known about the sources of the wave energy in the ocean, but some ideas about the relative effectiveness of generating waves at various levels can be obtained using an extension of the above arguments. For given external boundary conditions, it can be shown that σ^2 defined by (10.3.1) is an increasing function of N, so that the probability of breakdown is increased where the density

Period (hours)

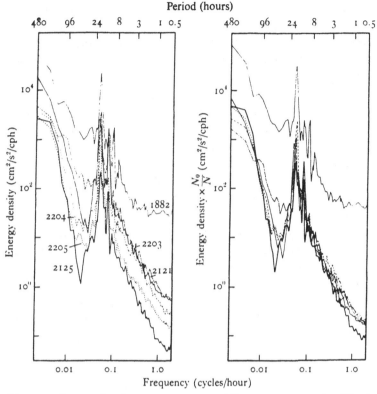

Fig. 10.7. Energy density spectra computed from measurements of horizontal velocities at various depths in the ocean, as reported by Fofonoff and Webster (1971). On the left are the directly measured values and on the right are the same data normalized using the local value of the buoyancy frequency. Note that the curves for all depths below the surface layer are superimposed in the range of frequencies above the inertial peak. The record numbers correspond to the following depths: 1882, 7 m; 2121, 50 m; 2203, 106 m; 2204, 511 m; 2205, 1013 m; and 2125, 1950 m.

gradient is large (cf. §4.3.2, where a similar result was quoted for interfaces). This means that, if energy is added at the top of a thermocline where the density gradients are a maximum, no instability can be produced at greater depths unless 'breaking' is already present nearer the surface. The observations suggest that a wind-mixed surface layer (§9.2.1) can have immediately below it a gradient region in which intermittent mixing is driven by wave energy propagating downwards from the interface (see, for example,

Kitaigordskii and Miropol'kii 1970). The thickness of this layer, which is indirectly but simultaneously mixed by the surface processes, must be strictly limited, however, since so much energy will be lost near the source that the amplitude of the propagating waves will fall below the level necessary to produce instability lower down. Wave energy can nevertheless be fed in this way from the surface into the deeper ocean, to interact later with that originating in other regions and produce intermittent mixing which cannot be identified in space or time with particular sources.

On the other hand, waves generated on the ocean bottom or at the edge of a shelf, which are regions of minimum N^2, will tend to become more unstable as they propagate upwards. Some fraction of the energy can be absorbed at each level due to breaking, while still allowing enough through to make the higher layers unstable in their turn. No such difficulty arises in the discussion of wave generation in the atmosphere, since the major wave generating processes (at the surface) are naturally associated with the weakest density gradients. Note again too that the increasing instability with increasing height (and increasing N^2) will be enhanced in the atmosphere by the decrease in air density (see §4.3.5).

10.3.4. Waves and turbulence in large scale flows

The interpretations of geophysical phenomena suggested above have of course been deliberately simplified, so that the physical processes could be identified and discussed individually, and always with buoyancy forces taken as the major controlling factor. We will now consider the relation between the various phenomena described, and comment briefly on some of the effects which have been omitted.

It is apparent that 'clear air turbulence' (and its equivalent in the ocean) can encompass a great variety of physically distinct phenomena, each of which may become important if the circumstances are right. Sometimes it can be associated with a particular part of the boundary (where separation occurs behind an obstacle, for example, or an internal hydraulic jump is generated downstream). Strong enough shears, especially across density interfaces, will lead to a local instability and the production of turbulence. With the generation of waves, the effects of boundaries can be propagated far into the interior of a stratified fluid. In different conditions, stable wave motions of large amplitude, those which are in the process of over-turning, and those which have actually broken down to give smaller scale turbulence may have similar effects on aircraft, but the various alternatives must be distinguished if one is to obtain a proper understanding of what is observed.

This raises the more general question of the criteria which can be used, in principle and in practice, to decide whether motions in a stably stratified fluid should be called 'waves' or 'turbulence'. It is clearly inefficient to lump everything together. We must learn to recognize the characteristic features of the two kinds of motion so that we can design observations more intelligently and analyse them in the most suitable way. As pointed out by Stewart (1969),† the distinction is not always clearcut. Internal waves of large amplitude produce turbulent patches (§§4.3, 10.3.2) and turbulence can generate waves (§§7.3.4, 9.2.4), and when such non-linear effects can continuously exchange energy between the two kinds of motion it is difficult to draw a firm line between them.

Many phenomena can, however, be put unambiguously in one category or the other, and the distinctive features can be listed. The most important property of turbulence is its ability to produce mixing, thereby transporting scalar quantities such as heat or salt both along and across surfaces of constant density. Associated with this is the fact that turbulence is a strong interaction phenomenon, and highly dissipative. There is little connection between the motion at well separated points in the fluid, and energy propagates slowly, with the speed of the fluid motion. Waves, on the other hand, may distort the density distribution to produce an apparent, transient layering (§ 10.1), but they cannot permanently change the strati-fication unless they 'break' to produce turbulence. Waves are characterized by a dispersion relation, a definite relation between their speed and scale. They transfer energy more rapidly through the fluid, varying only gradually as they do so, and preserve phase relations over several wavelengths at least.

Both observation and theory support the view that in the interior of strongly stratified fluids, such as the atmosphere and ocean,

† Presented at a Colloquium on 'Spectra of Meteorological Variables', the whole of which is recommended to those interested in this field.

turbulence as defined in this way can occur only sporadically and in isolated patches. Through most of the volume, wave theories should therefore be the more appropriate, whereas inside these patches one should change to a description in terms of turbulence (cf. §5.2). Sometimes, by using tracer techniques for instance (§4.3.3), it may be possible to pick out these regions directly, but more often the state of motion will have to be diagnosed from temperature and velocity records like those of fig. 10.2. Analysis of the cross spectra of these variables seems the most promising method so far proposed. Wave motions will give a high coherence between temperature and vertical velocity, but since the fluctuations are 90° out of phase (§2.2), no vertical transport is implied. (See, for example, Axford (1971), who has analysed observations of internal gravity waves in this way.) Phase angles near zero or 180°, implying a much stronger co-spectrum than quadrature spectrum and a non-zero vertical transport of heat, will be indicative of motions which are mostly turbulence.

Patches of turbulence in the ocean or atmosphere can arise as a result of the superposition of motions from many sources and on many scales. A completely deterministic theory is therefore un-likely, and detailed forecasting of clear air turbulence will always be very difficult, but there are obvious ways in which the discussion given here can be extended and made more realistic. The mean shears certainly contribute to the internal breakdown mechanism, and one must also include the larger scale, quasi-horizontal motions associated with inertial waves. A full description of even the smaller scale motions will strictly involve both stratification and rotation, with the theory of inertio–gravitational waves replacing that of pure gravity waves. It seems likely, for instance, that inertial waves could lead to the critical conditions for turbulence being achieved simultaneously everywhere in a horizontally elongated patch, rather than in a more localized region which then spreads sideways.

More detailed measurements will be needed before we can even properly describe the transport mechanisms in the atmosphere and ocean. Already in recent years there has been a notable change in attitude to small scale observations and the associated theoretical and laboratory work, and each new technique has led to further advances. The characteristic layered and patchy structure, and the importance of density interfaces, have become widely appreciated, and more information will come with increased resolution. There is still a great need, however, for more systematic surveys, designed not only to learn more about individual mixing events but also to determine how these are distributed in space and time, and how they are related to the local gradients and the energy inputs. Remote sensing techniques based on radar and acoustic sounding offer the best hope for rapid progress, and an example of the kind of measurement which can already be made is given in fig. 10.8 pl. XXII. (See also Beran, Little and Willworth 1971.) More attention should also be given to the problem of identification of significant events on a record, before embarking on routine analyses (such as the calculation of spectra) which may sometimes be quite inappropriate.

The detailed description of the ocean and atmosphere will undoubtedly change as more information becomes available and additional effects, especially those introduced by rotation, are added to those considered here. We can be confident, however, that microphysical processes will always be important, and that they will be described using concepts closely related to those which have been discussed. It is only by building systematically on our theoretical and laboratory knowledge of individual physical processes that we can hope first to recognize, and then to understand, the fascinating array of more complex natural phenomena.

BIBLIOGRAPHY AND AUTHOR INDEX

Numbers in square brackets refer to the pages on which the reference is cited.

Abbott, M. R. (1960). Salinity effects in estuaries. *J. Marine Res.* **18**, 101–11. [p. 163.]

Abraham, G. and Eysink, W. D. (1969). Jets issuing into fluid with a density gradient. *J. Hydraulic Res.* **7**, 145–75. [p. 198.]

Anati, D. A. (1971). On the mechanism of the deep mixed layer formation during Medoc '69. *Cahiers Oceanographique*, **23**, 427–43. [p. 304.]

Atlas, D., Metcalf, J. I., Richter, J. H. and Gossard, E. E. (1970). The birth of 'CAT' and microscale turbulence. *J. Atmos. Sci.* **27**, 903–13. [p. 106.]

Axford, D. N. (1971). Spectral analysis of an aircraft observation of gravity waves. *Quart. J. Roy. Met. Soc.* **97**, 313–21. [p. 336.]

Baines, P. G. (1969). Waves and stability in rotating or stratified fluids. *Ph.D. thesis.* University of Cambridge. [p. 30.]

Baines, P. G. and Gill, A. E. (1969). On thermohaline convection with linear gradients. *J. Fluid Mech.* **37**, 289–306. [pp. 253, 256.]

Baines, W. D. (1975). Entrainment by a plume or jet at a density interface. *J. Fluid Mech.* **68**, 309–20. [pp. 296, 307.]

Baines, W. D. and Turner, J. S. (1969). Turbulent buoyant convection from a source in a confined region. *J. Fluid Mech.* **37**, 51–80. [p. 232.]

Bakke, P. and Leach, S. J. (1965). Turbulent diffusion of a buoyant layer at a wall. *Appl. Sci. Res.* **15**, 97–136. [p. 159.]

Ball, F. K. (1957). The katabatic winds of Adelie Land and King George V Land. *Tellus*, **9**, 201–8. [p. 69.]

Ball, F. K. (1960). Control of inversion height by surface heating. *Quart. J. Roy. Met. Soc.* **86**, 483–94. [p. 306.]

Ball, F. K. (1964). Energy transfer between external and internal gravity waves. *J. Fluid Mech.* **19**, 465–78. [p. 41.]

Barr, D. I. H. (1967). Densimetric exchange flow in rectangular channels. III. Large scale experiments. *La Houille Blanche*, **22**, 619–32. [p. 71.]

Batchelor, G. K. (1952). Diffusion in a field of homogeneous turbulence. *Proc. Camb. Phil. Soc.* **48**, 345–62. [p. 205.]

Batchelor, G. K. (1953*a*). The conditions for dynamical similarity of motions of a frictionless perfect-gas atmosphere. *Quart. J. Roy. Met. Soc.* **79**, 224–35. [p. 12.]

Batchelor, G. K. (1953*b*). *The theory of homogeneous turbulence.* Cambridge University Press. [pp. 141, 146.]

Batchelor, G. K. (1954*a*). Heat convection and buoyancy effects in fluids. *Quart. J. Roy. Met. Soc.* **80**, 339–58. [pp. 166, 170, 194.]

Batchelor, G. K. (1954b). Heat transfer by free convection across a closed cavity between vertical boundaries at different temperatures. *Quart. Appl. Maths.* **12**, 209–33. [p. 246.]

Batchelor, G. K. (1967). *An introduction to fluid dynamics.* Cambridge University Press. [p. 3.]

Bénard, H. (1901). Les tourbillons cellulaires dans une nappe liquide transportant de la chaleur par convection en régime permanent. *Ann. Chim. Phys.* (7) **23**, 62–144. [pp. 208, 216.]

Benjamin, T. B. (1963). The threefold classification of unstable disturbances in flexible surfaces bounding inviscid flows. *J. Fluid Mech.* **16**, 436–50. [pp. 92, 94.]

Benjamin, T. B. (1966). Internal waves of finite amplitude and permanent form. *J. Fluid Mech.* **25**, 241–70. [pp. 53, 55, 57, 67.]

Benjamin, T. B. (1967). Internal waves of permanent form in fluids of great depth. *J. Fluid Mech.* **29**, 559–92. [pp. 54, 67.]

Benjamin, T. B. (1968). Gravity currents and related phenomena. *J. Fluid Mech.* **31**, 209–48. [pp. 70, 73.]

Benjamin, T. B. (1970). Upstream influence. *J. Fluid Mech.* **40**, 49–79. [pp. 57, 62.]

Beran, D. W., Little, C. G. and Willworth, B. C. (1971). Acoustic Doppler measurements of vertical velocities in the atmosphere. *Nature*, **230**, 160–2. [p. 337.]

Betchov, R. and Criminale, W. O. (1967). *Stability of parallel flows.* New York: Academic Press. [pp. 92, 107.]

Birikh, R. V., Gershuni, G. Z. and Zhukhovitshii, E. M. (1969). Stability of the steady convective motion of a fluid with a longitudinal temperature gradient. *J. Appl. Math. and Mech.* **33**, 937–47. [p. 249.]

Booker, J. R. and Bretherton, F. P. (1967). The critical layer for internal gravity waves in a shear flow. *J. Fluid Mech.* **27**, 513–39. [p. 38.]

Boussinesq, J. (1903). *Théorie analytique de la chaleur*, vol. 2. Paris: Gathier-Villars. [p. 9.]

Bretherton, F. P. (1964). Resonant interactions between waves: The case of discrete oscillations. *J. Fluid Mech.* **20**, 457–79. [p. 41.]

Bretherton, F. P. (1966). The propagation of groups of internal gravity waves in a shear flow. *Quart. J. Roy. Met. Soc.* **92**, 466–80. [p. 37.]

Bretherton, F. P. (1967.) The time-dependent motion due to a cylinder moving in an unbounded rotating or stratified fluid. *J. Fluid Mech.* **28**, 545–70. [p. 80.]

Bretherton, F. P. (1969a). Momentum transport by gravity waves. *Quart. J. Roy. Met. Soc.* **95**, 213–43. [p. 327.]

Bretherton, F. P. (1969b). On the mean motion induced by internal gravity waves. *J. Fluid Mech.* **36**, 785–803. [p. 327.]

Bretherton, F. P. (1969c). Waves and turbulence in stably stratified fluids. *Radio Science*, **4**, 1279–87. [p. 330.]

Bretherton, F. P. and Garrett, C. J. R. (1968). Wave trains in inhomogeneous moving media. *Proc. Roy. Soc.* A **302**, 529–54. [pp. 37, 38.]

Briggs, G. A. (1969). *Plume rise.* U.S. Atomic Energy Commission Critical Review Series. [pp. 193, 198.]

340 BUOYANCY EFFECTS IN FLUIDS

Browning, K. A. (1971). Structure of the atmosphere in the vicinity of large-amplitude Kelvin–Helmholtz billows, *Quart. J. Roy. Met. Soc.* **97**, 283–99. [p. 318.]

Browning, K. A. and Watkins, C. D. (1970). Observations of clear air turbulence by high power radar. *Nature*, **227**, 260–3. [p. 106.]

Brunt, D. (1952). *Physical and dynamical meteorology.* Cambridge University Press. [pp. 12, 216.]

Businger, J. A. (1969). Note on the critical Richardson number(s). *Quart. J. Roy. Met. Soc.* **95**, 653–4. [p. 323.]

Busse, F. H. (1967). On the stability of two-dimensional convection in a layer heated from below. *J. Math. and Phys.* **46**, 140–50. [p. 212.]

Busse, F. H. (1969). On Howard's upper bound for heat transport by turbulent convection. *J. Fluid Mech.* **37**, 457–77. [p. 215.]

Busse, F. H. and Whitehead, J. A. (1971). Instabilties of convection rolls in a high Prandtl number fluid. *J. Fluid Mech.* **47**, 305–20. [p. 218.]

Cabelli, A. and Davis, G. de Vahl (1971). A numerical study of the Bénard cell. *J. Fluid Mech.* **45**, 805–29. [p. 223.]

Carstens, T. J. (1964). Stability of shear flow near the interface of two fluids. *Doctoral Dissertation.* University of California. [p. 122.]

Chandrasekhar, S. (1961). *Hydrodynamic and hydromagnetic stability.* Oxford: Clarendon Press. [pp. 211, 216.]

Charnock, H. (1967). Flux-gradient relations near the ground in unstable conditions. *Quart. J. Roy. Met. Soc.* **93**, 97–100. [pp. 133, 135.]

Chen, C. F., Briggs, D. G. and Wirtz, R. A. (1971). Stability of thermal convection in a salinity gradient due to lateral heating. *Int. J. Heat and Mass Trans.* **14**, 57–65. [p. 268.]

Chen, M. M. and Whitehead, J. A. (1968). Evolution of two-dimensional periodic Rayleigh convection cells of arbitrary wavenumbers. *J. Fluid Mech.* **31**, 1–15. [p. 217.]

Claus, A. J. (1964). Large-amplitude motion of a compressible fluid in the atmosphere. *J. Fluid Mech.* **19**, 267–89. [pp. 11, 59.]

Corby, G. A. and Wallington, C. E. (1956). Airflow over mountains: the lee-wave amplitude. *Quart. J. Roy. Met. Soc.* **82**, 266–74. [p. 35.]

Craik, A. D. D. (1968). Resonant gravity-wave interactions in a shear flow. *J. Fluid Mech.* **34**, 531–49. [p. 124.]

Craya, A. (1949). Theoretical research on the flow of non-homogeneous fluids. *La Houille Blanche*, **4**, 44–55. [p. 76.]

Cromwell, T. (1960). Pycnoclines created by mixing in an aquarium tank. *J. Mar. Res.* **18**, 73–82. [p. 289.]

Dake, J. M. K. and Harleman, D. R. F. (1966). An analytical and experimental investigation of thermal stratification in lakes and ponds. *M.I.T. Hydrodynamics Lab. Rep. no. 99.* [p. 306.]

Davis, R. E. (1969). The two-dimensional flow of a stratified fluid over an obstacle. *J. Fluid Mech.* **36**, 127–43. [pp. 62, 119.]

Davis, R. E. and Acrivos, A. (1967a). Solitary internal waves in deep water. *J. Fluid Mech.* **29**, 593–607. [p. 54.]

Davis, R. E. and Acrivos, A. (1967b). The stability of oscillatory internal waves. *J. Fluid Mech.* **30**, 723–36. [pp. 43, 123.]

Deardorff, J. W. (1964). A numerical study of two-dimensional parallel-plate convection. *J. Atmos. Sci.* **21**, 419–38. [p. 222.]

Deardorff, J. W. (1965). A numerical study of pseudo three-dimensional parallel-plate convection. *J. Atmos. Sci.* **22**, 419–35. [p. 223.]

Deardorff, J. W. (1966). The counter-gradient heat flux in the lower atmosphere and the laboratory. *J. Atmos. Sci.* **23**, 503–6. [p. 234.]

Deardorff, J. W. and Willis, G. E. (1965). The effect of two-dimensionality on the suppression of thermal turbulence. *J. Fluid Mech.* **23**, 337–53. [p. 221.]

Deardorff, J. W. and Willis, G. E. (1967). Investigation of turbulent thermal convection between horizontal plates. *J. Fluid Mech.* **28**, 675–704. [p. 221.]

Deardorff, J. W., Willis, G. E. and Lilly, D. K. (1969). Laboratory investigation of non-steady penetrative convection. *J. Fluid Mech.* **35**, 7–31. [pp. 235, 306].

Debler, W. R. (1959). Stratified flow into a line sink. *J. Eng. Mech. Div., Proc. Am. Soc. Civil Eng.* **85**, 51–65. [p. 77.]

Defant, A. (1961). *Physical oceanography*, vols. I and II. London: Pergamon Press. [pp. 17, 125, 306.]

Degens, E. T. and Ross, D. A. (eds.) (1969). *Hot brines and recent heavy metal deposits in the Red Sea.* Berlin: Springer–Verlag. [p. 271.]

Dore, B. D. (1970). Mass transport in layered fluid systems. *J. Fluid Mech.* **40**, 113–26. [p. 51.]

Drazin, P. G. (1958). The stability of a shear layer in an unbounded heterogeneous inviscid fluid. *J. Fluid Mech.* **4**, 214–24. [p. 99.]

Drazin, P. G. (1962). On stability of parallel flow of an incompressible fluid of variability density and viscosity. *Proc. Camb. Phil. Soc.* **58**, 646–61. [p. 107.]

Drazin, P. G. (1969). Non-linear internal gravity waves in a slightly stratified atmosphere. *J. Fluid Mech.* **36**, 433–46. [p. 11.]

Drazin, P. G. (1970). Kelvin–Helmholtz instability of finite amplitude. *J. Fluid Mech.* **42**, 321–35. [p. 114.]

Drazin, P. G. and Howard, L. N. (1966). Hydrodynamic stability of parallel flow of inviscid fluid. *Advanc. Appl. Mech.* **9**, 1–89. [pp. 92, 100.]

Drazin, P. G. and Moore, D. W. (1967). Steady two-dimensional flow of fluid of variable density over an obstacle. *J. Fluid Mech.* **28**, 353–70. [p. 59.]

Dubreil–Jacotin, M. L. (1937). Sur les théorèmes d'existence relatifs aux ondes permanentes périodiques à deux dimensions dans les liquides hétérogènes. *J. Math. Pures Appl.* (9), **16**, 43–67. [p. 55.]

Dyer, A. J. (1965.) The flux-gradient relation for turbulent heat transfer in the lower atmosphere. *Quart. J. Roy. Met. Soc.* **91**, 151–7. [p. 135.]

Eckert, E. R. G. and Carlson, W. O. (1961). Natural convection in an air layer enlosed between two vertical plates with different temperatures. *Int. J. Heat and Mass Trans.* **2**, 106–20. [p. 246.]

Ekman, V. W. (1904). On dead water. *Sci. Results Norwegian N. Polar Exp. Vol. 5 No. 15* (Christiania). [pp. 1, 19.]

Elder, J. W. (1965a). Laminar free convection in a vertical slot. *J. Fluid Mech.* **23**, 77–98. [p. 246.]

Elder, J. W. (1965b). Turbulent free convection in a vertical slot. *J. Fluid Mech.* **23**, 99–111. [pp. 246, 249.]

Elder, J. W. (1967). Steady free convection in a porous medium heated from below. *J. Fluid Mech.* **27**, 29–48. [p. 225.]

Elder, J. W. (1968). The unstable thermal interface. *J. Fluid Mech.* **32**, 69–96. [pp. 227, 229.]

Elder, J. W. (1969a). The temporal development of a model of high Rayleigh number convection. *J. Fluid Mech.* **35**, 417–37. [pp. 224, 237.]

Elder, J. W. (1969b). Numerical experiments with thermohaline convection. *Phys. of Fluids*, **12**, Suppl. II, 194–7. [p. 261.]

Ellison, T. H. (1957). Turbulent transport of heat and momentum from an infinite rough plane. *J. Fluid Mech.* **2**, 456–66. [pp. 135, 137, 148, 160.]

Ellison, T. H. (1959). Turbulent diffusion. *Sci. Prog.* **47**, 495–506. [p. 140.]

Ellison, T. H. (1961). Stratified flows. *Sci. Prog.* **49**, 57–67. [p. 67.]

Ellison, T. H. (1966). A note on the influence of density stratification on turbulence. Unpub. manuscript, quoted by Yaglom (1969). [pp. 148, 150.]

Ellison, T. H. and Turner, J. S. (1959). Turbulent entrainment in stratified flows. *J. Fluid Mech.* **6**, 423–48. [pp. 177, 179, 182, 185, 187, 297.]

Ellison, T. H. and Turner, J. S. (1960). Mixing of dense fluid in a turbulent pipe flow. *J. Fluid Mech.* **8**, 514–44. [pp. 159, 161.]

Fischer, H. B. (1971). The dilution of an undersea sewage cloud by salt fingers. *Water Research*, **5**, 909–15. [p. 272.]

Fischer, H. B. (1972). Mass transport mechanisms in partially stratified estuaries. *J. Fluid Mech.* **53**, 671–87. [p. 162.]

Fjeldstad, J. E. (1933). Interne Wellen. *Geofys. Publ.* **10**, no. 6, 1–35. [p. 21.]

Fofonoff, N. P. and Webster, T. F. (1971). Current measurements in the western Atlantic. *Phil. Trans.* A **270**, 423–36. [p. 333.]

Fohl, T. (1967). Optimization of flow for forcing stack wastes to high altitudes. *J. Air. Poll. Control. Assoc.* **17**, 730–3. [p. 202.]

Fortescue, G. E. and Pearson, J. R. A. (1967.) On gas absorption into a turbulent liquid. *Chem. Eng. Sci.* **22**, 1163–76. [p. 297.]

Foster, M. R. and Saffman, P. G. (1970). Drag of a body moving transversely in a confined stratified fluid. *J. Fluid Mech.* **43**, 407–18. [p. 87.]

Foster, T. D. (1965). Onset of convection in a layer of fluid cooled from above. *Phys. Fluids*, **8**, 1770–4. [p. 229.]

Foster, T. D. (1969). The effect of initial conditions and lateral boundaries on convection. *J. Fluid Mech.* **37**, 81–94. [p. 217.]

Foster, T. D. (1971a). A convective model for the diurnal cycle in the upper ocean. *J. Geophys. Res.* **76**, 666–75. [p. 306.]

Foster, T. D. (1971b). Intermittent convection. *Geophys. Fluid Dyn.* **2**, 201–17. [p. 229.]

Fromm, J. E. (1965). Numerical solutions of the nonlinear equations for a heated fluid layer. *Phys. Fluids*, **8**, 1757–69. [p. 222.]

Gage, K. S. (1971). The effect of stable stratification on the stability of viscous parallel flows. *J. Fluid Mech.* **47**, 1–20. [p. 109.]

Gage, K. S. and Reid, W. H. (1968). The stability of thermally stratified plane Poiseuille flow. *J. Fluid Mech.* **33**, 21–32. [pp. 109, 211.]

Garrett, C. and Munk, W. (1971). Internal wave spectra in the presence of fine structure. *Phys. Oceanog.* **1**, 196–202. [p. 21.]

Gebhart, B. (1969). Natural convection flow, instability, and transition. *J. Heat Transfer*, **91**, 293–309. [p. 242.]

Gill, A. E. (1966). The boundary layer régime for convection in a rectangular cavity. *J. Fluid Mech.* **26**, 515–36. [pp. 243, 247.]

Gill, A. E. and Davey, A. (1969). Instabilities of a buoyancy-driven system. *J. Fluid Mech.* **35**, 775–98. [pp. 242, 248.]

Gill, A. E. and Turner, J. S. (1969). Some new ideas about the formation of Antarctic bottom water. *Nature*, **224**, 1287–8. [p. 286.]

Gille, J. (1967). Interferometric measurement of temperature gradient reversal in a layer of convecting air. *J. Fluid Mech.* **30**, 371–84. [p. 221.]

Globe, S. and Dropkin, D. (1959). Natural convection heat transfer in liquids confined by two horizontal plates and heated from below. *J. Heat Trans.* **81**, 156–65. [pp. 219, 223.]

Goldstein, S. (1931). On the stability of superposed streams of fluids of different densities. *Proc. Roy. Soc.* A **132**, 524–48. [p. 97.]

Görtler, H. (1943). Uber eine Schwingungserscheinung in Flüssigkeiten mit stabiler Dichteschichtung. *Z. agnew. Math. Mech.* **23**, 65–71. [p. 24.]

Gough, D. O., Spiegel, E. A. and Toomre, J. (1975). Modal equations for cellular convection. *J. Fluid Mech.* **68**, 695–719. [p. 225.]

Graebel, W. P. (1969). On the slow motion of bodies in stratified and rotating fluids. *Q. J. Mech. Appl. Maths.* **22**, 39–54. [p. 85.]

Grant, H. L., Moilliet, A. and Vogel, W. M. (1968). Some observations of the occurrence of turbulence in and above the thermocline. *J. Fluid Mech.* **34**, 443–8. [p. 316.]

Grant, H. L., Stewart, R. W. and Moilliet, A. (1962). Turbulence spectra from a tidal channel. *J. Fluid Mech.* **12**, 241–68. [p. 141.]

Grigg, H. R. and Stewart, R. W. (1963). Turbulent diffusion in a stratified fluid. *J. Fluid Mech.* **15**, 174–86. [pp. 143, 202, 289.]

Hanna, S. R. (1969). The thickness of the planetary boundary layer. *Atmos. Envir.* **3**, 519–36. [p. 309.]

Hansen, D. V. and Rattray, M. (1965). Gravitational circulation in straits and estuaries. *J. Marine Res.* **23**, 104–22. [p. 163.]

Harleman, D. R. F. (1961). Stratified flow. In *Handbook of fluid mechanics* (ed. V. L. Streeter), ch. 26. New York: McGraw-Hill. [p. 75.]

Harleman, D. R. F. and Ippen, A. T. (1967). Two-dimensional aspects of salinity intrusion in estuaries. *Comm. on Tidal Hydr. U.S. Army Corps of Eng. Tech. Bull. no. 23.* [p. 158.]

Harleman, D. R. F., Jordan, J. M. and Lin, J. D. (1959). The diffusion of two fluids of different density in a homogeneous turbulent field. *M.I.T. Hydr. Lab. Tech. Report no. 31.* [p. 162.]

Hart, J. E. (1971 a). Stability of the flow in a differentially heated inclined box. *J. Fluid Mech.* **47**, 547–76. [p. 249.]

Hart, J. E. (1971b). On sideways diffusive instability. *J. Fluid Mech.* **49**, 279–88. [p. 269.]

Hart, J. E. (1971c). A possible mechanism for boundary layer mixing and layer formation in a stratified fluid. *J. Phys. Oceanog.* **1**, 258-62.[p.126].

Hasselman, K. (1966). Feynman diagrams and interaction rules of wave–wave scattering processes. *Rev. Geophys.* **4**, 1–32. [p. 41.]

Hasselman, K. (1967). A criterion for non-linear wave stability. *J. Fluid Mech.* **30**, 737–9. [p. 43.]

Hazel, P. (1967). The effects of viscosity and heat conduction on internal gravity waves at a critical level. *J. Fluid Mech.* **30**, 775–81. [p. 38.]

Hazel, P. (1972). Numerical studies of the stability of inviscid stratified shear flows. *J. Fluid Mech.* **51**, 39–61. [pp. 99, 101, 103.]

Herring, J. R. (1963). Investigation of problems in thermal convection. *J. Atmos. Sci.* **20**, 325–38. [pp. 224, 238.]

Herring, J. R. (1964). Investigation of problems in thermal convection: rigid boundaries. *J. Atmos. Sci.* **21**, 277–90. [pp. 224, 237.]

Hines, C. O. (1971). Generalizations of the Richardson criterion for the onset of atmospheric turbulence. *Quart. J. Roy. Met. Soc.* **97**, 429–39. [p. 323.]

Hinwood, J. B. (1967). The stability of a stratified fluid. *J. Fluid Mech.* **29**, 233–40. [p. 115.]

Hinwood, J. B. (1970). The study of density stratified flows up to 1945. 1. Nearly horizontal flows with stable interfaces. *La Houille Blanche*, **25**, 347–59 [p. 2.]

Hoare, R. A. (1966). Problems of heat transfer in Lake Vanda, a density stratified Antarctic Lake. *Nature*, **210**, 787–9. [p. 270.]

Houghton, D. and Woods, J. D. (1969). The slippery seas of Acapulco. *New Scientist*, January, 134–6. [p. 156.]

Howard, L. N. (1961). Note on a paper of John W. Miles. *J. Fluid Mech.* **10**, 509–12. [p. 100.]

Howard, L. N. (1963). Heat transport by turbulent convection. *J. Fluid Mech.* **17**, 405–32. [p. 215.]

Howard, L. N. (1964). Convection at high Rayleigh number. *Proc. eleventh Int. Congress Applied Mechanics, Münich* (ed. H. Görtler), pp. 1109–15. Berlin: Springer–Verlag [pp. 214, 226, 229, 277.]

Huber, D. G. (1960). Irrotational motion of two fluid strata towards a line sink. *J. Eng. Mech. Div. Am. Soc. Civil Eng.* **86**, EM 4, 71–86. [p. 76.]

Hunt, J. N. (1961). Interfacial waves of finite amplitude. *La Houille Blanche*, **16**, 515–31. [p. 49.]

Huppert, H. E. (1968). Appendix to paper by J. W. Miles. *J. Fluid Mech.* **33**, 803–14. [p. 61.]

Huppert, H. E. (1971). On the stability of a series of double-diffusive layers. *Deep-Sea Res.* **18**, 1005–21. [pp. 275, 280.]

Huppert, H. E. and Miles, J. W. (1969). Lee waves in a stratified flow. Part 3. Semi-elliptical obstacle. *J. Fluid Mech.* **35**, 481–6. [p. 60.]

Hurle, D. T. J. and Jakeman, E. (1971). Soret-driven thermosolutal convection. *J. Fluid Mech.* **47**, 667–87. [p. 263.]

Hurley, D. G. (1969). The emission of internal waves by vibrating cylinders. *J. Fluid Mech.* **36**, 657–72. [p. 30.]

Hurley, D. G. (1970). Internal waves in a wedge-shaped region. *J. Fluid Mech.* **43**, 97-120. [p. 30.]

Imberger, J. (1972). Two dimensional sink flow of a stratified fluid contained in a duct. *J. Fluid Mech.* **53**, 329-49. [p. 89.]

Ippen, A. T. (1966). *Estuary and coastline hydrodynamics.* New York: McGraw-Hill. [p. 74.]

Ippen, A. T. and Harleman, D. R. F. (1952). Steady-state characteristics of subsurface flow. *Proc. NBS Symp. on Gravity Waves, Nat. Bur. Stand. Circ.* **521**, 79-93. [p. 112.]

Iselin, C. O'D. (1939). The influence of vertical and lateral turbulence on the characteristic of the waters at mid-depths. *Trans. Amer. Geophys. Union 1939*, 414-17. [p. 316.]

Izumi, Y. (1964). The evolution of temperature and velocity profiles during breakdown of a nocturnal inversion and a low-level jet. *J. Appl. Meteor.* **3**, 70-82. [p. 308.]

Jakob, M. (1957). *Heat transfer.* New York: John Wiley. [p. 250.]

Janowitz, G. S. (1968). On wakes in stratified fluids. *J. Fluid Mech.* **33**, 417-32. [p. 85.]

Jeffreys, H. (1926). The stability of a layer of fluid heated from below. *Phil. Mag.* **2**, 833-44. [p. 210.]

Jeffreys, H. (1928). Some cases of instability in fluid motion. *Proc. Roy. Soc.* A **118**, 195-208. [p. 210.]

Johns, B. and Cross, M. J. (1970). The decay and stability of internal wave modes in a multisheeted thermocline. *J. Marine Res.* **28**, 215-24. [p. 330.]

Johnson, M. A. (1963). Turbidity currents. *Sci. Prog.* **198**, 257-73. [p. 178.]

Kármán, T. von (1940). The engineer grapples with non-linear problems. *Bull. Am. Math. Soc.* **46**, 615-83. [pp. 71, 73.]

Kato, H. and Phillips, O. M. (1969). On the penetration of a turbulent layer into a stratified fluid. *J. Fluid Mech.* **37**, 643-55. [pp. 293, 297.]

Kelly, R. E. (1968). On the resonant interaction of neutral disturbances in two inviscid shear flows. *J. Fluid Mech.* **31**, 789-99. [p. 124.]

Keulegan, G. H. (1949). Interfacial instability and mixing in stratified flows. *J. Res. Nat. Bur. Stand.* **43**, 487-500. [p. 107.]

Keulegan, G. H. (1953). Characteristics of internal solitary waves. *J. Res. Nat. Bur. Stand.* **51**, 133-40. [p. 53.]

Keulegan, G. H. (1957). Form characteristics of arrested saline wedges. *Nat. Bur. Stand. Rept. 5482.* [p. 75.]

Keulegan, G. H. (1958). The motion of saline fronts in still water. *Nat. Bur. Stand. Rept. 5831.* [p. 72.]

Keulegan, G. H. and Carpenter, L. H. (1961). An experimental study of internal progressive oscillatory waves. *Nat. Bur. Stand. Rept. 7319.* [pp. 44, 50, 123.]

Kitaigorodskii, S. A. (1960). On the computation of the thickness of the wind-mixing layer in the ocean. *Bull. Acad. Sci. U.S.S.R. Geophys. Ser.* **3**, 284-7. [p. 301.]

Kitaigorodskii, S. A. and Miropol'kii, Yu. Z. (1970). On the theory of the open-ocean active layer. *Bull. Acad. Sci. U.S.S.R. Atmos. and Oceanic Phys.* **6**, 97-102. [p. 334.]

Klug, W. (1968). Diffusion in the atmospheric surface layer: comparison of similarity theory with observations. *Quart. J. Roy. Met. Soc.* **94**, 555–62. [p. 140.]

Koh, R. C. Y. (1966). Viscous stratified flow towards a sink. *J. Fluid Mech.* **24**, 555–75. [pp. 87, 89.]

Koschmieder, E. L. (1967). On convection under an air surface. *J. Fluid Mech.* **30**, 9–15. [p. 216.]

Kraichnan, R. H. (1962). Turbulent thermal convection at arbitrary Prandtl number. *Phys. Fluids*, **5**, 1374–89, [p. 214.]

Kraus, E. B. and Rooth, C. G. H. (1961). Temperature and steady state vertical heat flux in the ocean surface layers. *Tellus*, **13**, 231–8. [p. 305.]

Kraus, E. B. and Turner, J. S. (1967). A one-dimensional model of the seasonal thermocline. II. The general theory and its consequences. *Tellus*, **19**, 98–106. [p. 302.]

Krishnamurti, R. (1968). Finite amplitude convection with changing mean temperature. *J. Fluid Mech.* **33**, 445–63. [p. 217.]

Krishnamurti, R. (1970). On the transition to turbulent convection. *J. Fluid Mech.* **42**, 295–320. [pp. 218, 220.]

La Fond, E. C. (1962). Internal waves. *The Sea* (ed. M. N. Hill.) vol. 1, pp. 731–51. New York: Interscience. [p. 19.]

La Fond, E. C. (1966). Internal waves. *Encyclopedia of Oceanography* (ed. R. W. Fairbridge), pp. 402–8. New York: Reinhold. [p. 50.]

Laikhtman, D. L. (1961). *Physics of the boundary layer of the atmosphere.* English trans., U.S. Dept. Commerce, Washington. [p. 309.]

Lamb, H. (1932). *Hydrodynamics*, 6th ed. Cambridge University Press. [pp. 9, 15, 48, 94.]

Larsen, L. H. (1969a). Internal waves incident upon a knife edge barrier. *Deep-Sea Res.* **16**, 411–19. [p. 30.]

Larsen, L. H. (1969b). Oscillations of a neutrally bouyant sphere in a stratified fluid. *Deep-Sea Res.* **16**, 587–603. [pp. 31, 202.]

Lighthill, M. J. (1962). Physical interpretation of the mathematical theory of wave generation by wind. *J. Fluid Mech.* **14**, 385–98. [p. 93.]

Lighthill, M. J. and Whitham, G. B. (1955). On kinematic waves. *Proc. Roy. Soc.* A **229**, 281–317. [p. 66.]

Lilly, D. K. (1964). Numerical solutions for the shape-preserving two-dimensional thermal convection element. *J. Atmos. Sci.* **21**, 83–98. [p. 193.]

Lilly, D. K. (1968). Models of cloud-topped mixed layers under a strong inversion. *Quart. J. Roy. Met. Soc.* **94**, 292–309. [p. 309.]

Lin, S. P. (1969). Finite-amplitude stability of a parallel flow with a free surface. *J. Fluid Mech.* **36**, 113–26. [p. 113.]

Linden, P. F. (1971a). Salt fingers in the presence of grid-generated turbulence. *J. Fluid Mech.* **49**, 611–24. [p. 285.]

Linden, P. F. (1971b). The effect of turbulence and shear on salt fingers. *Ph.D. thesis.* University of Cambridge. [pp. 278, 281, 296.]

Lloyd, J. R. and Sparrow, E. M. (1970). On the instability of natural convection flow on inclined plates. *J. Fluid Mech.* **42**, 465–70. [p. 243.]

Lock, R. C. (1951). The velocity distribution in the laminar boundary layer between parallel streams. *Quart. J. Mech. Appl. Math.* **4**, 42–63. [p. 112.]

Lofquist, K. (1960). Flow and stress near an interface between stratified liquids. *Phys. Fluids.* **3**, 158–75. [pp. 72, 182.]

Long, R. R. (1953a). Some aspects of the flow of stratified fluids. I. A theoretical investigation, *Tellus*, **5**, 42–57. [p. 55.]

Long, R. R. (1953b). A laboratory model resembling the 'Bishop-wave' phenomenon. *Bull. Amer. Met. Soc.* **34**, 205–11. [pp. 68, 125.]

Long, R. R. (1954). Some aspects of the flow of stratified fluids. II. Experiments with a two-fluid system. *Tellus*, **6**, 97–115. [pp. 65, 68.]

Long, R. R. (1955). Some aspects of the flow of stratified fluids. III. Continuous density gradients. *Tellus*, **7**, 342–57. [p. 58.]

Long, R. R. (1959). The motion of fluids with density stratification. *J. Geophys. Res.* **64**, 2151–63. [p. 84.]

Long, R. R. (1962). Velocity concentrations in stratified fluids. *J. Hydr. Div., Amer. Soc. Civil Eng.*, **88**, HY1, 9–26. [p. 88.]

Long, R. R. (1965). On the Boussinesq approximation and its role in the theory of internal waves. *Tellus*, **17**, 46–52. [p. 53.]

Long, R. R. (1970). A theory of turbulence in stratified fluids. *J. Fluid Mech.* **42**, 349–65. [p. 317.]

Ludlam, F. H. (1967). Characteristics of billow clouds and their relation to clear-air turbulence. *Quart. J. Roy. Met. Soc.* **93**, 419–35. [p. 106.]

Lumley, J. L. (1964). The spectrum of nearly inertial turbulence in a stably stratified fluid. *J. Atmos. Sci.* **21**, 99–102. [p. 144.]

Lumley, J. L. and Panofsky, H. A. (1964). *The structure of atmospheric turbulence.* New York: Interscience. [p. 139.]

Lyra, G. (1943). Theorie der stationären Leewellenstromung in freier Atmosphäre. *Z. agnew. Math. Mech.* **23**, 1–28. [p. 34.]

Macagno, E. O. and Rouse, H. (1961). Interfacial mixing in stratified flow. *J. Eng. Mech. Div., Amer. Soc. Civil Eng.* **87**, EM 5, 55–81. [p. 156.]

Malkus, W. V. R. (1954a). Discrete transitions in turbulent convection. *Proc. Roy. Soc.* A **225**, 185–95. [p. 219.]

Malkus, W. V. R. (1954b). The heat transport and spectrum of thermal turbulence. *Proc. Roy. Soc.* A **225**, 196–212. [pp. 214, 220.]

Malkus, W. V. R. and Veronis, G. (1958). Finite amplitude cellular convection. *J. Fluid Mech.* **4**, 225–60. [pp. 212, 238.]

Martin, S. (1966). The slow motion of a finite flat plate through a viscous stratified fluid. *Johns Hopkins, Dept. of Mechanics ONR Tech. Rept. no. 21.* [pp. 83, 86.]

Martin, S. and Long, R. R. (1968). The slow motion of a plate through a viscous stratified fluid. *J. Fluid Mech.* **31**, 669–88. [pp. 82, 85.]

Martin, S., Simmons, W. F. and Wunsch, C. (1969). Resonant internal wave interactions. *Nature*, **224**, 1014–15. [p. 45.]

Maslowe, S. A. and Thompson, J. M. (1971). Stability of a stratified free shear layer. *Phys. Fluids*, **14**, 453–8. [p. 104.]

McEwan, A. D. (1971). Degeneration of resonantly-excited standing internal gravity waves. *J. Fluid. Mech.* **50**, 431–48. [pp. 46, 328, 330.]

McIntyre, M. E. (1972). On Long's hypothesis of no upstream influence in uniformly stratified or rotating flow. *J. Fluid Mech.* **52**, 209–43. [p. 57.]

Middleton, G. V. (1966). Experiments on density and turbidity currents. 1. Motion of the head. *Canad. J. Earth Sci.* **3**, 523–46. [p. 70.]

Miles, J. W. (1957). On the generation of surface waves by shear flows. *J. Fluid Mech.* **3**, 185–204. [p. 93.]

Miles, J. W. (1961). On the stability of heterogeneous shear flows. *J. Fluid Mech.* **10**, 496–508. [p. 100.]

Miles, J. W. (1963). On the stability of heterogeneous shear flows. Part 2. *J. Fluid Mech.* **16**, 209–27. [p. 100.]

Miles, J. W. (1968a). Lee waves in a stratified flow. Part 1. Thin barrier. *J. Fluid Mech.* **32**, 549–68. [p. 59.]

Miles, J. W. (1968b). Lee waves in stratified flow. Part 2. Semicircular obstacle. *J. Fluid Mech.* **33**, 803–14. [p. 59.]

Miles, J. W. (1969). Waves and wave drag in stratified flows. In *Proc. twelfth Int. Congress Applied Mechanics, Stanford* (eds. M. Hetenyi and W. G. Vincenti). Berlin: Springer–Verlag. [p. 59.]

Miles, J. W. and Howard, L. N. (1964). Note on a heterogeneous shear flow. *J. Fluid Mech.* **20**, 331–6. [p. 97.]

Miles, J. W. and Huppert, H. E. (1969). Lee waves in a stratified flow. Part 4. Perturbation approximations. *J. Fluid Mech.* **35**, 497–525. [pp. 59, 62.]

Mittendorf, G. H. (1961). The instability of stratified flow. *M.Sc. thesis.* State University of Iowa. [p. 156.]

Monin. A. S. (1959). Smoke propagation in the surface layer of the atmosphere. *Advances in Geophysics*, **6**, 331–44. [p. 140.]

Monin, A. S. (1962a). Empirical data on turbulence in the surface layer of the atmosphere. *J. Geophys. Res.* **67**, 3103–9. [pp. 132, 139.]

Monin, A. S. (1962b). On the turbulence spectrum in a stratified atmosphere. *Bull. Acad. Sci. U.S.S.R. Geophys. Series*, **3**, 266–71. [p. 141.]

Monin, A. S. (1965). On the symmetry properties of turbulence in the surface layer of air. *Bull. Acad. Sci. U.S.S.R. Atmos. and Oceanic Phys.* **1**, 25–30. [pp. 148, 150.]

Monin, A. S. and Obukov, A. M. (1954). Basic laws of turbulent mixing in the ground layer of the atmosphere. *Acad. Sci. U.S.S.R. Leningrad Geophys. Inst.* **24**, 163–87. [p. 131.]

Moore, M. J. and Long, R. R. (1971). An experimental investigation of turbulent stratified shearing flow. *J. Fluid Mech.* **49**, 635–55. [p. 295.]

Mortimer, C. H. (1952). Water movements in lakes during summer stratification: evidence from the distribution of temperature in Windermere. *Phil. Trans. Roy. Soc.* B **236**, 355–404. [p. 21.]

Morton, B. R. (1956). Buoyant plumes in a moist atmosphere. *J. Fluid Mech.* **2**, 127–44. [p. 196.]

Morton, B. R. (1959a). Forced plumes. *J. Fluid Mech.* **5**, 151–63. [pp. 173, 200.]

Morton, B. R. (1959b). The ascent of turbulent forced plumes in a calm atmosphere. *Int. J. Air Poll.* **1**, 184–97. [p. 173.]

Morton, B. R. (1960). Laminar convection in uniformly heated vertical pipes. *J. Fluid Mech.* **8**, 227-40. [p. 240.]

Morton, B. R. (1968). Turbulent structure in cumulus models. *Int. Conf. on Cloud Physics, Toronto Aug. 1968.* [p. 171.]

Morton, B. R., Taylor, Sir Geoffrey and Turner, J. S. (1956). Turbulent gravitational convection from maintained and instantaneous sources. *Proc. Roy. Soc.* A **234**, 1-23. [pp. 171, 197, 200.]

Mowbray, D.E. and Rarity, B.S.H. (1967). A theoretical and experimental investigation of the phase configuration of internal waves of small amplitude in a density stratified liquid. *J. Fluid Mech.* **28**, 1-16. [pp. 24, 29.]

Munk, W. H. (1966). Abyssal recipes. *Deep-Sea Res.* **13**, 707-30. [pp. 316, 331.]

Musman, S. (1968). Penetrative convection. *J. Fluid Mech.* **31**, 343-60. [p. 238.]

Myrup, L. O. (1968). Atmospheric measurements of the buoyant subrange of turbulence. *J. Atmos. Sci.* **25**, 1160-4. [p. 144.]

Neal, V. T., Neshyba, S. and Denner, W. (1969). Thermal stratification in the Arctic Ocean. *Science,* **166**, 373-4. [p. 270.]

Nichol, C. I. H. (1970). Some dynamical effects of heat on a turbulent boundary layer. *J. Fluid Mech.* **40**, 361-84. [p. 155.]

Nield, D. A. (1964). Surface tension and buoyancy effects in cellular convection. *J. Fluid Mech.* **19**, 341-52. [p. 217.]

Nield, D. A. (1967). The thermohaline Rayleigh–Jeffreys problem. *J. Fluid Mech.* **29**, 545-58. [pp. 253, 263.]

Ogura, Y. (1963). The evolution of a moist convective element in a shallow, conditionally unstable atmosphere: a numerical calculation. *J. Atmos. Sci.* **20**, 407-24. [p. 196.]

Okamoto, M. and Webb, E. K. (1970). The temperature fluctuations in stable stratification. *Quart. J. Roy. Met. Soc.* **96**, 591-600. [p. 155.]

Oke, T. R. (1970). Turbulent transport near the ground in stable conditions. *J. Appl. Met.* **9**, 778-86. [p. 137.]

Okubo, A. (1968). Some remarks on the importance of the 'shear effect' on horizontal diffusion. *J. Ocean. Soc. Japan,* **24**, 60-9. [p. 317.]

Okubo, A. (1970). Oceanic mixing. *Chesapeake Bay Inst. Tech. Rept. no. 62,* Johns Hopkins University. [p. 317.]

Orlanski, I. and Bryan, K. (1969). Formation of the thermocline step structure by large-amplitude internal gravity waves. *J. Geophys. Res.* **74**, 6975-83. [p. 329.]

Ostrach, S. (1964). Laminar flows with body forces. In *Theory of laminar flows* (ed. F. K. Moore). Princeton University Press. [p. 242.]

Ozmidov, R. V. (1965a). Energy distribution between oceanic motions of different scales. *Bull. Acad. Sci. U.S.S.R. Atmos. and Oceanic Phys.* **1**, 257-61. [p. 142.]

Ozmidov, R. V. (1965b). On the turbulent exchange in a stably stratified ocean. *Bull. Acad. Sci. U.S.S.R. Atmos. and Oceanic. Phys.* **1**, 493-7. [p. 143.]

Pao, Y-H. (1968a). Laminar flow of a stably stratified fluid past a plate. *J. Fluid Mech.* **34**, 795-808. [p. 85.]

Pao, Y-H. (1968 b). Undulance and turbulence in stably stratified media. *Symposium on clear air turbulence and its detection*, Seattle, Washington. [p. 151.]

Pao, Y-H. (1969). Inviscid flows of stably stratified fluids over barriers. *Quart. J. Roy. Met. Soc.* **95**, 104–19. [p. 59.]

Pearce, R. P. and White, P. W. (1967). Lee wave characteristics derived from a three layer model. *Quart. J. Roy. Met. Soc.* **93**, 155–65. [p. 35.]

Pearson, J. R. A. (1958). On convection cells induced by surface tension. *J. Fluid Mech.* **4**, 489–500. [p. 217.]

Pellew, A. and Southwell, R. V. (1940). On maintained convective motion in a fluid heated from below. *Proc. Roy. Soc.* A **176**, 312–43. [p. 211.]

Phillips, A. C. and Walker, G. T. (1932). The forms of stratified clouds. *Quart. J. Roy. Met. Soc.* **58**, 23–30. [p. 216.]

Phillips, O. M. (1960). On the dynamics of unsteady gravity waves of finite amplitude. *J. Fluid Mech.* **9**, 193–217. [p. 41.]

Phillips, O. M. (1966 a). *The dynamics of the upper ocean.* Cambridge University Press. (2nd ed. 1977.) [pp. 2, 6, 17, 46, 120, 141, 310, 332.]

Phillips, O. M. (1966 b). On turbulent convection currents and the circulation of the Red Sea. *Deep-Sea Res.* **13**, 1149–60. [pp. 163, 287.]

Phillips, O. M. (1968). The interaction trapping of internal gravity waves. *J. Fluid Mech.* **34**, 407–16. [p. 45.]

Phillips, O. M. (1970). On flows induced by diffusion in a stably stratified fluid. *Deep-Sea Res.* **17**, 435–43. [p. 244.]

Phillips, O. M. (1971). On spectra measured in an undulating layered medium. *J. Phys. Oceanog.* **1**, 1–6. [p. 21.]

Phillips, O. M., George, W. K. and Mied, R. P. (1968). A note on the interaction between internal gravity waves and currents. *Deep-Sea Res.* **15**, 267–73. [p. 45.]

Pillow, A. F. (1952). The free convection cell in two dimensions. *Rep. Aero. Res. Lab., Melbourne*, A79. [p. 246.]

Pingree, R. D. (1971). Analysis of the temperature and salinity small-scale structure in the region of the Mediterranean influence in the N.E. Atlantic. *Deep-Sea Res.* **18**, 485–91. [pp. 273, 317.]

Polymeropoulos, C. E. and Gebhart, B. (1967). Incipient instability in free convection laminar boundary layers. *J. Fluid Mech.* **30**, 225–39. [p. 242.]

Prandtl, L. (1952). *Essentials of fluid dynamics.* London: Blackie. [pp. 7, 23, 73, 129, 240, 243.]

Priestley, C. H. B. (1953). Buoyant motion in a turbulent environment. *Aust. J. Phys.* **6**, 279–90. [p. 203.]

Priestley, C. H. B. (1956). A working theory of the bent-over plume of hot gas. *Quart. J. Roy. Met. Soc.* **82**, 165–76. [p. 205.]

Priestley, C. H. B. (1959). *Turbulent transfer in the lower atmosphere.* University of Chicago Press. [pp. 127, 199, 203, 230, 234.]

Priestley, C. H. B. and Ball, F. K. (1955). Continuous convection from an isolated source of heat. *Quart. J. Roy. Met. Soc.* **81**, 144–57. [p. 199.]

Proudman, J. (1953). *Dynamical Oceanography.* London: Methuen. [p. 21.]

Prych, E. A., Harty, F. R. and Kennedy, J. F. (1964). Turbulent wakes in density stratified fluids of finite extent. *MIT Hydrodynamics Laboratory Report no. 65.* [p. 152.]

Queney, P. (1941). Ondes de gravité produites dans un courant aérien par une petite chaîne de montagnes. *Comptes Rendus,* 213, 588–91. [pp. 34, 36.]

Queney, P. (1948). The problem of air flow over mountains: a summary of theoretical studies. *Bull. Amer. Meteor. Soc.* 29, 16–26. [p. 34.]

Rarity, B. S. H. (1967). The two-dimensional wave pattern produced by a disturbance moving in an arbitrary direction in a density stratified liquid. *J. Fluid Mech.* 30, 329–36. [p. 31.]

Rayleigh, Lord (1916). On convection currents in a horizontal layer of fluid when the higher temperature is on the under side. *Phil. Mag.* (6) 32, 529–46. [p. 210.]

Richards, J. M. (1963). Experiments on the motions of isolated cylindrical thermals through unstratified surroundings. *Intern. J. Air Water Poll.* 7, 17–34. [p. 192.]

Richardson, L. F. (1920). The supply of energy from and to atmospheric eddies. *Proc. Roy. Soc.* A 97, 354–73. [p. 12.]

Ricou, F. P. and Spalding, D. B. (1961). Measurements of entrainment by axisymmetric turbulent jets. *J. Fluid Mech.* 11, 21–32. [p. 173.]

Riehl, H., Yeh, T. C., Malkus, J. S. and La Seur, N. E. (1951). The northeast trade of the Pacific ocean. *Quart. J. Roy. Met. Soc.* 77, 598–626. [p. 309,]

Rosenhead, L. (1931). The formation of vortices from a surface of discontinuity. *Proc. Roy. Soc.* A 134, 170–92. [p. 96.]

Rossby, H. T. (1969). A study of Bénard convection with and without rotation. *J. Fluid Mech.* 36, 309–35. [pp. 218, 229.]

Rouse, H. (1956). Seven exploratory studies in hydraulics 1. Development of the non-circulatory waterspout. *J. Hydr. Div. ASCE,* 82, 1038-3 to 1038-7. [p. 77.]

Rouse, H. and Dodu, J. (1955). Turbulent diffusion across a density discontinuity. *La Houille Blanche,* 10, 530–2. [pp. 289, 295.]

Rouse, H., Yih, C.-S. and Humphreys, H. W. (1952). Gravitational convection from a boundary source. *Tellus,* 4, 201–10. [pp. 168, 177.]

Sandstrom, H. (1969). Effect of topography on propagation of waves in stratified fluids. *Deep-Sea Res.* 16, 405–10. [p. 30.]

Sani, R. L. (1965). On finite amplitude roll cell disturbances in a fluid layer subjected to a heat and mass transfer. *Am. Inst. Chem. Eng. J.* 11, 971–80. [p. 259.]

Sawyer, J. S. (1960). A numerical calculation of the displacements of a stratified airstream crossing a ridge of small height. *Quart. J. Roy. Met. Soc.* 86, 326–45. [p. 36.]

Schlichting, H. (1935). Turbulenz bei Wärmeschichtung. *Z. angew. Math. Mech.* 15, 313–38. [p. 110.]

Schlüter, A., Lortz, D. and Busse, F. (1965). On the stability of steady finite amplitude convection. *J. Fluid Mech.* 23, 129–44. [p. 213.]

Schmidt, R. J. and Milverton, S. W. (1935). On the instability of a fluid when heated from below. *Proc. Roy. Soc.* A 152, 586–94. [p. 216.]

Schmidt, R. J. and Saunders, O. A. (1938). On the motion of a fluid heated from below. *Proc. Roy. Soc.* A **165**, 216–28. [p. 216.]

Schooley, A. H. and Stewart, R. W. (1963). Experiments with a self-propelled body submerged in a fluid with a vetical density gradient. *J. Fluid Mech.* **15**, 83–96. [p. 151.]

Schwarzschild, M. and Härm, R. (1958). Evolution of very massive stars. *Astrophys. J.* **128**, 348–60. [p. 279.]

Scorer, R. S. (1949). Theory of waves in the lee of mountains. *Quart. J. Roy. Met. Soc.* **75**, 41–56. [p. 35.]

Scorer, R. S. (1954). Theory of airflow over mountains. III. Airstream characteristics. *Quart. J. Roy. Met. Soc.* **80**, 417–28. [p. 35.]

Scorer, R. S. (1957). Experiments on convection of isolated masses of buoyant fluid. *J. Fluid Mech.* **2**, 583–94. [p. 188.]

Scorer, R. S. and Wilson, S. D. R. (1963). Secondary instability in steady gravity waves. *Quart. J. Roy. Met. Soc.* **89**, 532–9. [p. 125.]

Scotti, R. S. and Corcos, G. M. (1969). Measurements on the growth of small disturbances in a stratified shear layer. *Radio Science*, **4**, 1309–13. [p. 102.]

Shirtcliffe, T. G. L. (1967). Thermosolutal convection: observation of an overstable mode. *Nature*, **213**, 489–90. [p. 262.]

Shirtcliffe, T. G. L. (1969*a*). An experimental investigation of thermosolutal convection at marginal stability. *J. Fluid Mech.* **35**, 677–88. [p. 262].

Shirtcliffe, T. G. L. (1969*b*). The development of layered thermosolutal convection. *Int. J. Heat. Mass Transfer*, **12**, 215–22. [p. 265.]

Shirtcliffe, T. G. L. (1972). Colour–Schlieren observations of double-diffusive interfaces. *J. Fluid Mech.* (in the press). [p. 282.]

Shirtcliffe, T. G. L. and Turner, J. S. (1970). Observations of the cell structure of salt fingers. *J. Fluid Mech.* **41**, 707–19. [pp. 282, 284.]

Silveston, P. L. (1958). Wärmedurchgang in waagerechten Flüssigkeits-schichten. *Forsch. a.d. Geb. des. Ingen.* **24**, 29–32 and 59–69. [pp. 219, 223.]

Simmons, W. F. (1969). A variational method for weak resonant wave interactions. *Proc. Roy. Soc.* A **309**, 551–75. [p. 41.]

Simpson, J. E. (1969). A comparison between laboratory and atmospheric density currents. *Quart. J. Roy. Met. Soc.* **95**, 758–65. [pp. 74, 106.]

Simpson, J. H. and Woods, J. D. (1970). Temperature microstructure in a fresh water thermocline. *Nature*, **226**, 832–5. [p. 315.]

Slawson, P. R. and Csanady, G. T. (1967). On the mean path of buoyant, bent-over chimney plumes. *J. Fluid Mech.* **28**, 311–22. [p. 205.]

Sparrow, E. M., Husar, R. B. and Goldstein, R. J. (1970). Observations and other characteristics of thermals. *J. Fluid Mech.* **41**, 793–800. [p. 229.]

Spiegel, E. A. (1969). Semiconvection. *Comments on Astrophys. Space Phys.* **1**, 57–61. [p. 279.]

Spiegel, E. A. and Veronis, G. (1960). On the Boussinesq approximation for a compressible fluid. *Astrophys. J.* **131**, 442–7. [p. 9.]

Squires, P. and Turner, J. S. (1962). An entraining jet model for cumulo-nimbus updraughts. *Tellus*, **14**, 422–34. [p. 196.]

Stern, M. E. (1960). The 'salt fountain' and thermohaline convection. *Tellus.* **12**, 172–5. [pp. 253, 258.]

Stern, M. E. (1969). Collective instability of salt fingers. *J. Fluid Mech.* **35**, 209–18. [p. 284.]

Stern, M. E. and Turner, J. S. (1969). Salt fingers and convecting layers. *Deep-Sea Res.* **16**, 497–511. [pp. 267, 281.]

Stevenson, T. N. (1968). Some two-dimensional internal waves in a stratified fluid. *J. Fluid Mech.* **33**, 715–20. [p. 31.]

Stewart, R. W. (1959). The problem of diffusion in a stratified fluid. *Adv. in Geophysics*, **6**, 303–11. [p. 146.]

Stewart, R. W. (1969). Turbulence and waves in a stratified atmosphere. *Radio Science*, **4**, 1269–78. [pp. 137, 316, 320, 335.]

Stokes, G. G. (1847). On the theory of oscillatory waves. *Trans. Camb. Phil. Soc.* **8**, 441–55. [p. 1.]

Stommel, H. (1962). On the smallness of sinking regions in the ocean. *Proc. Nat. Acad. Sci. Washington*, **48**, 766–72. [p. 231.]

Stommel, H. and Fedorov, K. N. (1967). Small scale structure in temperature and salinity near Timor and Mindanao. *Tellus*, **19**, 306–25. [p. 273.]

Stommel, H., Arons, A. B. and Blanchard, D. (1956). An oceanographical curiosity: the perpetual salt fountain. *Deep-Sea Res.* **3**, 152–3. [p. 252.]

Tait, R. I. and Howe, M. R. (1968). Some observations of thermohaline stratification in the deep ocean. *Deep-Sea Res.* **15**, 275–80. [p. 271.]

Tait, R. I. and Howe, M. R. (1971). Thermohaline staircase. *Nature*, **231**, 178–9. [p. 272.]

Taylor, G. I. (1931a). Effect of variation of density on the stability of superposed streams of fluid. *Proc. Roy. Soc.* A **132**, 499–523. [p. 97.]

Taylor, G. I. (1931b). Internal waves and turbulence in a fluid of variable density. *Conseil Perm. Int. pour. l'Expl. de la Mer., Rapp. et Proc-Verb*, **76**, 35–42. [pp. 157, 161.]

Taylor, G. I. (1954). The dispersion of matter in turbulent flow through a pipe. *Proc. Roy. Soc.* A **223**, 446–68. [p. 162.]

Taylor, G. I. (1958). Flow induced by jets. *J. Aero./Space Sci.* **25**, 464–5. [p. 170.]

Telford, J. W. (1966). The convective mechanism in clear air. *J. Atmos. Sci.* **23**, 652–66. Also discussion by B. R. Morton, and author's reply, *ibid.* **25**, 135–9. [p. 171.]

Telford, J. W. (1970). Convective plumes in a convective field. *J. Atmos. Sci.* **27**, 347–58. [pp. 171, 231.]

Thomas, D. B. and Townsend, A. A. (1957). Turbulent convection over a heated horizontal surface. *J. Fluid Mech.* **2**, 473–92. [p. 220.]

Thompson, S. M. (1969). Turbulent interfaces generated by an oscillating grid in a stably stratified fluid. *Ph.D. thesis*, University of Cambridge. [p. 290.]

Thorpe, S. A. (1966a). Internal gravity waves, *Ph.D. thesis*. University of Cambridge. [p. 125.]

Thorpe, S. A. (1966b). On wave interactions in a stratified fluid. *J. Fluid. Mech.* **24**, 737–51. [pp. 42, 44.]

Thorpe, S. A. (1968a). On standing internal gravity waves of finite amplitude. *J. Fluid Mech.* **32**, 489–528. [pp. 24, 30, 49, 123.]

Thorpe, S. A. (1968b). A method of producing a shear flow in a stratified fluid. *J. Fluid Mech.* **32**, 693–704. [p. 103.]

Thorpe, S. A. (1968c). On the shape of progressive internal waves. *Phil. Trans.* A **263**, 563–614. [pp. 44, 49, 52, 55.]

Thorpe, S. A. (1969a). Neutral eigensolutions of the stability equation for stratified shear flow. *J. Fluid Mech.* **36**, 673–83. [p. 115.]

Thorpe, S. A. (1969b). Experiments on the instability of stratified shear flows: immiscible fluids. *J. Fluid Mech.* **39**, 25–48. [p. 104.]

Thorpe, S. A. (1969c). Experiments on the stability of stratified shear flows. *Radio Science*, **4**, 1327–31. [p. 103.]

Thorpe, S. A. (1971). Experiments on the instability of stratified shear flows: miscible fluids. *J. Fluid Mech.* **46**, 299–319. [pp. 103, 104.]

Thorpe, S. A., Hall, A. and Crofts, I. (1972). The internal surge in Loch Ness. *Nature*, **237**, 96–8. [pp. 20, 68.]

Thorpe, S. A., Hutt, P. K. and Soulsby, R. (1969). The effect of horizontal gradients on thermohaline convection. *J. Fluid Mech.* **38**, 375–400. [p. 268.]

Townsend, A. A. (1956). *The structure of turbulent shear flow.* Cambridge University Press. (2nd ed. 1976.) [p. 127.]

Townsend, A. A. (1957). Turbulent flow in a stably stratified atmosphere. *J. Fluid Mech.* **3**, 361–72. [p. 154.]

Townsend, A. A. (1958). The effects of radiative transfer on turbulent flow of a stratified fluid. *J. Fluid Mech.* **4**, 361–75. [p. 146.]

Townsend, A. A. (1959). Temperature fluctuations over a heated horizontal surface. *J. Fluid Mech.* **5**, 209–41. [pp. 214, 220, 226, 250.]

Townsend, A. A. (1962). Natural convection in the earth's boundary layer. *Quart. J. Roy. Met. Soc.* **88**, 51–6. [pp. 134, 136, 214.]

Townsend, A. A. (1964). Natural convection in water over an ice surface. *Quart. J. Roy. Met. Soc.* **90**, 248–59. [pp. 236, 238.]

Townsend, A. A. (1965). Excitation of internal waves by a turbulent boundary layer. *J. Fluid Mech.* **22**, 241–52. [pp. 237, 311.]

Townsend, A. A. (1966). Internal waves produced by a convective layer. *J. Fluid Mech.* **24**, 307–19. [pp. 237, 311.]

Townsend, A. A. (1967). Wind and the formation of inversions. *Atmos. Envir.* **1**, 173–5. [p. 155.]

Townsend, A. A. (1968). Excitation of internal waves in a stably stratified atmosphere with considerable wind shear. *J. Fluid Mech.* **32**, 145–71. [pp. 237, 311.]

Townsend, A. A. (1970). Entrainment and the structure of turbulent flow. *J. Fluid Mech.* **41**, 13–46. [pp. 167, 171.]

Tritton, D. J. (1963). Transition to turbulence in the free convection boundary layers on an inclined heated plate. *J. Fluid Mech.* **16**, 417–35. [pp. 113, 243.]

Tritton, D. J. and Zarraga, M. N. (1967). Convection in horizontal layers with internal heat generation. Experiments. *J. Fluid Mech.* **30**, 21–31. [p. 217.]

Trustrum, K. (1964). Rotating and stratified fluid flow. *J. Fluid Mech.* **19**, 415–32. [p. 79.]

Tsang, G. (1970). Laboratory study of two-dimensional starting plumes. *Atmos. Envir.* **4**, 519–44. [p. 194.]

Tully, J. P. and Giovando, L. F. (1963). Seasonal temperature structure in the eastern subarctic Pacific ocean. *Marine distributions* (ed. M. J. Dunbar), pp. 10–36. Toronto: University Press. [p. 304.]

Turner, J. S. (1957). Buoyant vortex rings. *Proc. Roy. Soc.* A **239**, 61–75. [p. 189.]

Turner, J. S. (1960a). A comparison between buoyant vortex rings and vortex pairs. *J. Fluid Mech.* **7**, 419–32. [pp. 193, 202.]

Turner, J. S. (1960b). Intermittent release of smoke from chimneys. *Mech. Eng. Science*, **2**, 97–100. [p. 202.]

Turner, J. S. (1962). The starting plume in neutral surroundings. *J. Fluid Mech.* **13**, 356–68. [p. 192.]

Turner, J. S. (1963a). Model experiments relating to thermals with increasing buoyancy. *Quart. J. Roy. Met. Soc.* **89**, 62–74. [p. 195.]

Turner, J. S. (1963b). The motion of buoyant elements in turbulent surroundings. *J. Fluid Mech.* **16**, 1–16. [p. 204.]

Turner, J. S. (1964a). The flow into an expanding spherical vortex. *J. Fluid Mech.* **18**, 195–208. [p. 189.]

Turner, J. S. (1964b). The dynamics of spheroidal masses of buoyant fluid. *J. Fluid Mech.* **19**, 481–90. [p. 188.]

Turner, J. S. (1965). The coupled turbulent transports of salt and heat across a sharp density interface. *Int. J. Heat and Mass Trans.* **8**, 759–67. [pp. 274, 276.]

Turner, J. S. (1966). Jets and plumes with negative or reversing buoyancy. *J. Fluid Mech.* **26**, 779–92. [p. 175.]

Turner, J. S. (1967). Salt fingers across a density interface. *Deep-Sea Res*, **14**, 599–611. [p. 281.]

Turner, J. S. (1968a). The behaviour of a stable salinity gradient heated from below. *J. Fluid Mech.* **33**, 183–200. [p. 264.]

Turner, J. S. (1968b). The influence of molecular diffusivity on turbulent entrainment across a density interface. *J. Fluid Mech.* **33**, 639–56 [pp. 290, 296.]

Turner, J. S. (1969a). Buoyant plumes and thermals. *Annual Rev. Fluid Mech.* **1**, 29–44. [p. 165.]

Turner, J. S. (1969b). A note on wind mixing at the seasonal thermocline. *Deep-Sea Res.* Suppl. to vol. **16**, 297–300. [p. 300.]

Turner, J. S. (1972). On the energy deficiency of self-preserving convective flows. *J. Fluid Mech.* **53**, 217–26. [p. 189.]

Turner, J. S. and Kraus, E. B. (1967). A one-dimensional model of the seasonal thermocline. I. A laboratory experiment and its interpretation. *Tellus*, **19**, 88–97. [p. 302.]

Turner, J. S. and Yang, I. K. (1963). Turbulent mixing at the top of stratocumulus clouds. *J. Fluid Mech.* **17**, 212–24. [p. 237.]

Turner, J. S., Shirtcliffe, T. G. L. and Brewer, P. G. (1970). Elemental variations of transport coefficients across density interfaces in multiple-diffusive systems. *Nature*, **228**, 1083–4. [p. 278.]

356   BUOYANCY EFFECTS IN FLUIDS

Veronis, G. (1963). Penetrative convection. *Astrophys. J.* **137**, 641–63. [p. 238.]

Veronis, G. (1965). On finite amplitude instability in thermohaline convection. *J. Mar. Res.* **23**, 1–17. [pp. 253, 259.]

Veronis, G. (1966). Large amplitude Bénard convection. *J. Fluid Mech.* **26**, 49–68. [p. 225.]

Veronis, G. (1967). Analogous behaviour of homogeneous rotating fluids and stratified, non-rotating fluids. *Tellus*, **19**, 326–36. [p. 243.]

Veronis, G. (1968). Effect of a stabilizing gradient of solute on thermal convection. *J. Fluid Mech.* **34**, 315–36. [pp. 259, 277.]

Walin, G. (1964). Note on the stability of water stratified by both salt and heat. *Tellus*. **16**, 389–93. [p. 253.]

Walin, G. (1971). Contained non-homogeneous flow under gravity, or how to stratify a fluid in the laboratory. *J. Fluid Mech.* **48**, 647–72. [p. 249.]

Ward, P. R. B. and Fischer, H. B. (1971). Some limitations on use of the one-dimensional dispersion equation. *Water Resources Res.* **7**, 215–20. [p. 162.]

Warner, J. (1970). On steady-state one-dimensional models of cumulus convection. *J. Atmos. Sci.* **27**, 1035–40. [p. 196.]

Warner, J. and Telford, J. W. (1967). Convection below cloud base. *J. Atmos. Sci.* **24**, 374–82. [pp. 234, 307.]

Warren, F. W. G. (1960). Wave resistance to vertical motion in a stratified fluid. *J. Fluid Mech.* **7**, 209–29. [p. 202.]

Webb, E. K. (1970). Profile relationships: the log-linear range, and the extension to strong stability. *Quart. J. Roy. Met. Soc.* **96**, 67–90. [pp. 131, 136.]

Webster, C. A. G. (1964). An experimental study of turbulence in a density-stratified shear flow. *J. Fluid Mech.* **19**, 221–45. [p. 160.]

Webster, T. F. (1971). On the intensity of horizontal ocean currents. *Deep-Sea Res.* **18**, 885–93. [p. 332.]

Whitehead, J. A. and Chen, M. M. (1970). Thermal instability and convection of a thin fluid layer bounded by a stable region. *J. Fluid. Mech.* **40**, 549–70. [p. 238.]

Wilkinson, D. L. and Wood, I. R. (1971). A rapidly varied flow phenomenon in a two-layer flow. *J. Fluid Mech.* **47**, 241–56. [p. 182.]

Wood, I. R. (1966). Studies in unsteady self-preserving flows. Univ. of N.S.W. *Water Research Lab. Rept. no. 81.* [p. 73.]

Wood, I. R. (1968). Selective withdrawal from a stably stratified fluid. *J. Fluid Mech.* **32**, 209–23. [pp. 74, 76.]

Wood, I. R. (1970). A lock exchange flow. *J. Fluid Mech.* **42**, 671–87. [p. 76.]

Woods, J. D. (1968a). An investigation of some physical processes associated with vertical flow of heat through the upper ocean. *Met. Mag.* **97**, 65–72. [p. 325.]

Woods, J. D. (1968b). Wave-induced shear instability in the summer thermocline. *J. Fluid Mech.* **32**, 791–800. [pp. 122, 325.]

Woods, J. D. (1969). On Richardson's number as a criterion for laminar-turbulent-laminar transition in the atmosphere and ocean. *Radio Science*, **4**, 1289–98. [p. 316.]

Woods, J. D. and Wiley, R. L. (1972). Billow turbulence and ocean microstructure. *Deep-Sea Res.* **19**, 87–121. [pp. 317, 325.]

Wu, J. (1969). Mixed region collapse with internal wave generation in a density stratified medium. *J. Fluid Mech.* **35**, 531–44. [p. 152.]

Wunsch, C. (1969). Progressive internal waves on slopes. *J. Fluid Mech.* **35**, 131–44. [pp. 28, 30, 45, 125.]

Wunsch, C. (1970). On oceanic boundary mixing. *Deep-Sea Res.* **17**, 293–301. [pp. 244, 273.]

Yaglom, A. M. (1969). Horizontal turbulent transport of heat in the atmosphere and the form of the eddy diffusivity tensor. *Polish Acad. Sci., Fluid Dynamics Trans.* **4**, 801–12, [p. 150.]

Yih, C-S. (1955). Stability of two-dimensional parallel flows for three-dimensional disturbances. *Quart. Appl. Math.* **12**, 434–5. [p. 94.]

Yih, C-S. (1958). On the flow of a stratified fluid. *Proc. 3rd U.S. Nat. Cong. Appl. Mech.* 857–61. [p. 10.]

Yih, C-S. (1960). Exact solutions for steady two-dimensional flow of a stratified fluid. *J. Fluid Mech.* **9**, 161–74. [pp. 56, 58.]

Yih, C-S. (1965). *Dynamics of nonhomogeneous fluids*. New York and London: Macmillan. [pp. 6, 56, 68, 77, 79, 113.]

Zenk, W. (1970). On the temperature and salinity structure of the Mediterranean water in the north-east Atlantic. *Deep-Sea Res.* **17**, 627–32. [p. 273.]

The following references, with brief notes on their relation to the subjects discussed in the text, were added in proof to the first printing.

Arya, S. P. S. (1972). The critical condition for the maintenance of turbulence in stratified flows. *Quart. J. Roy. Met. Soc.* **98**, 264–73. Presents wind tunnel measurements and theoretical arguments which support the idea of a critical value of K_H/K_M in stable stratification (see §§5.2.3 and 5.3.3).

Gargett, A. E. and Hughes, B. A. (1972). On the interaction of surface and internal waves. *J. Fluid Mech.* **52**, 171–91. Gives a new explanation for the appearance of 'slicks' (fig. 2.3 pl. IV). Suggests that the increase in reflectivity may be due not to surface contaminants, but to short steep waves formed by resonant interaction (§2.4).

Gregg, M. C. and Cox, C. S. (1972). The vertical microstructure of temperature and salinity. *Deep-Sea Res.* **19**, 355–76. Ocean microstructure measurements extending into the millimetre range. These contain evidence of both mechanical (§4.3) and double-diffusive (§8.2) mixing processes.

Huppert, H. E. and Turner, J. S. (1972). Double-diffusive convection and its implications for the temperature and salinity structure of the ocean and Lake Vanda. *J. Phys. Oceanog.* **2**, 456–61. It is suggested that one should be cautious about applying laboratory data to poorly defined situations in the ocean. A confirmation of the relation (8.3.3) can be obtained on a larger scale, however, using observations in an Antarctic Lake.

Kullenberg, G. (1972). Apparent horizontal diffusion in a stratified vertical shear flow. *Tellus*, **24**, 17–28.

Field observations of the spread of patches of dye are interpreted using a theoretical model which combines vertical diffusion and an oscillating vertical current shear (see §5.3.4).

Martin, S., Simmons, W. and Wunsch, C. (1972). The excitation of resonant triads by single internal waves. *J. Fluid Mech.* **53**, 17–44.

An extended and up to date description of interaction experiments using travelling internal waves (§2.4.3).

Morton, B. R. (1971). The choice of conservation equations for plume models. *J. Geophys. Res.* **76**, 7409–16.

A comparison of various theories of turbulent plumes (see p. 199) which lends support to the formulation based on flux equations for mass and vertical momentum (§6.1.2).

Pochapsky, T. E. (1972). Internal waves and turbulence in the deep ocean. *J. Phys. Oceanog.* **2**, 96–103.

The author concludes on the basis of ocean measurements that most of the kinetic energy in the fluctuating components is associated with non-mixing or internal wave type motions (see §10.3.4).

Schooley, A. H. and Hughes, B. A. (1972). An experimental and theoretical study of internal waves generated by the collapse of a two-dimensional mixed region in a density gradient. *J. Fluid Mech.* **51**, 159–76.

The internal wave amplitude generated by a collapsing, initially turbulent region is well predicted using a highly idealized linear model (§5.3.1).

Scotti, R. S. and Corcos, G. M. (1972). An experiment on the stability of small disturbances in a stratified shear layer. *J. Fluid Mech.* **52**, 499–528.

This gives a more detailed description of the experiment discussed in §4.1.4. It also shows photographs of the breakdown of a subcritical flow as it passes through a contraction (see p. 96).

Simpson, J. E. (1972). Effects of the lower boundary on the head of a gravity current. *J. Fluid Mech.* **53**, 759–67.

A laboratory study of the three-dimensional structure, which supports an explanation based on the gravitational instability of the lighter fluid overrun by the nose (see fig. 3.14 pl. VI).

Spiegel, E. A. (1971). Convection in stars. I. Basic Boussinesq convection. *Ann. Rev. Astron. & Astrophys.* **9**, 323–52.

An excellent review of convection, emphasizing the theory and especially the recent numerical work referred to in §7.2.3. Part II. 'Special effects', *ibid.* **10**, 261–304, includes a discussion of double-diffusive convection.

Thomas, N. H. and Stevenson, T. N. (1972). A similarity solution for viscous internal waves. *J. Fluid Mech.* **54**, 495–506.

A theoretical description, supported by detailed experiments, of internal waves generated by a localized disturbance in a viscous fluid. (cf. figs. 2.7 and 2.11).

RECENT PUBLICATIONS (1979 REPRINT)

Briscoe, M. G. (1975). Internal waves in the ocean: 1975. *Rev. Geophys. Space Phys.* **13**, 591–8.
A good review of the subject, with a comprehensive bibliography of recent work.

Farmer, D. M. (1975). Penetrative convection in the absence of mean shear. *Quart. J. Roy. Met. Soc.* **101**, 869–91.
A clearcut series of observations on convective deepening in an ice-covered lake, where there can be no mechanical energy inputs (§§7.3.4, 9.2.3).

Fedorov, K. N. (1976). *The thermohaline finestructure of the ocean.* Leningrad: Gidrometeiozdat.
Gives a good summary of recent ocean observations, especially of layering associated with two-dimensional double-diffusive processes (§8.2.3). In the English translation (Pergamon 1978) extra references (to mid-1977) have been added.

Fischer, H. B. (1976). Mixing and dispersion in estuaries. *Ann. Rev. Fluid. Mech.* **8**, 107–33.
An excellent review of a wide range of processes, including the effects of stratification and lateral mixing (§5.3.4).

Gargett, A. E. (1978). Microstructure and finestructure in an upper ocean frontal regime. *J. Geophys. Res.* **83**, 5123–34.
Documents the importance of double-diffusive processes near oceanic fronts. Two-dimensional effects must be considered, as well as the one-dimensional processes described in §8.2.4.

Garrett, C. J. R. and Munk, W. (1975). Space-time scales of internal waves: a progress report. *J. Geophys. Res.* **80**, 291–7.
The authors have proposed a universal wave number-frequency spectrum, based on observations, which has had a great influence on work in this field.

Gregg, M. C. (1975). Microstructure and intrusions in the California Current. *J. Phys. Oceanogr.* **5**, 253–78.
Shows that the regions of most intense activity are the upper and lower boundaries of intrusions produced by interleaving (§8.2,3).

Hopfinger, E. J. and Toly, J-A. (1976). Spatially decaying turbulence and its relation to mixing across density interfaces. *J. Fluid Mech.* **78**, 155–75.
Substantial confirmation of the results described in §9.1, derived from larger-scale laboratory experiments.

Huppert, H. E. and Moore, D. A. (1976). Nonlinear double-diffusive convection. *J. Fluid Mech.* **78**, 821–54.
A numerical study of large-amplitude motions (§8.1.4) with wider implications for other convection problems. The minimum value of *Ra* for which time-independent motions can occur can be less than the minimum critical value predicted by linear theory.

Huppert, H. E. and Turner, J. S. (1978). On melting icebergs. *Nature*
  271, 46–8.
The two-dimensional system of layers produced by cooling and
melting at a vertical ice surface in a stratified fluid (§8.2.3) is shown
to have important oceanographic implications.

Imberger, J., Thompson, R. and Fandry, C. (1976). Selective withdrawal
  from a finite rectangular tank. *J. Fluid Mech.* **78**, 489–512.
A unified treatment of the various flow regimes described in §3.3. It
also deals with the related intrusion problems.

Joyce, T. M., Zenk, W. and Toole, J. M. (1978). The anatomy of the
  Antarctic polar front zone in the Drake Passage. *J. Geophys. Res.* **83**,
  6093–113.
Includes observations of interleaving structures which are interpreted
in terms of double-diffusive processes (§8.2.3).

Katsaros, K. B., Liu, W. T., Businger, J. A. and Tillman, J. E. (1977).
  Heat transport and thermal structure in the interfacial boundary
  layer measured in an open tank of water in turbulent free convection.
  J. Fluid Mech. **83**, 311–35.
Gives an excellent summary of the background, as well as experimental
results at high Ra which support the $Nu \propto Ra^{1/3}$ relation (pp. 213, 220).

Kotsovinos, N. E. and List, E. J. (1977). Plane turbulent buoyant jets.
  Part I. Integral properties. *J. Fluid Mech.* **81**, 25–44.
This paper (and Part II on the turbulence structure) gives a sub-
stantially improved description of the entrainment process (§6.1.4).

Kraus, E. B. (Ed.) (1977). *Modelling and prediction of the upper layers of
  the ocean.* Pergamon, Oxford.
The proceedings of a conference at which the leading workers in the
field surveyed various aspects of the problems discussed in §9.2.

Launder, B. E. (1976). Heat and mass transport. In: *Turbulence*, ed. P.
  Bradshaw, pp. 231–87. Berlin: Springer-Verlag.
Summarizes methods of calculating turbulent structure and fluxes,
based on closure schemes which are developments of those described
in §5.2.3. Other articles in the volume are also relevant.

Lighthill, Sir James (1978). *Waves in fluids.* Cambridge University Press.
An authoritative monograph covering all types of wave motion,
including linear and non-linear internal waves. There is a helpful
annotated bibliography.

Maxworthy, T. and Browand, F. K. (1975). Experiments in rotating and
  stratified flows: oceanographic application. *Ann. Rev. Fluid Mech.*
  7, 273–305.
Reviews in particular experimental methods, intrusions (§3.3), and
instabilities due to internal waves (§§4.1.4, 10.3).

McEwan, A. D. and Robinson, R. M. (1975). Parametric instability of
  internal gravity waves. *J. Fluid Mech.* **67**, 667–87.
The previously puzzling instabilities described on p. 328 and shown in
fig. 10.5 have now been explained in terms of another resonance
mechanism.

Niiler, P. P. (1977). One-dimensional models of the seasonal thermocline. In: *The Sea* Volume 6, ed. E. D. Goldberg *et al.*, pp. 97–115. London, New York: Wiley Interscience.
A good summary of the various approaches to this problem (§9.2) developed over the past ten years.

Plate, E. J. (ed.) (1979). *Engineering meteorology*. Amsterdam: Elsevier. This multi-author volume ranges widely over topics related to most chapters in this book.

Pollard, R. T. (1977). Observations and theories of Langmuir circulations and their role in near surface mixing. In: *A Voyage of Discovery*, ed. M. Angel, pp. 235–51. Oxford: Pergamon.
A good summary of the development of this field, and of the currently preferred theoretical explanation in terms of wave-current interactions (p. 300). Other articles in the volume are also of interest.

Sherman, F. S., Imberger, J. and Corcos, G. M. (1978). Turbulence and mixing in stably stratified waters. *Ann. Rev. Fluid Mech.* **10**, 267–88.
Up-to-date review of double-diffusive convection, especially recent theories (§8.1); mixing driven by shear or waves (§§4.1.3, 10.3) and mixing due to external energy inputs (§§4.3.1, 9.1).

Stern, M. E. (1975). *Ocean circulation physics*. New York, London: Academic Press.
A more theoretical discussion of various subjects treated in this book. In particular the chapter on 'Thermohaline convection' complements our chapter 8.

Thorpe, S. A. (1973). Turbulence in stably stratified fluids: a review of laboratory experiments. *Boundary Layer Meteorology* **5**, 95–119.
A useful discussion of the experimental work in this field.

Thorpe, S. A. (1978). On internal gravity waves in an accelerating shear flow. *J. Fluid Mech.* **88**, 623–39.
Numerical and laboratory experiments on the interaction between a mean flow and a train of internal gravity waves. The early experiments in this elegant series are described in §4.1.4 and in the above review.

Tritton, D. J. (1977). *Physical Fluid Dynamics*. New York: Van Nostrand Reinhold.
An excellent textbook which discusses a wide range of fluid phenomena from a physical viewpoint.

Turner, J. S. (1974). Double-diffusive phenomena. *Ann. Rev. Fluid Mech.* **6**, 37–56.
A review of theory, experiments and applications which extends the subject matter of chapter 8.

Turner, J. S. (1978). Double-diffusive intrusions into a density gradient. *J. Geophys. Res.* **83**, 2887–901.
Laboratory experiments which demonstrate the importance of double-diffusive processes in driving intrusive motions when there are strong horizontal property gradients.

Williams, A. J. (1975). Images of ocean microstructure. *Deep-Sea Res.* **22**, 811–29.
The first direct observations of salt fingers in the ocean, using an optical technique. They thus fill the gap mentioned on p. 272.

SUBJECT INDEX